T0134105

The Biometric Computing

The Biometric Computing
Recognition and Registration

Edited by
Karm Veer Arya and Robin Singh Bhadoria

CRC Press
Taylor & Francis Group
Boca Raton London New York

CRC Press is an imprint of the
Taylor & Francis Group, an **informa** business

A CHAPMAN & HALL BOOK

CRC Press
Taylor & Francis Group
52 Vanderbilt Avenue,
New York, NY 10017

© 2020 by Taylor & Francis Group, LLC
CRC Press is an imprint of Taylor & Francis Group, an Informa business

No claim to original U.S. Government works

Printed on acid-free paper

International Standard Book Number-13: 978-0-8153-9364-1 (Hardback)

Visit the Taylor & Francis Web site at
http://www.taylorandfrancis.com

and the CRC Press Web site at
http://www.crcpress.com

Contents

Part I   The Biometric Computing – Fundamentals & Definitions

Part II   The Biometric Computing – Algorithms & Methodologies

Part III The Biometric Computing – Futuristic Research & Case Studies

Preface

The word 'biometrics' is derived from the Greek words *'bios'* and *'metric'*, which means life and measurement, respectively. Biometrics computing defines the processing of biometric data using computer and communication technology. This book focuses on various aspects of biometric computing using human characteristics like the face, fingerprint, palm print, iris, etc., that eventually works for person authentication in real-time systems. This book also provides the platform and paradigm for the development of a surveillance system in the identification and access control for security measures. The different chapters in this book demonstrate the methods that help in feature extraction, feature matching, and classification using various pattern recognition techniques.

Another aim of this book is to focus on analyzing the data generated by the biometric system with the help of different devices. The purpose of this is to define the person identification and human gesture recognition. The design and architecture issues are deeply discussed in this book, which not only reveals fundamental principles but also leverages the study by taking different use cases on different biometric computing models.

The key factor of this book is to exhibit on application-specific techniques using pattern recognition methods that are implemented through an efficient hardware platform for recognizing human characteristics as input data. Such biometric data are categorized by several features that affect the acceptability of input and parameters like distinctiveness, reliability, permanency, contestability, and universality must be perceived.

This book is divided into three major parts and each part contains 4–5 chapters. Part I discusses the fundamentals of data acquisition and computation for biometric computation and processing. It includes the standards and identification for biometrics along with data acquisition methodologies. In the second part, the different methodologies and algorithms for human identification using biometric traits, such as the face, iris, fingerprint, palm print, voiceprint, etc., have been discussed, and exclusive chapters have been dedicated on each of the above-mentioned traits with the aid of different case studies. Part III of this book opens up the interdisciplinary research as it includes chapters on sharing a common platform of biometrics. This book also includes the chapters on trending topics like Internet of Biometric Things (IoBT), newborn biometrics, use of deep neural networks, and artificial intelligence for biometric applications.

This book is an edited volume by prominent industry professionals and academic researchers in the area of biometrics computing. It will be a very good reference book for undergraduate students, young researchers, and practitioners who are willing to carry out the carrier in the field of biometric recognition and security.

Editors
Karm Veer Arya
Robin Singh Bhadoria

MATLAB® is a registered trademark of The MathWorks, Inc. For product information, please contact:

The MathWorks, Inc.
3 Apple Hill Drive
Natick, MA 01760-2098
USA Tel: 508-647-7000
Fax: 508-647-7001
Email: info@mathworks.com
Web: www.mathworks.com

Editors

Prof. Karm Veer Arya received his Master's degree in Electrical Engineering from the Indian Institute of Science (IISc), Bangalore, India, and the Ph.D. degree in Computer Science & Engineering from the Indian Institute of Technology Kanpur (IITK), Kanpur, India. His research area includes image processing, biometrics, information security, and reliability. He is currently working as a professor in the Department of Information and Communication Technology at the ABV-Indian Institute of Information Technology and Management, Gwalior, India. He has published more than 150 papers in international journals and peer-reviewed conferences around the world. He has guided 10 Ph.D. and 80 Master's theses students. He is on the editorial board of many international journals. He is a member of the organizing committee and programme committee for various national and international conferences and workshops. He has edited the Proceedings of International Conference on Industrial & Information Systems (ICIIS2014) held on 15th–17th, December 2014 at Gwalior, India. Prof. Arya has co-authored a book titled *Security in Mobile Ad-hoc Networks: A Clustering based Approach* (LAP-Lambert Academic Publishing). He is Senior Member of IEEE (USA), Fellow of IETE (INDIA), Fellow of Institution of Engineers (India), Member of ACM and Life Member of ISTE (INDIA).

Current Affiliation: Professor, Department of Information and Communication Technology, ABV-Indian Institute of Information Technology & Management, Gwalior, India

Dr. Robin Singh Bhadoria received his Ph.D. degree from the Indian Institute of Technology (IIT) Indore, India, in 2018. His research interests include data mining, frequent pattern mining, cloud computing, service-oriented architecture, wireless sensor network and Internet of Things. He received his Bachelor and Master of Technology degrees in Computer Science & Engineering from Rajiv Gandhi Technological University, Bhopal, India. He has published more than 60 articles in international journals and peer-reviewed conferences around the world with reputed publishers like IEEE, Elsevier, and Springer, including book chapters. He is currently working as an assistant professor in the Department of Computer Science & Engineering at the Indian Institute of Information Technology (IIIT) Bhopal, India, and also an associate editor in the *International Journal of Computing, Communications and Networking* (IJCCN) ISSN 2319–2720. He is also serving as an editorial board member for different journals around the globe. He is a professional member for different professional research bodies like IEEE (USA), IAENG (Hong-Kong), Internet Society, Virginia (USA), and IACSIT (Singapore).

Current Affiliation: Department of Computer Science & Engineering, Indian Institute of Information Technology (IIIT), Bhopal, Madhya Pradesh, India

Contributors

Akella Amarendra Babu
St. Martin's Engineering College
Hyderabad, India

Asish Bera
Department of Computer Science and
 Engineering
Haldia Institute of Technology
West Bengal, India

Debotosh Bhattacharjee
Department of Computer Science and
 Engineering
Jadavpur University
Kolkata, India

Ankit A. Bhurane
Department of Electronics &
 Communication Engineering
Indian Institute of Information Technology
 (IIIT) Nagpur
Nagpur, India

Zoran S. Bojkovic
Engineering Academy of Serbia
University of Belgrade
Beograd, Serbia

Jaime S. Cardoso
Faculty of Engineering
University of Porto
and
INESC TEC
Porto, Portugal

Mireya Saraí García Vázquez
Instituto PolitécnicoNacional–CITEDI
Tijuana, México

Mario Graff
CONACyT—Consejo Nacional de Ciencia y
 Tecnología, Dirección de Cátedras
Ciudad de México, México
and
INFOTEC Centro de Investigación e
 Innovación en Tecnologías de la
 Información y Comunicación
Aguascalientes, Mexico

Phalguni Gupta
Department of Computer Science and
 Engineering
National Institute of Technical Teachers'
 Training and Research
Kolkata, India

Puneet Gupta
Department of Computer Science and
 Engineering, Indian Institute of
 Technology
Embedded system and Robotics
TCS Research Kolkata
Kolkata, India
and
Indian Institute of Technology Indore
Indore, India

D. Jude Hemanth
Department of Electrical Sciences
Karunya Institute of Technology and
 Science
Coimbatore, India

Lawrence Henessey
Department of Systems and Software
 Engineering
Blekinge Institute of Technology
Karlskrona, Sweden

Chiung Ching Ho
Department of Computing and
    Informatics
Multimedia University
Cyberjaya, Malaysia

Gaurav Jaswal
Department of Electrical Engineering
National Institute of Technology
Hamirpur, India

Amit Kaul
Department of Electrical Engineering
National Institute of Technology
Hamirpur, India

Abhinav Kumar
Computer Science and Engineering
    Department
IIT (BHU)
Varanasi, India

Pradeep Kundu
Department of Mechanical Engineering
Indian Institute of Technology (IIT) Delhi
Delhi, India

Eduardo Garea Llano
Advanced Technologies Application
    Center–CENATAV
Havana, Cuba

André Lourenço
CardioID Technologies, Instituto Superior
    de Engenharia de Lisboa (ISEL)
and
Instituto de Telecomunicações (IT)
Lisbon, Portugal

Dragorad A. Milovanovic
Engineering Academy of Serbia
University of Belgrade
Beograd, Serbia

Sabino Miranda-Jiménez
CONACyT—Consejo Nacional de Ciencia y
    Tecnología, Dirección de Cátedras
Ciudad de México, México
and
INFOTEC Centro de Investigación e
    Innovación en Tecnologías de la
    Información y Comunicación
Aguascalientes, México

Daniela Moctezuma
CONACyT—Consejo Nacional de Ciencia y
    Tecnología, Dirección de Cátedras
Ciudad de México, México
and
Centro de Investigación en Ciencias de
    Información Geoespacial
Aguascalientes, México

Mita Nasipuri
Department of Computer Science and
    Engineering
Jadavpur University
Kolkata, India

Ravinder Nath
Department of Electrical Engineering
National Institute of Technology
Hamirpur, India

João Ribeiro Pinto
Faculty of Engineering
University of Porto
and
INESC TEC
Porto, Portugal

Shyam Singh Rajput
Department of Computer Science &
    Engineering
Bennett University
Greater Noida, India

Yellasiri Ramadevi
Department of Computer Science &
   Engineering
Chaitanya Bharati Institute of Technology
Hyderabad, India

Alejandro Álvaro Ramírez Acosta
MIRAL R&D&I
San Diego, California

Manish Rawat
Discipline of Mechatronics Engineering
Manipal University
Jaipur, India

K. Martin Sagayam
Department of Electrical Sciences
Karunya Institute of Technology and
   Science
Coimbatore, India

Poonam Sharma
Department of Computer Science &
   Engineering
Visvesvaraya National Institute of
   Technology Nagpur
Nagpur, India

Sanjay Kumar Singh
Computer Science and Engineering
   Department
IIT (BHU)
Varanasi, India

Sankirthana Suresh
Department of Electrical Sciences
Karunya Institute of Technology and
   Science
Coimbatore, India

Eric S. Tellez
CONACyT—Consejo Nacional de Ciencia y
   Tecnología, Dirección de Cátedras
Ciudad de México, Mexico
and
INFOTEC Centro de Investigación e
   Innovación en Tecnologías de la
   Información y Comunicación
Aguascalientes, México

Sridevi Tumula
Department of Computer Science &
   Engineering
Chaitanya Bharati Institute of Technology
Hyderabad, India

Juan Miguel Colores Vargas
Universidad Autónomade Baja
   California–ECITEC
Mexicali, México

Part I

The Biometric Computing – Fundamentals & Definitions

1

Acquisition and Computation for Data in Biometric System

Poonam Sharma

Visvesvaraya National Institute of Technology Nagpur

Karm Veer Arya

ABV-Indian Institute of Information Technology & Management

CONTENTS

1.1 Introduction

Biometrics refers to a method for measurement of human characteristics. Biometrics is basically used for authentication and identification of a human being. Multiple applications of biometrics also include identification of a person in groups that are under surveillance. Different traits used for authentication and identification are human physical appearance, habits and behavioral aspects. For a specific application, selection of the particular trait or multiple traits depends upon the accuracy desired and time constraint. For any application, a single biometric does not meet the requirement. Physiological biometrics is used in recognition system in which biometric data such as signature, face, speech, fingerprint, iris, retina, gait, hand and ear geometry, etc. are acquired from person and compared from the stored biometric data. There are many application of biometrics such as security purpose, identification of individual, crime prevention and airport security, in defense, smart cards, law enforcement and surveillance. Depending on the applications, it basically operates in two ways: verification and identification.

Biometrics is automated methods for identifying a person using his biological traits or behavioral aspects or characteristics. It involves distinctive features of a person stored in the form of a binary code for subsequently recognizing that person by automated method.

Sample characteristics are taken from the person asking authentication to compare the similarity to biometric references previously stored from known persons.

In present day applications, identification of a person reliably and conveniently is a major challenge. This is mainly due to explosive growth in internet connectivity and human mobility. Thus a number of different biometric methods including face, fingerprint and multimodal methods have been developed. Fingerprint recognition is used to identify a person by matching fingerprint with that of the fingerprint stored in the database. It has very high accuracy, most economical biometric, most developed biometrics, easy to use and standardized. But it has a disadvantage that the accuracy reduces due to with the dryness in fingers or dirty skin of the finger, it also reduces due to changes in fingerprints with the age and large memory requirement.

Face recognition is preferred due to its non-intrusive capability, ease to capture and it doesn't need user cooperation. But, biometric images are affected by a number of variations due to data acquisition methods or due to aging, lighting, pose, etc., and thus recognition method must be adaptive to the changes in pose, expression, lighting, occlusion, aging, etc.

Multimodal biometric systems data acquisition recognition accuracy of many applications can be enhanced in comparison to unimodal biometric systems, which is done using multiple sensors. In multimodal biometrics, data acquisition can be done using different sensors for the same trait or data information can be collected using sensors for different biometrics. Multimodal biometric systems can be implemented by a sequential, parallel, hierarchical or combination of different unimodal biometrics.

The rest of this chapter is organized as follows. In Section 1.2, elements of biometric system including enrollment, image acquisition, feature extraction, database development and classification have been described. Section 1.3 describes the performance measures for biometric systems. The detailed applications of biometric systems have been presented in Section 1.4. Section 1.5 includes different biometric systems. Limitations of biometric systems have been described in Section 1.7, which concludes the chapter.

1.2 Elements of Biometric System

Biometric system is an application where a system automatically identifies or verifies an individual using image or video of different biometric properties of a person. It includes two basic parts: Biometric detection where a trait is detected from a cluttered scene by positioning and detecting the edges by discarding the image background. Recognition methods are used to verify or identify one or more traits using stored database of trait image. Biometric system may be divided into four steps: In the first step, image is acquired from a camera. Second step includes detection from the acquired image. Third step includes feature extraction from the detected part which act as the image signature. Fourth step includes the recognition method which leads to verification or identification of the person. The elements of biometric system are shown in Figure 1.1.

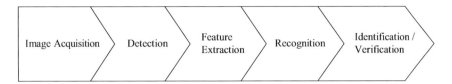

FIGURE 1.1
Elements of biometric system.

1.2.1 Image Acquisition

At their most basic level, biometric methods use pattern recognition using either image acquisition devices, like scanners or cameras for face, fingerprint or iris recognition methods, or sound or movement acquisition devices, like microphones or platens for voice recognition or signature recognition methods, to collect the biometric features. The characteristics of the acquired samples, which are unique for each user, are extracted and converted to a unique code as a reference or template, which represents a person's biometric feature. These features are stored in the form of a database and used for comparison when recognition is required.

The output of the acquisition method is used as an input data to extract features using different methods. The feature extracted is given to any of the classification methods for further classification. The efficiency of the system depends on different parameters like biometric measure used, method of storage of data, feature extraction and classification, etc. The results may also vary due to change in environment where the data is collected.

1.2.2 Feature Extraction

To detect the representations that store invariant and discriminatory information in the input obtained from image acquisition is the most challenging problem in biometric methods. This representation has a large implication on the design of the rest of the system. The unprocessed measurement values are not invariant over the time of capture, and thus the salient features of the input measurement need to be determined which both discriminate between the identities as well as remain invariant for a given individual. Thus, the problem of representation is to determine a measurement (feature) space, which is invariant (less variant) for the input signals belonging to the same identity and which differs maximally for those belonging to different identities (high interclass variation and low interclass variation). To determine the discriminatory power of an image obtained after acquisition and obtain an effective feature is a challenging task. From raw acquisition inputs, automatically extracting the features is an extremely difficult problem, especially when input acquisition is accompanied with noise. Today's biometric systems need feature extraction to be automatic and must not require human intervention. But for most of the biometric method, it is not easy to reliably detect these features using state-of-the-art image processing techniques. With the development of various applications that require biometric features, the need for fully automatic feature extraction method has increased in pattern recognition research. Traditionally, the feature extraction methods followed sequential architecture and were comparatively less efficient especially for noisy images. A number of rigorous models have been developed for feature representations, especially, in noisy situations. Although determining an appropriate model for the features is still challenging. The selection of the prominent and discriminative features for dimension reduction is a crucial issues of pattern recognition. The aim of feature extraction is to reduce intra-class variation and increase inter-class variation. Thus discriminatory information needs to be retained. Since the dimensionality of the feature vector is reduced after the feature extraction step, extraction will yield savings in memory and time consumption.

As we reduce the dimensionality, uniqueness of face image get lost. It is more critical with the variation in pose, illumination, scaling and unconstrained environment. Feature extraction methods are basically divided into two parts: holistic and local feature extraction. Holistic method uses lexicographic ordering of pixel value and whole image is converted to a single vector. The whole image is considered as a single two-dimensional matrix. This results in large dimensionality, which needs to be reduced to classify the image [1].

Local feature extraction considers only a local region of face image extracted by some method and feature vector generated is unique. Gottumukkal and Asari [2] proposed local PCA where an image is divided into grid and each region of the grid is extracted as an individual vector. Zou et al. [3] proposed local feature extraction and is invariant to intrinsic image properties. Feature extraction by including only significant points using adaboost was developed by Wu et al. [4]. Liu et al. [5] used Gabor Kernel PCA to reduce the dimensionality. Cardinaux et al. [6] used modified discrete cosine transform (DCT) and generated a feature vector with dimensionality of 18. Shan et al. [7] proposed local binary pattern (LBP) to represent image as a single feature vector. Zhang et al. [8] used Local Gabor binary pattern histogram sequence (LGBPHS). Initially images are converted to Gabor images and LBP is applied to each image. Feature vector is obtained by taking histogram of the LBP image. Mikolajczyk and Schmid [9] proposed log-polar location grid. In this method three bins are considered in radial direction and eight in angular direction. Then PCA is used for dimensionality reduction. Sarfraz and Hellwich [10] proposed feature extraction using gradient location orientation histogram [11] and PCA for feature extraction. Scale adaptive spatial gradients are calculated and partitioned into eight parts on a grid in polar coordinates. Then histogram of each part is calculated and results are concatenated to form a single feature vector.

Cootes et al. [11] proposed active shape model (ASM). In this model, 68 landmark points are considered and are located on the facial regions and a shape model is generated by concatenating all the vectors obtained from individual landmark. The PCA is applied to get the shape free face patches.

1.2.3 Recognition

The recognition or classification is the method of assigning the feature vector obtained after feature extraction to a predefined class. The output of the classifier is a selection of one of the predefined classes. The complexity of the classification method depends on the similarity metrics defined between the patterns of different classes [13,14]. Therefore, the success of classification is significantly affected by the method feature extraction.

1.3 Performance Measures

Parameters like false acceptance rate (FAR), false rejection rate (FRR), equal error rate (EER) curve and receiver operating characteristic (ROC) curve are basic face recognition evaluation methods. In biometrics, there are two types of errors considered for recognition. When the individual tries to get identified even when his/her identity is not in the trained data but similarity score is higher than the given threshold, this is called a false accept. False acceptance rate (FAR) is the percentage that number of times that a false accept occurs. The second type of error is false reject this is when there is a proper claim as to identity, and yet the returned similarity score is lower than preset threshold. False reject rate (FRR) is the percentage that false reject occurs across all individuals. Subtracting this rate from 100% (100% − false reject rate) gives us the *probability of verification*. FAR and FRR are inversely proportional to each other and both depend on the acceptance threshold. High threshold results in low FAR but high FRR. Thus threshold is set according to desired FAR and FRR. Receiver operating characteristic (ROC) curve is used for effective comparison of different biometrics systems. It is used to visualize and compare the effectiveness of the different face recognition methods. Ideal ROC has zero value at zero false acceptance rate and one at all other values of false acceptance rate. ROC curve represents a plot between the genuine acceptance rate and false acceptance rate, as shown in Figure 1.2.

To evaluate the performance of a face recognition system with a threshold, equal error rate (EER) is calculated. It is calculated by finding a common value of false acceptance and false rejection errors on the receiver operating curve. EER point can be calculated on error versus threshold graph on which FAR and FRR probability density is plotted. For an efficient face recognition system, EER is desired to be very low as shown in Figure 1.3. The best method of evaluating the value of threshold is by trial method till the optimization of results for a particular application. If false acceptance can be tolerated then the threshold is set high to minimize the value of FRR.

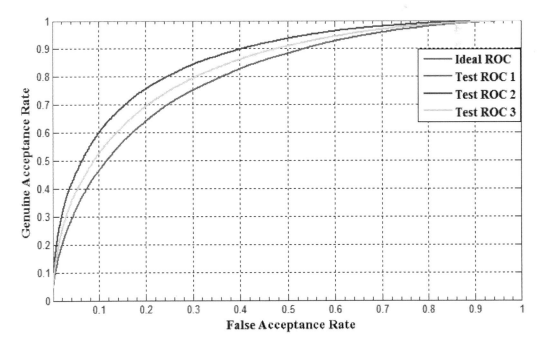

FIGURE 1.2
The receiver operating characteristic curve.

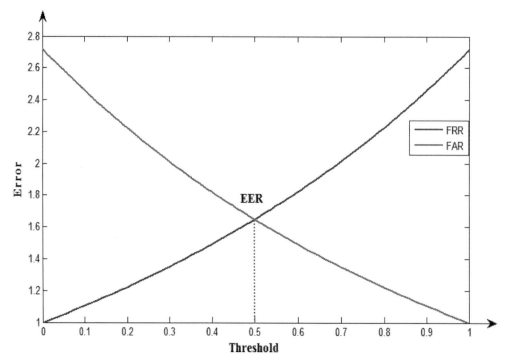

FIGURE 1.3
The equal error rate curve.

1.4 Applications of Face Recognition

Face recognition system is very effective and reliable among different biometric technologies. Major areas of applications for face recognition/registration are:

- **Security:** For law enforcement and surveillance at airport, railway stations, portal control, tracking of criminals, traffic control, finding fugitives and fighting against terrorism.

- **Personal Information Access:** Smart cards including ATM cards, immigration, passport, voter identity cards, driving license and national identity cards, entry at defense areas, sporting event, home access or any other application where personal identification is required.

- **Human–Machine Interaction:** Human–computer interaction, gaming, virtual reality, etc.

- **Commercial Applications:** Surveillance in stores, shopping, etc.

- **Information Security:** Computer or electronic gadget login, database administrator login, intranet and intranet login, online banking, etc.

Although numerous methods have been developed for different applications, the available methods may not produce efficient results due to wide variation in lighting pose and other factors.

1.5 Image Acquisition Methods

1.5.1 Fingerprint Recognition Methods

Fingerprint recognition technology mostly deployed in the biometric system. Recently, for the security reason, many organizations have deployed a fingerprint-based biometric system for identification. The fingerprint recognition carried out in four step: data acquisition, preprocessing, feature extraction and template matching. Data acquisition is a process to gather good quality of fingerprints from a human being. There are two ways to collect fingerprints from a human being: off-line and online. In off-line acquisition, smear ink on a fingertip and put an impression on white paper. Then these fingerprints get digitized by high-resolution digital camera or scanner, this process is called rolled fingerprint. Another off-line acquisition is a latent fingerprint, it is collected by a forensic expert. A latent fingerprint is bad quality and difficult to extract discriminative feature than a rolled fingerprint. In online acquisition, fingerprint is collected by using the scanner where a person has to put his/her finger on the flat surface sensor. Based on sensing technology the electronic sensor categories in three families: silicon sensor, optical and ultrasound, these sensors also be recognized by the touch sensor. Recently in mobile sweep sensor has been used for fingerprint identification. There are several challenges in fingerprint recognition for data acquisition: skin condition, sensor condition, poor quality of images, size of the scanner. The important step of the fingerprint recognition is a feature extraction and classification (template matching). When the fingerprint image partially corrupted or overlapped, this issue is not explored properly. The SIFT (scale invariant feature transform) is another method for partial fingerprints SIFT but it generates a large number of key points which increases computational complexity and increases time consumption. Aravindan et al. [15] have used wavelet SIFT descriptor in fingerprint recognition to resolve the problem of overlapped, corrupted, rotation and scaling of the images. An author used SIFT on decomposed images of the wavelet to get the less number of key points to recognize the fingerprints. The fingerprint biometric system used for security reason but the system is vulnerable to face or spoof fingerprint. This problem is overcome by Kim et al. [16] who proposed liveness detection of a fingerprint using deep belief network (DBN). In this method, a restricted Boltzmann machine (RBM) is applied to extract features using training set of artificial fingerprints. The DBN provides a conditional probability to check whether a test fingerprint image can be identified from the training set and also to test the liveness of fingerprint using the output of multiple patches of DBN. Ali et al. [17] proposed fingerprint verification using minutiae features. In this method, the author used minutiae extractor algorithm to extract a bifurcation and ridge ending and used minutiae matching algorithm with Euclidean distance as the similarity measure. Similarly, Yuan et al. [18] discussed about testing the liveness fingerprint using local phase quantity (LPQ) followed by principal component analysis (PCA) for dimensionality reduction. The support vector machine (SVM) has used to train the model with an ample number of train samples. After training the model used to detect the liveness of fingerprint. Balaji et al. [19] proposed spoof detection algorithm using white light, this approach has taken two things to identify the artificial fingerprint. When the light passed through the system, it captures a fingerprint and photo of a fingernail. Then process the photo of a fingernail and extract key descriptor using SIFT and compare with the reference database. This approach calculates the correlation value between them and decides fake fingerprint on the basis of the correlation value. The latent fingerprint is partially corrupted and difficult to generate a discriminative feature vector for its recognition. Cao et al. [20] developed a method to recognize latent fingerprint. This algorithm has used convolution neural network to represent the latent fingerprint using one texture template and two minutiae templates. Retrieve a similar candidate list by comparing score between the latent and gallery reference database. The latent fingerprint contains more nonlinear deformation, which can affect the performance of recognition. Medina-Pérez et al. [21] proposed latent iris recognition based on deformable minutiae clustering. This method finds multiple overlapping clusters to overcome the problem of nonlinear variation of a latent fingerprint. Al-Ani [22] has proposed method based on double fingerprint thumb. In this method, author used 2D-DWT to extract the feature from the fingerprint at different levels 1, 2, and 3 and used simple correlation approach to measure the similarity between fingerprints. The matching fingerprints produce more similarity than mismatching fingerprints. Su et al. [23] extract pore for high-resolution fingerprint recognition. Here deep learning (CNN) model used to extract pore and also present affine Fourier moment matching method which fuses three features like pore, ridge pattern and minutiae to find similarity.

Fingerprints of each person have a permanent uniqueness. Thus, fingerprints are widely used for identification and forensic investigations of a person for a long time. It consists of many ridges and furrows, which can be compared in each small local window based on parallelism and average width.

However, intensive research on fingerprint recognition shows that these are compared or identified using two most significant minutiae, called termination, immediate ending of a ridge, and bifurcation, point on the ridge from which two branches derive.

Fingerprint recognition approaches are of two types based on the representation of fingerprints. The first approach, is based on minutia, where fingerprint is represented by its local features including terminations and bifurcations. This approach has been intensively studied, also is the backbone of the current available fingerprint recognition products. The second approach is based on image methods [13,14], which is based on the global features of a complete fingerprint image and is an advanced method for fingerprint recognition.

These minutiae consider a variety of properties as features for identification including their position and orientations. In Ref. [24], a novel matching method to detect the matched minutiae pairs is done incrementally. The final matching score of two fingerprints is the maximum score computed.

Menotti et al. [25] proposed two methods. Convolution network architectures are used for verification in the first method, and the weights are updated using back propagation neural network in the second method. Results were tested on different biometric spoofing test images along with real samples. Deep representations for each image are obtained by combining the results from the two learning approaches. This provides promising results as compared to existing methods with approximately eight recognitions out of nine spoofing images. The results show that the proposed method using convolutional neural networks is robust to different methods of spoofing.

Chavan et al. [26] proposed fingerprint authentication using Gabor filter bank for feature extraction. Initially a filter bank is created for five scales and eight orientations. Then image is convolved with all these filters, and final feature vector is obtained by combining features from all the filters. Identification and recognition can be done by comparing the feature vector with the feature vector stored in the database. Classification is done using Euclidean distance between the test and train image feature vector. The results are tested on two databases FVC2000 and DBIT and proved the performance and efficiency of the algorithm.

In biometric system, [27] security is not maintained as all the training data is stored in the form of feature vector on single server. Visual cryptography can be combined with biometrics to handle the problem of data security. Multimodal biometrics is another solution to deal with data security. Fingerprint and iris data can be combined together to improve the recognition accuracy. Features obtained from fingerprints and iris when combined provides a unique data for each person thus improving the accuracy. This also reduces the requirement of large training data.

Most of the methods consider identification of a single fingerprint where the feature vector of fingerprint is compared with that of the existing fingerprint image stored in the database [28]. Proposed a method based on composite fingerprints (CFs), CFs are generated by combining feature vector of many fingerprint into a single feature vector. The training data of fingerprints is as an unordered binary search tree in this approach, the internal node in the tree represents a fingerprint which is composited from all the fingerprints beneath that node. The results are compared with different search strategies. The results obtained using this approach have good accuracy with lower computational cost.

In Ref. [29] Donida et al. presented a fully touchless fingerprint recognition system using computational 3D models. It uses less constrained acquisition setup as compared to other existing methods and simultaneously captures multiple images while the finger is moving. IT takes the shape of the figure also into consideration. Also, comparison is done using computation of multiple touch compatible images. Thus the algorithm is evaluated using different aspects of the biometric system including acceptability, usability, recognition performance, robustness to environmental conditions and finger misplacements, and compatibility and interoperability with touch based technologies and results provided satisfactory accuracy, the system is robust to environmental conditions and finger rotations.

Kumar and Kwong [30] proposed a matching strategy to match popular minutiae features in 3D space. This strategy uses 3D representation of 2D minutiae features and it recovers and includes minutiae height and its 3D orientation. 3D fingerprint identification systems are complex in terms of space and computation, which is due to the use of structured lighting system or multiple cameras. The single camera based 3D fingerprint identification systems have been presented so as to reduce the complexity. A generalized 3D minutiae matching model have been used to obtain extended 3D fingerprint features from the 3D fingerprints. 2D fingerprint images are simultaneously used to improve the efficiency. Database of 240 clients' 3D fingerprints is used for the evaluation of the method, and thus proved that the proposed method has the discriminate power of 3D minutiae representation and matching to achieve performance improvement.

1.5.2 Face Recognition

The performance of many face recognition algorithms could be effected by the data acquisition module, which is the very first module in any algorithm. Data can fall in any of the categories like text, image, audio or a video clip. In case of face recognition, an image acts as the data. Image acquisition can be achieved by performing digital scan on an already available photographs or by making use of an electro-optical camera in order to take pictures of the individual. The face images can also be constructed from a video. The existing facial recognition systems make use of a single camera. The images taken under unconstrained environment, with respect to pose, illumination and expression, results in the low recognition rates. As the pose angle of the subject goes high, greater than 30°, the recognition rate drops down. Some algorithms sustain the illumination variation problem, but not the same with PCA. The problem can be solved by generating the face images with frontal view, less face expression and same illumination.

Since the movement of the face in video surveillance system cannot be restrained, it is very rare to achieve the perfect face that is straight to the camera in the picture by using only a single camera. Hence, there is a need for deploying multiple cameras as stated in the paper [8] to capture face images from different views. Then, the face detection and face postures estimation have to be performed on every captured face images. And then, the face images which are near to frontal view are proceeded for recognition purpose. The deployment of multiple cameras, which gets effected by the human head posture during movement, plays the key role in obtaining acceptable images. The deployment can be done according to the horizontal rotation, since it is usually changeable. The possible rotation angle of face for identification is less than ±15°.

The most easiest way to build complete face image logs is by applying the existing face detectors to the each and every video sequences, and then adding all the faces that are detected to the log. The number of faces to be detected per individual per video sequence is restricted to one. So, a camera that records 15 video sequences per unit time results in 900 face images per individual per minute. The drawback of this approach, person oriented face log construction [31] is that the constructed logs of the faces are lengthy because there are chunks of faces to be logged corresponding to a single individual. Thus, by considering only a few best images per individual, the above-mentioned approach attempts to lessen the count of faces that are to be logged. The faces need to be not only detected but also tracked in order to achieve the proper results. Thus, a face image track can be collated for each subject. If a person goes out of the scene, then the high-quality images can be picked up from the person's face history, and then added to the face log. More faces are chosen if the tracker doesn't give good outcomes. Hence, the construction of person oriented face logs avoids the high computational cost issue in many face recognition technologies. The key factor to be considered while building a database of faces is to limit the number of duplicates per individual such that it comprises of key information about the video sequence face. The authors of Ref. [32] introduced a module that selects only a few number of high-quality images as local maxima in constructing a database of faces. After experimenting on the various datasets that include faces that are taken under unconstrained environments, this module has given the satisfying results.

Modern face recognition technologies consider the processing speeds of the best quality frames to obtain acceptable performance in the recognition rate as compared to conventional recognition systems. Anantharajah et al. has come up with an approach [33] that uses each and every detail of the face including the distance between the lips of an individual for the selection of the high-quality faces from a frame. A neural network has been used to fuse the feature scores that are standardized. And, the local Gabor binary pattern histogram sequence-based face recognition technologies make use of the images that have high-quality scores. This method resulted in achieving good recognition rates as it picks up the top quality face images.

To deal with the problems caused due to the unconstrained environments like pose, size and illumination variations, a system [34] which uses pan-tilt-zoom parameters of a camera to capture complete face. And a module that considers the quality of the image has been employed for logging the high-quality faces from the pictures that are already been clicked. Prior to the construction of the log, the functions for detecting the faces, controlling the camera, face tracking and face quality assessment have been performed. This system has gained an accuracy of around 85%.

The author, S. Parupati et al. [35] designed an image acquisition service to obtain face images of individuals from multiple views and at the same time to index each obtained image into its respective pose. If the non-frontal faces are detected, then they are very challenging as compared to the frontal faces. Thus resulting in introducing the additional image processing step. To decrease the computational cost and also to maintain a high detection accuracy, camera that can view in all directions and can communicate to other cameras is used. Hence, a high capture rate for both front side viewed faces and other poses has been achieved. This system is processed on the group of cameras that have a 1.6 GHz intel atom processor which resulted in reliable front side viewed faces at 11 fps and the other angled faces at 10 fps on the captured images that have the size of 640 × 480.

Another impact of low-quality face images is the low recognition rate of dynamic scene algorithms. Since, their inability to efficiently obtain features on the targets that are far away in the video. A framework that has been presented in the paper [36] analyzes in all aspects and controls the camera angle in order to explore all the scenes while making note of the performance measures such as tracking accuracy, best click and image size through dynamic camera-to-target assignment and effective feature acquisition

The important parameters as stated in Ref. [37] with respect to image acquisition that helps in improving the performance of face recognition algorithms are sensor memory, tasking threshold and number of cameras used. In the case of sensor memory, since caching capacity of the sensor is limited the useful images can be lost. Thus, the choice lies between the performance and the cost of the sensor. The tasking threshold gives the option to choose between the energy consumption and the performance. And, if more number of cameras are used, then low risk of occlusion is observed while capturing the images. But this is achieved when the performance is chosen rather than the deployment cost.

Many techniques have been developed for face recognition [38–40]. Different methods of face recognition for still images are categorized into three main groups such as holistic method, feature based method and hybrid method. The entire face region is considered as input data in this method. In feature based method, input data for classifier is the segmented local feature on face such as nose and eyes. Hybrid methods include both local features and global features to counteract the drawbacks of both the holistic and feature based methods. Hybrid methods are developed by the proper combination of already existing holistic or feature based methods. Most of the methods discussed below are hybrid methods because local or global features alone cannot be used to develop a highly efficient face recognition method. Kalocsai et al. [41] model uses multi-scale and multi-orientation Gabor kernels to calculate the similarity of face images. Weighted contribution of each element (1,920 kernels) is taken into consideration for the representation of face images. Also the change in features due to changes in orientation (horizontal and vertical), expression, illumination and background have been used to calculate the overall variance in the Gabor kernel activations. Caucasian and Japanese image sets have been used in the representation to improve their discriminative power so as to improve its recognition performance.

Zhong and Zhang [42] paper presented an enhanced local directional patterns (ELDP) to face recognition, which uses local edge gradient information to represent face images. The face is divided into sub-blocks, and each pixel of sub-block obtains eight edge response values by convolving the local 3 * 3 neighborhood with eight Kirsch masks. The ELDP dominant patterns (ELDPd) are generated by testing the occurrence rates of the ELDP codes in a sub-block of face images. Global concatenated histogram with ELDP is used to represent the face descriptor. Andreu et al. [43] worked on neutral and distorted faces for gender classification methodologies. In this approach, global and local representations are used, types of features like gray levels, PCA and LBP are used and comparison by 1-NN,

SVM and PCA + LDA classifiers is done. They used different combinations of these methods and obtained efficiencies above 90%.

Heusch and Marcel [44] paper presented a generative model to describe a face and is applied to the face authentication task. The presented model proposed to encode relationships between features using a static Bayesian Network. Hidden Markov Models (HMM) and Gaussian Mixture Models (GMM) are used for feature extraction. XM2VTS database and BANCA database are used for experiment, it gives 95% confidence.

Seo and Park [45] proposed a robust face recognition method by introducing a statistical learning process for local features, which represents a facial image as a set of local feature descriptors such as Scale Invariant Feature Transform (SIFT). Local feature use feature extraction on the feature applies weight distance measure for the classification purpose. It used AR database for the results.

Wan et al. [46] proposed a technique which fuses geometric features and algebraic features of the original image face recognition method for the revolution of feature. Random weight network (RWN) is proposed as efficient classifier. RWN is applied to classify fused features to improve the recognition rate and speed. Geometric features are extracted using fast discrete curvelet transform and two-dimensional principal component analysis and the algebraic features are extracted by quasi-singular value decomposition method to create relationship of each image.

Varadarajan et al. [47] proposed Chirp Z-Transform and Goertzel algorithms as preprocessing, block-based feature extraction and exponential binary particle swarm optimization for feature selection. For illumination normalization, combined approach of CZT and Goertzel algorithm is used. In Ref. [48], two techniques, anisotropic diffusion-based preprocessing and Gabor filter-based feature extraction, have been proposed. Anisotropic diffusion is used to preserve the edges for smoothing and enhancement. To capture facial features of specific angles Gabor filter is used. To search the feature space for the optimal feature subset, a binary particle swarm optimization-based feature selection algorithm is used. Thus improvement in efficiency is observed by using anisotropic diffusion. Varun et al. [49] developed a feature extraction method based on Hough Transform peaks. For efficient feature extraction, Block-wise Hough Transform peaks are used, and to search the feature space for the optimal feature subset, a Binary Particle Swarm Optimization (BPSO)-based feature selection algorithm is used.

Li et al. [50] proposed a dense feature extraction method for face recognition which consists of two steps: first is an encoding scheme that compresses high dimensional dense features and second is to develop an adaptive feature matching algorithm. Adaptive matching is used which requires small subset of training samples and provides better accuracy. Lenc et al. [38] proposed a SIFT-based approach for the face recognition. The databases used for comparison of efficiency are FERET, AR and LFW. A corpus creation algorithm is used to extract the faces from the database, and for classification, two supervised confidence measure methods are used and that is based on a posterior class probability and a multilayer perceptron.

Ma et al. [39] proposed combination of a simple effective dimensionality increasing (DI) method and an information preserving dimensionality reduction (DR) method. To extract the features rectangle filters in DI method sums up the pixel values within a randomized rectangle window on the face image, which in turn reduces the computational complexity and it also preserves the informative features. But DR is based on compressed sensing theory. It has the partial based and data-independent properties. The proposed method shows better robustness to variation in occlusion and disguise.

A feature extraction method for robust face recognition [40] based on local patterns of gradients (LPOG) is developed. LPOG uses block-wised elliptical local binary pattern, a refined variant of ELBP and local phase quantization operators for capturing local as well as global features. For dimension reduction, Whitened Principal Component Analysis is used, and for classification, weighted angle-based distance is used. This model of LPOG WPCA system is robust to illumination, expression, occlusion, pose, time-lapse variations and low resolution. This approach is faster than some of the advanced feature extraction algorithms.

1.5.3 Iris Recognition

Iris is a circular, flat, colored membrane behind the cornea of the eye that is all around the pupil in the center of the eye. It is unique for each and every individual for the entire lifetime and can be used as an efficient way for recognizing people that can be used in biometric systems. Image Acquisition forms the very first step in every recognition-based system and it is of great significance as well. It plays a very important role in the recognition system because overall accuracy of the system depends on the quality of the captured information.

The basic architecture of a usual iris image acquisition system consists of image sensor, optical lens, illuminator and access control point. The features which are essential to design a suitable lens for iris image acquisition system are field of view, depth of field and focal length.

Particularly for the purpose of iris recognition, this step of image acquisition basically deals with capturing high-resolution images of iris from the subject being considered using cameras and specific sensors with good sharpness. Pupil and iris part of the eye should be clearly captured by all these iris images, and then some preprocessing operations such as noise removal, histogram equalization, filtering, etc. may be applied to upgrade the image quality.

As discussed, accurate iris recognition much more depends on high-resolution iris images. However, it is very hard to capture accurate images of iris with high resolution because human iris is very small with approximately 11 mm diameter and near infrared (NIR) of about 700–900 nm illumination is needed in order to clearly illustrate texture details of iris.

Various factors should be considered while acquiring data for iris recognition:

- First, images of the iris which are with appropriate resolution and sharpness are required for iris recognition.
- Second, sufficient amount of contrast should be there in the interior iris pattern that would be constrained by operator comfort.
- Third, these iris images should be well framed (i.e., centered) without the need of the operator to make any use of either an eye piece, chin rest or other such things that would be invasive.

Among the early iris image acquisition systems proposed were Daugman [51] and Wildes et al. [53]. The Daugman system captures images of iris from a distance of 15–46 cm where the diameter is roughly between 100 and 200 pixels with the help of a 330-mm lens. Similarly, the Wildes et al. [53] proposed a system where iris images are captured from a distance of 20 cm with approximately 256 pixels diameter using an 80-mm lens.

Auto-focus lens and pan-tilt-zoom (PTZ) units are being used in order to enlarge the volume of capture. For capturing scene image, these systems consist of a wide field of view camera, and for capturing iris images, a narrow field of view camera mounted on a PTZ unit is deployed. To detect the 2D position of face, the wide field of view camera is unified with depth data which adjusts the narrow field of view camera toward iris. Even though by making use of auto-focus lens and pan-tilt-zoom units, the capture volume is significantly extended but these are still too slow to trace the human movement because the cameras used here have to mechanically adjusted. Narayanswamy et al. [54] proposed a system for extending the field depth of an iris imaging system by making use of a camera with fixed focus, in absence of a zoom lens whereas Zhang et al. [55] increased depth of field of the iris imaging system through a simple architecture composed by light field cameras. McCloskey et al. [56] made use of the flutter shutter technique in order to avoid motion blur.

A long standoff distance is also an appealing feature along with large capture volume for an efficient iris imaging system. The goal here is to acquire images of iris where the subject walks at normal speed through an access control point. The first ever study was proposed by Fancourt et al. [57] for iris recognition from a certain distance. In the proposed system, iris images are captured with an infrared camera and telescope from a distance of up to 10 m with sufficient quality. There was no performance degradation while capturing iris images from this much distance, but this system requires the head of the subject to be placed in a chin rest. Matey et al. [58] proposed a system where they deployed two narrow field of view cameras having fixed focal length and stacked them vertically to provide a larger capture volume for developing Iris-On-the-Move system which can identify subjects while they walk at a normal speed. This proposed system can capture iris images from a 3 m distance with the advantages of high-resolution cameras and high power NIR strobe illumination. Further, Eagle-Eyes Bashir et al. [59] proposed an iris recognition system that can capture accurate iris images at 3–6 m distance using a laser illuminator.

Current researchers [60] interest is mainly on the analysis of the quality of the data for iris recognition systems. Accurate and efficient iris image acquisition methods are still a challenging task for an iris recognition system. Great efforts and in-depth research are needed to develop innovative iris imaging systems that can accurately and quickly acquire good quality iris images from a long distance and in a large capture volume.

Data acquisition is a process to collect a good quality of the iris images. A system captures a number of iris images by the digital camera but it should care about noise and environmental effect. The performance of iris recognition is depended on the quality of images. The iris recognition system divided into five steps: data acquisition, data segmentation, data normalization, feature extraction and classification or template matching. Data acquisition is the first step of iris recognition. The iris component is a small part of the face, with a diameter approximately equals to 11 mm. Most iris acquisition system used near-infrared (NIR) band. Near-infrared (NIR) sensor is used to capture iris images with a 700–900 nm range of wavelength. The data acquisition classified into two parts: controlled mode (conventional acquisition) and uncontrolled mode. In conventional acquisition system, a person is stationary during capturing the images. The person should stand near the camera at a fixed place, and look straight to the camera. The eyes have to be open in front of camera and eyes should be center of the image. In the process of data acquisition, the quality of images depends on the focus of the camera, limbic contrast, pupillary, illumination and iris resolution. This factor can change the quality of the iris images. In less constrained acquisition, all the quality factors are not possible to maintain on the person. In uncontrolled mode distance and on-the-move plays a vital role in the acquisition process. Several methods have been developed for iris recognition, such as Eagle-Eyes [59] operate 3–6 m distance. IOM (Iris-On-the-Move) system [58], working with moving object and operate at 3 m distance, stand-off system with 1.5 m distance [61] and 3 m distance [62], pan-tilt-zoom camera operates with 1.5–3 m distance [63] and long-range iris acquisition system used in video surveillance for 8–12 m [64] and 30 m [65]. Conventional iris system has several drawbacks which are not possible to maintain real-time scenario.

1. The distance between person and camera should be 50 cm.
2. A person should be static (not movable).
3. The time-consuming process to maintain all the condition.

To overcome the above-mentioned problem, researchers have made acquisition process in realistic conditions, they mostly target the distance and move issue. These problems have been resolved by Fancourt et al. [57] by proposing two iris recognition systems at 5 and 10 m. In this acquisition, process person has to seat with the fixed head at in front of the camera. In this process, a person should not move. To make more unconstrained Matey et al. [58] proposed Iris-On-the-Move (IOM) system that provides both flexibilities: move and capture an image at a distance, but the person has to move in normal speed. Iris-On-the-Move system can capture the images at a 3 m distance and with 100 pixels across the iris diameter. The IOM system camera is fixed and cannot work properly for people with different heights due to the small area of capturing field and also face the problem of low resolution. This problem has been resolved by Yoon et al. [102] by proposing pan-tilt-zoom camera to capture images at a long distance. This camera provides large area to capture an iris image and resolve height issue also. The system required two steps to capture images: firstly it detects the face and secondly moves toward eyes region and zoom automatically until camera captures clear iris image in order to generate a high-resolution image. Similarly, Wheeler et al. [61] proposed a stand-off iris recognition system, this system covers only 1.5 m distance and a person has to cooperate with some constraint. The main intention is to design a system for different height people who look toward the camera. This camera capture approximately 200 pixels across the iris diameter. First time Bashir et al. [59] proposed Eagle-Eyes system which is used in video surveillance techniques to acquire multiple biometric. Eagle-Eyes system can recognize the person at a stand-off distance of 3–6 m by their irises. The Eagle-Eyes system used multiple cameras: (i) Fixed scene camera for human tracking. (ii) Face camera for detection of the face. (iii) NIR camera used to capture iris of eyes. Eagle-Eyes system produces 100–200 pixel across the iris diameter. Like Eagle-Eyes system, Dong [62] build a new self-adaptive (stand-off) system for iris images for acquiring a high-quality image at a 3 m distance. Still above all mentioned system did not cover more than 5 m distance but De Villar et al. [65] proposed long-range iris acquisition and recognition system. Long-range system is able to identify a person at a distance of 30 m. This system covers a large area to capture iris images and used a view camera to locate face and eyes and then a narrow field of view camera attached to a telescope is pointed to the detected eye. The long-range system generates an iris image with 190 pixels across the iris diameter. A person can be stationary or walk toward the system camera. If a person is moving toward the camera, the person speed is calculated in order to tune the focus of the system to the person motion and then a set of high-resolution images are acquired at a desired checkpoint based on the estimated speed, the focus is continuously adjusted while the person is moving toward the system. The iris image quality depends on the system and environmental condition during capturing the images.

The acquisition system should have a high-resolution camera and good NIR illumination intensity. He [66] proposed a fast evaluation method which is based on weighted entropy. In this method, author evaluates the quality of images. A localization method is used for segmentation of the iris images, different weight distributed to the different division. Then the author evaluate the quality of image based on sum of weighted entropy. In such wavelength, iris images highlight the intricate texture information of the iris even for dark eyes. Lu [67] proposed quality assessment scheme for iris images which deals with eyelash occlusion, eyelid occlusion, defocus and motion blur. Iris recognition becomes difficult when data acquisition is noisy. Iris segmentation method is used to identify pixels of eyes belongs to iris region. Frucci [68] proposed WIRE (watershed-based iris recognition) method which overcomes the issues of existing segmented methods. The preprocessing step of WIRE removes the illumination effect from iris images using watershed transform. Umer et al. [69] proposed a method based on morphology using multiscale morphologic feature for recognition. In the proposed method, a feature of iris images is represented by the sum of dissimilarity residues obtained by applying the morphologic transform and used SVM as a classifier. Naseem et al. [70] used sparse representation-based classification in iris recognition. SRC test image represents a linear combination of the dictionary atom and achieves state-of-the-art result. In this method, author used class-specific component of the training sample and corporate class-specific component into the dictionary as an atom. Ramya et al. [71] developed an approach using Bernstein polynomial and Fourier transform. These methods extract the feature from iris images and are converted to a continuous information of a specific region of the target image. Zhang et al. [63] proposed an application of convolution neural network to extract an automatic feature from iris images. In this method, iris region is divided into eight rectangular component and applied CNN on each component to extract discriminative feature. Vyas et al. [72] discussed iris recognition using discriminative features extracted using GLCM (gray level co-occurrence matrices) and 2D Gabor filter. GLCM extracts a haralick feature from iris images and used a neural network for classification. Kavosi et al. [73] used Ripplet transform feature in iris recognition; in this method, first, extract iris component from the face image, select diagonal Ripplet component and then generate iris code using BPSK (binary phase shift keying symbols); this iris code can be used for recognition.

1.5.4 Multi-Biometrics

Proliferation of criminal activities and security breaches in public and private areas questions the reliability of prevailing surveillance systems and urges to more definitive identification and verification of a person. Earlier physical identity proofs like identity cards and several cryptographic techniques involving passwords, PINs, email ids, etc. were put into use for person authentication, but these faced some serious challenges like spoofing, impersonation, passwords thefts and many more. They also burdened the person with remembering such personal information or carrying the identity proofs along with them all the time.

Later researchers [74] recommended utilizing person's unique physiological and behavioral characteristics, commonly referred to as biometrics, for verification and authorization tasks. Such characteristics are difficult to forge and less susceptive to fraudulent acts. Different types of biometrics (described in Ref. [75]) can be listed as physiological attributes (iris, fingerprints, etc.), behavioral attributes (voice, signatures, etc.), soft attributes (gender, color, height, etc.) and medico-chemical attributes (ECG, DNA, etc.). Biometric identification corresponds to a one-to-many process where test sample obtained from a sensor is compared to a pre-stored database of similar attribute for its existence, producing either a pre-stored image with highest resemblance or a rejection indicating the absence of similar sample. While, biometric verification corresponds to a one-to-one process between the test image and the trained images of the person he claims to be.

Initially, unimodal biometric identification technique was deployed, where a single biometric attribute was used for person authentication. But these unimodal techniques [76] had certain limitations that resulted in increased false acceptance rate (FAR) and false rejection rate (FRR). Major challenges include (i) noisy data generation which occurs due to different environmental conditions, fingerprint with a scar, voice affected due to cough and cold, (ii) Inter-class similarities where features space corresponding to various users overlap, (iii) Inter-class variations which might happen due to wrong interaction between user and sensor resulting in generation of dissimilar test and sample data, (iv) spoof attacks that are likely to occur in case of forged signatures or voice, etc., (v) non-universality where relevant information couldn't be captured due to inaccurate sensor, etc.

Hence, data from multiple biometrics, commonly referred to as multimodal biometrics, can be deployed to achieve more precise human identification and verification. Such systems have higher reliability as they are resistant to issues discussed above to a great extent. Multiple biometrics allow wider population coverage by increasing feature space, as forging multiple biometrics becomes quite difficult. In short, shortcoming of one biometric can be overcome with the help of other biometrics. Employing multimodal biometrics for person identification involves information fusion obtained from various sensors which can be done at different steps involved in the system. A basic multimodal biometric method mainly consists of following the four steps:

1. Acquisition step: The first step consists of acquiring biometric data of a person through interaction with various sensors deployed. For face or iris recognition, camera can act as an acquisition tool, scanners can be used for palm of fingerprints, microphones for recording voice, etc.

2. Feature extraction step: Once raw data is obtained, it is pre-processed for noise removal to obtain a feature vector that is capable of accurately and uniquely defining a person in such a way that feature values extracted for a person under different scenarios are relatively similar while that of for a different person under similar situations are fairly dissimilar. It's a compact and more precise form of information that's easier to deal with. Information fusion in feature extraction level is referred to as feature-level fusion.

3. Matching attributes: In this step, feature vectors pertaining to an individual is matched against a pre-stored template database producing corresponding degree of similarity in such a way that obtained matching score is fairly high for features of similar person and fairly low for that of dissimilar ones. Usually this step is a tough pattern recognition issue as it may involve large intra-class variations (caused due to bad acquisition, different environmental scenarios, noise, distortions, etc.) and huge inter-class similarity (i.e. differentiating identical twins is quite tough in face recognition). Integration of match scores is known as match-level information fusion.

4. Decision making step: In this step, the person is identified or verified based upon the obtained matching score. Usually the matching score is compared against a set threshold for decision making, and this threshold can be decided based upon the security level of the application. Here, final outputs from different classifiers are integrated for information fusion, called as decision-level information fusion.

Biometric systems involving information fusion at feature level are considered to be more accurate as feature sets contain richer information as compared to match scores but is practically infeasible due to incompatibility of feature sets. Information present at decision level is very less, resulting in a rigid information fusion. Hence, practically match score-level information fusion is preferred for developing a multimodal biometric system.

Sources of information:

Based upon the number of sensors, scenarios and feature sets used for data acquisition, multimodal biometrics can be broadly divided into the following five categories.

1. Multi-sensor systems: Raw data pertaining to a common attribute and generated by different sensors is accumulated for pre-processing. Fusion of information extracted from raw data can be carried out at any step. Chang et al. [77] improved the accuracy of face recognition by capturing both 2D and 3D images of faces and by deploying information fusion at both feature level and match score level. Kumar et al. [78] used geometric features of hands along with palm prints for developing person verification system. Information fusion at match score level produced more efficient results as compared to information fusion at feature level probably due to high dimensional feature data generated.

2. Multi-algorithm systems: Raw data generated by deploying a single sensor is given as input to different classifiers which in turn either extract their own feature sets for further processing or operate on a similar feature set. Ross et al. [79] generated an enhanced person identification system by combining the match score of texture-based fingerprint system with minutiae-based fingerprint system. Jain et al. [80] integrated the data generated by three different fingerprint systems at match score level by using logistic function. All the three fingerprint matchers operated on same minutiae sets. Lu et al. [81] used PCA, ICA and LDA techniques to obtain three different feature sets from a single face image and further integrated the corresponding outputs at match score level.

3. Multi-instance systems: This refers to a scenario where a single camera is deployed to obtain images pertaining to multiple similar units, for instance, both irises, prints of multiple fingers, etc. It's a cheaper way to strengthen the biometric system as no additional sensor or classifier is required.

4. Multi-modal systems: In this type of multimodal system, different types of sensors are deployed to obtain raw data pertaining to different attributes of a person. Such attributes are fairly independent of each other, thereby, guaranteeing the increased efficiency of the system. Bigun et al. [82] integrated the speech data with face data of a person by using Bayesian statistical framework. Hong and Jain [83] proposed a method where face features were used for preliminary identification followed by fingerprint verification to narrow down to a specific identity. They also employed separate confidence measures to each matcher during the information fusion. BioID [84], a commercially used system, uses face features, lip motion and voice for user identification.

1.5.5 Hand Geometry

Human hand biometrics is an ensemble of techniques leading to the establishment of a person's identity based upon hand geometry and hand silhouette. Human hands are considerably capable of uniquely identifying a person by extracting the palm surface, length and width of fingers, etc. Hand geometry biometric systems deploy a camera or a scanner as a sensor for capturing hand images for further processing. This image information is further matched against the sample hand information in a pre-stored database for human identification and verification. Hand geometry biometric systems are not sufficiently accurate for a large population coverage where security issues are of utmost importance, but still these systems are in existence from many decades in average sized institutions for attendance purposes, person validation, etc. because of the following advantages offered by them to users:

1. Low-resolution images are used to extract hand information resulting in the efficient storage of hand templates in database. Many commercial systems use a 9 byte template to store hand information.
2. Sensors deployed are relatively inexpensive and provide a user-friendly environment for data acquisition.
3. It is commonly acceptable to public as it is free from criminal connotation.
4. Fingerprints and palm prints can easily be used in addition to hand geometry biometrics for more reliable human identification.

Every human hand has certain characteristics that make it unique. Researchers derived total 30 hand geometry features [85] in 2002, which are very precise. These include hand's inner surface which helps in extracting creases, minutiae, length of thumb finger, middle finger, index finger, ring finger, pinkie finger. Inner surface of every finger has three separate creases, namely dorsal phalanx, middle phalanx and proximal phalanx. Hand's upper surface contains dorsal while on its right border side we have thenar eminence, and on left border side we have hypothenar eminence. 'N' number of features reside in palm region which can be effectively used for activity recognition. Research has been carried out solely on finger features but many researchers also considered palmar regions [86] to fulfill this purpose. Radio metacarpal joint, metacarpal bones, styloid process can be used to demonstrate structure of bones and joints that can distinguish hands of people from different age groups [87]. 3D image and contours of human hand can be helpful in cloning a human hand [88]. Eight bones can also be used for feature vector [89]. A hand contains 19 bones, 5 metacarpals in palm and distal phalanx in fingertips [90].

Based upon contact of hand with the sensors, hand biometrics can be divided into the following categories, namely pose invariant contact-based approach but it depends on the placement of the hand with the help of pegs, pose variant contact based approach in which hands are simply placed on a flat surface without guiding the user about hand position and contact-free approach where no guidance or instruction is provided and hand images are clicked using CCTV cameras or webcams. Sensors obtain multiple samples of hands for further matching and decision-making process. Pegs employed in many scanners provide us axes for better feature extraction. Many variants of hand geometry biometrics have been proposed till date. Peter Varchol et al. [91] proposed by deploying a simple scanner on a flat surface guided by pegs for required hand position. In total, they extracted 21 features for each hand. Three separate recognition metrics were employed, namely Gaussian mixture model, Euclidean distance and Hamming distance.

Using pegs led to capture deformed structure of hands and also hindered hand movement, hence Miguel Adan et al. [92] used natural hand layout for person recognition. They captured images of both the hands using two CCD cameras and extracted features based upon the polar representation of hand contours. Joan Fabregas et al. [93] proposed method for different resolution of images. They lowered the resolution from 120 to 24 dpi and concluded that resolution can be lowered to 72 dpi without losing the features and recognition rate. Bahareh Agile et al. [94] proposed a pose variant and rotation insensitive hand recognition method that took into account scanning of four fingers. They used 24 features from four fingers and calculated similarity using Euclidean distance as a measure. Hand geometry was also combined with other biometrics like palm recognition and fingerprint recognition for better recognition rate.

Miguel A. Ferrer et al. [95] integrated fingerprint and palm print gesture along with hand geometry biometrics in which 10 images of right hand of a person were obtained and total 15 features were extracted. Wei-Chang Wang et al. [96] combined hand geometry features with palm prints and utilized Voronoi diagrams and morphological information for feature extraction. Coarse recognition followed by a fine recognition was deployed for better accuracy of the system.

Nowadays, some methodologies even use 90 features for hand geometry recognition. Many multimodal scenarios club hand geometry features with fingerprints and iris recognition for enhanced recognition rate.

Behavioral attributes in biometrics focus on day to day behavior and actions of a person like speech recognition, signature recognition, gait recognition, etc. The person need not to pay specific attention toward the sensors capturing the raw data. These aspects are usually combined with physical attributes to achieve higher efficiency in person identification. Human gait analysis in general terms differs from the rest of the attributes. It is preferred many a times due to its unobtrusive characteristic which dictates that special attention of a person under observation is not required. This raw data can be fed using sensors placed at distant places without requiring physical interaction between the sensor and the subject. These are useful in cases where facial- and fingerprints-related data become confidential. Also it allows to work with low-resolution videos.

Gait-based recognition systems utilize static features along with dynamic features for classification. Static features here correspond to the head and neck (mostly above the waist) portion of human body while dynamic features involve leg and foot movements. It can be deployed with other biometrics when they face problems with low-resolution videos as every human has a unique style of walking. Walks are cyclic in nature and can be divided into a series of events which then can be fed to template matching systems.

Methods of data acquisition in gait recognition:

1. Gait recognition based on moving videos: Here video cameras are used to record gait information of a person, which can be placed at certain distance. Later features are extracted using basic image processing which is used for further processing.

2. Gait recognition based on floor sensors: In such systems, sensors or force plates are used on the floors. These act as source for recording motion-related data of the person walking on them. Possible features might include minimum time between two consecutive heel strokes, amplitude value of heel strike, etc.

3. Gait recognition based on wearable sensors: These systems ask the user to bodily wear sensors required to record the unique features. MR sensors can be placed at different body parts and can record the acceleration and speed of human gait. These motion traits can be used for person identification.

But these attributes are often affected by many external factors like lighting conditions, shoe type of the person, surface type, viewing angles, etc. and internal factors like foot injury, drunkenness, aging, sickness, etc. Also gait analysis can be done using two approaches, namely model-based approach, where motion of the person is represented as a model (like 2D stick figure) explaining the joint positions, body part movements, etc. Prior knowledge of human shape and body part positions is required by the system. Second approach is a model-free approach where human motion is taken completely as input in the form of body silhouette. This technique is considered to be faster and less computation involving. Many gait recognition techniques have been developed till date. In Ref. [97], researchers calculated the dimensions of the contour of the silhouette, i.e. the horizontal distance between rightmost and leftmost pixel, but these contours might get affected by noise and thus are unreliable. Researchers in Ref. [98] proposed to use radon transformation of binary silhouette and to keep speed constant for any gait motion. Background subtraction techniques are applied to obtain the silhouette which in turn is used to extract walk cycle that is later fed to radon transformation for further processing.

Method in Ref. [99] proposes to analyze the human gait based on the images obtained using stereo vision camera in a 3D plane. 3D contours are obtained by analyzing binary silhouette and are sent for contour matching. Later, stereo gait feature is obtained by calculating the normal stereo silhouette vector. In Ref. [100], researchers extracted the feature vector using image processing techniques from the binary silhouette of the human motion. The main features are length of step, center of mass and length of cycle and are used to train a neural network. Authors in Ref. [101] proposed accelerometer-based gait recognition technique that has a very low error rate. Raw data corresponding to gait cycles are used as feature vectors and different distance functions are deployed for classification.

1.6 Conclusion

The accuracy of different face recognition methods reduces due to the changes in lighting, scale, pose, aging and other factors. The main technical challenges faced are:

- **Large Variation in the Appearance:** There may be large variations in the images obtained from different sources for the same person due to illumination, aging, emotion, expression, occlusion, pose, spectacles and beard makes the system complex. Therefore, due to this variability, it is very cumbersome to extract the basic face information from their respective images. Among these the four main problems of biometrics are variation in scaling, unconstrained environment, pose and illumination. Scaling is one of the problems of the size of biometrics. Due to scaling position of biometric features and distance between them changes which makes the problem complex. Unconstrained environment is the change in the background. Most of the methods developed deal with the cropped image and thus efficiency reduces abruptly with change in the background. Illumination and pose variation are another challenging problems and need to be solved.

- **High Dimensionality:** The size of the image is high. It further increases due to the use of different multiresolution techniques for feature extraction. Most of the classification methods are not capable of handling high dimensional inputs. If the dimensionality is reduced, it loses its uniqueness and results in reduced recognition efficiency. Thus dimensionality needs to be reduced to maintain the uniqueness.

- **Small Sample Size:** Most of the methods developed till date work efficiently when the training samples are more but the efficiency reduces as the training samples reduce. In most of the applications it is hard to get a many training images, thus face recognition must be capable of performing efficient recognition with single training image.

- **Huge Database Size:** The size of the database becomes huge when face recognition is used for applications like surveillance, authentication, etc. Thus effective method of feature extraction needs to be developed. Storing extracted features will result in lesser memory storage requirement.

- **Large Recognition Time:** Recognition time is high for most of the methods developed as these methods deal in increasing the efficiency under different changes in pose, illumination, aging, etc. For real-time application recognition, time needs to be reduced.

- **Highly Complex Methods for Dimensionality Reduction and Classification:** Techniques performing dimensionality reduction like PCA, ICA, LDA are not capable of preserving the non-convex variations of face image, which are very important to differentiate among individuals. Metrics like Euclidean distance and Mahalanobis distance do not perform well for classification which effects the efficiency of face recognition system.

References

1. T. Zhou, D. Tao and X. Wu, Manifold elastic net: A unified framework for sparse dimension reduction, *Data Mining and Knowledge Discovery*, vol. 22, no. 3, 340–371, May 2011.
2. R. Gottumukkal and V. K. Asari, An improved face recognition technique based on modular PCA approach, *Pattern Recognition Letters*, vol. 25, no. 4, 429–436, March 2004.
3. J. Zou, Q. Ji and G. Nagy, A comparative study of local matching approach for face recognition, *IEEE Transactions on Image Processing*, vol. 16, no. 10, 2617–2628, 2007.
4. H. Y. Wu, Y. Yoshida and T. Shioyama, Optimal Gabor filters for high speed face identification, In *Proceedings of International Conference on Pattern Recognition*, Quebec City, vol. 1, 107–110, 11–15 August 2002.
5. D. H. Liu, K. M. Lam and L. S. Shen, Optimal sampling of Gabor features for face recognition, *Pattern Recognition Letters*, vol. 25, no. 2, 267–276, January 2004.
6. F. Cardinaux, C. Sanderson and D. S. Bengio, User authentication via adapted statistical models of face images, *IEEE Transactions on Signal Processing*, vol. 54, no. 1, 361–373, 2006.
7. C. Shan, S. Gong and P. W. McOwan, Facial expression recognition based on local binary patterns: A comprehensive study, *Image and Vision Computing*, vol. 27, no. 6, 803–816, May 2009.
8. W. Zhang, S. Shan, W. Gao, X. Chen and H. Zhang, Local Gabor binary pattern histogram sequence (LGBPHS): A novel non-statistical model for face representation and recognition, In *Proceedings of International IEEE Conference on Computer Vision*, Beijing, vol. 1, 786–791, 17–21 October 2005.
9. K. Mikolajczyk and C. Schmid, Performance evaluation of local descriptors, *IEEE Transactions on Pattern Analysis and Machine Intelligence*, vol. 27, no. 10, 1615–1630, 2005.
10. M. Sarfraz and O. Hellwich, Head Pose Estimation in Face Recognition Across Pose Scenarios, *3rd International Conference on Computer Vision Theory and Applications*, Funchal, Madeira, Portugal, 235–242, 2008.
11. T. F. Cootes, T. F. Edwards and C. J. Taylor, Active appearance models, *IEEE Transactions on Pattern Analysis and Machine Intelligence*, vol. 23, no. 6, 681–685, June 2001.

12. W. Zhao, R. Chellapa, P. J. Phillip and A. Rosenfeld, Face recognition: A literature survey, *ACM Computing Surveys (CSUR)*, vol. 35, no. 4, 399–458, 2003.

13. A. S. Tolba, A. H. El-baz and A. A. El-Harby, Face recognition: A literature review, *International Journal of Signal Processing*, vol. 2, no. 2, 88–103, 2005.

14. X. Zhang and Y. Gao, Face recognition across pose: A review, *Pattern Recognition*, vol. 42, 2876–2896, 2009.

15. A. Aravindan and S. M. Anzar, Robust partial fingerprint recognition using wavelet SIFT descriptors, *Pattern Analysis and Applications*, vol. 20, 1–17, 2017. doi:10.1007/s10044-017-0615-x.

16. S. Kim, B. Park, B. S. Song and S. Yang, Deep belief network based statistical feature learning for fingerprint liveness detection, *Pattern Recognition Letters*, vol. 77, 58–65, 2016, ISSN:0167-8655.

17. M. M. H. Ali, V. H. Mahale, P. Yannawar and A. T. Gaikwad, Fingerprint recognition for person identification and verification based on minutiae matching, In *2016 IEEE 6th International Conference on Advanced Computing (IACC)*, Bhimavaram, 332–339, 2016.

18. C. Yuan, X. Sun and R. Lv, Fingerprint liveness detection based on multi-scale LPQ and PCA, *China Communications*, vol. 13, no. 7, 60–65, July 2016. doi:10.1109/CC.2016.7559076.

19. A. Balaji, H. S. Varun and O. K. Sikha, Multimodal fingerprint spoof detection using white light, *Procedia Computer Science*, vol. 78, 330–335, 2016.

20. K. Cao and A. K. Jain. Automated latent fingerprint recognition, CoRR, abs/1704. 01925, 2017.

21. M. A. Medina-Pérez, A. M. Moreno, M. Á. F. Ballester, M. García-Borroto, O. Loyola-González and L. Altamirano-Robles, Latent fingerprint identification using deformable minutiae clustering, *Neurocomputing*, vol. 175, Part B, 851–865, 2016, ISSN:0925-2312.

22. M. S. Al-Ani, T. N. Muhamad, H. A. Muhamad and A. A. Nuri, Effective fingerprint recognition approach based on double fingerprint thumb, In *2017 International Conference on Current Research in Computer Science and Information Technology (ICCIT)*, Slemani, 75–80, 2017.

23. H. R. Su, K. Y. Chen, W. J. Wong and S. H. Lai, A deep learning approach towards pore extraction for high-resolution fingerprint recognition, In *2017 IEEE International Conference on Acoustics, Speech and Signal Processing (ICASSP)*, New Orleans, LA, 2057–2061, 2017.

24. A. C. Lomte, Biometric fingerprint authentication with minutiae using ridge feature extraction, In *IEEE Conference on Pervasive Computing*, Pune, 1–6, 2015.

25. D. Menotti, G. Chiachia and A. Pinto, et al., Deep representations for iris, face, and fingerprint spoofing detection, *IEEE Transactions on Information Forensics and Security*, vol. 10, 864–879, 2015.

26. S. Chavan, P. Mundada and D. Pal, Fingerprint authentication using Gabor filter based matching algorithm, In *International Conference on Technologies for Sustainable Development*, Mumbai, India, 1–6, 2015.

27. S. Patil, K. Bhagat, S. Bhosale and M. Deshmukh, Intensification of security in 2-factor biometric authentication system, In *IEEE Conference on Pervasive Computing*, Pune, India, 10–14, 2015.

28. S. Bayram, H. T. Sencar and N. Memon, Sensor fingerprint identification through composite fingerprints and group testing, *IEEE Transactions on Information Forensics and Security*, vol. 10, 597–612, 2015.

29. R. Donida Labati, A. Genovese, V. Piuri and F. Scotti, Toward unconstrained fingerprint recognition: A fully touchless 3-D system based on two views on the move, In *IEEE Transactions on Systems, Man, and Cybernetics: Systems*, vol. 46, no. 2, 202–219, Feb. 2016.

30. A. Kumar and C. Kwong, Towards contactless, low-cost and accurate 3D fingerprint identification, *IEEE Transactions on Pattern Analysis and Machine Intelligence*, vol. 37, 681–696, 2015.

31. A. Fourney and R. Laganiere, Constructing face image logs that are both complete and concise, In *Fourth Canadian Conference on Computer and Robot Vision, 2007. CRV '07*, Montreal, 488–494, 2007.

32. K. Nasrollahi and T. B. Moeslund, Complete face logs for video sequences using face quality measures, *IET Signal Processing*, vol. 3, no. 4, 289–300, July 2009.

33. K. Anantharajah, S. Denman, S. Sridharan, C. Fookes and D. Tjondronegoro, Quality based frame selection for video face recognition, In *2012 6th International Conference on Signal Processing and Communication Systems*, Gold Coast, QLD, 1–5, 2012.

34. M. A. Haque, K. Nasrollahi and T. B. Moeslund, Real-time acquisition of high quality face sequences from an active pan-tilt-zoom camera, In *2013 10th IEEE International Conference on Advanced Video and Signal Based Surveillance*, Krakow, 443–448, 2013.

35. S. Parupati, R. Bakkannagari, S. Sankar and V. Kulathumani, Collaborative acquisition of multi-view face images in real-time using a wireless camera network, In *2011 Fifth ACM/IEEE International Conference on Distributed Smart Cameras*, Ghent, 1–6, 2011.

36. C. Ding, B. Song, A. Morye, J. A. Farrell and A. K. Roy-Chowdhury, Collaborative sensing in a distributed PTZ camera network, *IEEE Transactions on Image Processing*, vol. 21, no. 7, 3282–3295, July 2012.

37. K. Heath and L. Guibas, Facenet: Tracking people and acquiring canonical face images in a wireless camera sensor network, In *2007 First ACM/IEEE International Conference on Distributed Smart Cameras*, Vienna, 117–124, 2007.

38. L. Lenc and P. Kral, Automatic face recognition system based on the SIFT features, *Computers and Electrical Engineering*, vol. 1, 1057–71157, 2015.

39. C. Ma, J. Jung, S. Kim and S. Ko, Random projection-based partial feature extraction for robust face recognition, *Neurocomputing*, vol. 149, 1232–1244, February 2015.

40. C. Nguyen, H. Tuan and A. Caplier, Local pattern of gradients (LPOG) for face recognition, *IEEE Transactions on Information Forensics and Security*, vol. 99, 214–220, 2015.

41. P. Kalocsai and C. von der Malsburg, Face recognition by statistical analysis of feature detectors, *Image and Vision Computing*, vol. 18, 273–278, 2000.

42. F. Zhong and J. Zhang, Face recognition with enhanced local directional patterns, *Neurocomputing*, vol. 119, 375–384, 2013.

43. Y. Andreu and P. Garcia-Sevilla, Face gender classification: A statistical study when neutral and distorted faces are combined for training and testing purposes, *Image and Vision Computing*, vol. 32, 27–36, 2014.

44. G. Heusch and S. Marcel, A novel statistical generative model dedicated to face recognition, *Image and Vision Computing*, vol. 28, 101–110, 2010.

45. J. Seo and H. Park, Robust recognition of face with partial variations using local features and statistical learning, *Neurocomputing*, vol. 129, 41–48, 2014.

46. W. Wan, Z. Zhou, J. Zhao and F. Cao, A novel face recognition method: Using random weight network and quasi-singular decomposition, *Neurocomputing*, vol. 151, 1180–1186, March 2015.

47. K. K. Varadarajan, P. R. Suhasini, K. Manikantan and S. Ramchandran, Face recognition using block-based feature extraction with CZT and Goertzel algorithm as a preprocessing technique, *Procedia Computer Science*, vol. 46, 1458–1467, 2015.

48. T. M. Abhishree, J. Latha, K. Manikantan, S. Ramchandran, Face recognition using Gabor filter based feature extraction with anisotropic diffusion as a preprocessing technique, *International Conference on Advance Computing Technologies and Science (ICACTA)*, *Procedia Computer Science*, vol. 45, 312–321, 2015.

49. R. Varun, Y. V. Kini, K. Manikantan and S. Ramachandran, Face recognition using Hough transform based feature extraction, *International Conference on Advance Computing Technologies and Science (ICACTA)*, *Procedia Computer Science*, vol. 46, 1491–1500, 2015.

50. Z. Li, D. Gong, X. Li and D. Tao, Learning compact feature descriptor and adaptive matching framework for face recognition, *IEEE Transactions on Image Processing*, vol. 24, 9, 2736–2745, 2015.

51. J. G. Daugman, High confidence visual recognition of persons by a test of statistical independence, *IEEE Transactions on Pattern Analysis and Machine Intelligence*, vol. 15, 11, 1148–1161, 1993. doi:10.1109/34.244676.

52. J. G. Daugman, U.S. Patent No. 5,291,560, U.S. Patent and Trademark Office, Washington, DC, 1994.

53. R. P. Wildes, J. C. Asmuth, S. C. Hsu, R. J. Kolczynski, J. R. Matey and S. E. McBride, Automated, noninvasive iris recognition system and method, U.S. Patent 5 572 596, 1996.

54. R. Narayanswamy, G. Johnson, P. Silveira and H. Wach, Extending the imaging volume for biometric iris recognition, *Applied Optics*, vol. 44, no. 5, 701–712, February 2005.

55. X. Zhang, Z. Sun and T. Tan, Texture removal for adaptive level set based iris segmentation. In *International Conference on Image Processing*, Hong Kong, China, 1729–1732, 2010. doi:10.1109/ICIP.2010.5652941.

56. S. McCloskey, W. Au and J. Jelinek, Iris capture from moving subjects using a fluttering shutter, In *Fourth IEEE International Conference on Biometrics: Theory Applications and Systems. BTAS 2010*, Washington, DC, 1–6, 2010.

57. C. Fancourt, L. Bogoni, K. Hanna, Y. Guo, R. Wildes, N. Takahashi and U. Jain, Iris recognition at a distance, In *Proceedings of 2005 IAPR Conference on Audio and Video Based Biometric Person Authentication*, Hilton Rye Town, NY, 1–13, July 2005.

58. J. R. Matey, O. Naroditsky, K. Hanna, R. Kolczynski, D. J. LoIacono, S. Mangru and W. Y. Zhao, Iris on the move: Acquisition of images for iris recognition in less constrained environments, In *Proceedings of the IEEE*, 94, 11, 1936–1947, 2006. doi:10.1109/JPROC.2006.884091.

59. F. Bashir, P. Casaverde, D. Usher and M. Friedman, Eagle-Eyes: A system for iris recognition at a distance. In *IEEE Conference on Technologies for Homeland Security*, Waltham, MA, 426–431, 2008.

60. K. Park and J. Kim, A real-time focusing algorithm for iris recognition camera, *IEEE Transactions on Systems, Man, and Cybernetics*, vol. 35, no. 3, 441–444, August 2005.

61. F. W. Wheeler, G. Abramovich, B. Yu and P. H. Tu, et al., Stand-off iris recognition system, In *2nd IEEE International Conference on Biometrics: Theory, Applications and Systems, 2008. BTAS 2008*, Washington, DC, 1–7, 2008.

62. W. Dong, Z. Sun, T. Tan and X. Qiu, Self-adaptive iris image acquisition system, In *Proceedings of SPIE 6944, Biometric Technology for Human Identification V, 694406*, Orlando, FL17 March 2008. doi:10.1117/12.777516.

63. W. Zhang, C. Wang and P. Xue, Application of convolution neural network in iris recognition technology, In *2017 4th International Conference on Systems and Informatics (ICSAI)*, Hangzhou, 1169–1174, 2017. doi:10.1109/ICSAI.2017.8248462.

64. S. Venugopalan, U. Prasad, K. Harun, K. Neblett, D. Toomey, J. Heyman and M. Savvides, Long range iris acquisition system for stationary and mobile subjects, In *2011 International Joint Conference on Biometrics (IJCB)*, Washington, DC, 1–8, 2011.

65. J. De Villar, R. W. Ives and J. R. Matey, et al., Design and implementation of a long range iris recognition system, In *2010 Conference Record of the Forty Fourth Asilomar Conference on Signals, Systems and Computers (ASILOMAR)*, Pacific Grove, CA, 1770–1773, 2010.

66. Y. He, T. Liu, Y. Hou and Y. Wang, A fast iris image quality evaluation method based on weighted entropy, In *Proceedings of SPIE 6623, International Symposium on Photoelectronic Detection and Imaging 2007: Image Processing, 66231*, Beijing, China 2008.

67. G. Lu, J. Qi and Q. Liao, A new scheme of iris image quality assessment, In *Third International Conference on Intelligent Information Hiding and Multimedia Signal Processing (IIH-MSP 2007)*, Kaohsiung, 147–150, 2007.

68. M. Frucci, M. Nappi, D. Riccio and G. S. di Baja, WIRE: Watershed based iris recognition, *Pattern Recognition*, vol. 52, 148–159, 2016.

69. S. Umer, B. C. Dhara and B. Chanda, Iris recognition using multiscale morphologic features, *Pattern Recognition Letters*, vol. 65, 67–74, 2015, ISSN:0167-8655,

70. I. Naseem, A. Aleem, R. Togneri and M. Bennamoun, Iris recognition using class-specific dictionaries, *Computers and Electrical Engineering*, vol. 62, 178–193, 2017, ISSN:0045-7906.

71. M. Ramya, V. Krishnaveni and K. S. Sridharan, Certain investigation on iris image recognition using hybrid approach of Fourier transform and Bernstein polynomials, *Pattern Recognition Letters*, vol. 94, 154–162, 2017. doi:10.1016/j.patrec.2017.04.009.

72. R. Vyas, T. Kanumuri, G. Sheoran and P. Dubey, Co-occurrence features and neural network classification approach for iris recognition, In *Fourth International Conference on Image Information Processing (ICIIP)*, Solan, India, 2017, 978-1-5090-6734-3.

73. D. Kavosi and A. Karimi, Iris recognition based on Ripplet transform feature extraction, vol. 17, No. 7, 231–234, 2005.

74. D. D. Zhang, *Automated Biometrics: Technologies and Systems*, Springer Science & Business Media, New York, 2013.

75. J. A. Unar, W. C. Seng and A. Abbasi, A review of biometric technology along with trends and prospects, *Pattern Recognition*, vol. 47, 2673–2688, 2014. doi:10.1016/j. patcog.2014.01.016.

76. A. Ross and A. K. Jain, Multimodal biometrics: An overview, In *12th European Signal Processing Conference 2004, IEEE*, Vienna, Austria, 1221–1224, 2004.

77. K. I. Chang, K. W. Bowyer and P. J. Flynn, Face recognition using 2D and 3D facial data, In *Proceedings of Workshop on Multimodal User Authentication*, Santa Barbara, CA, 25–32, December 2003.

78. A. Kumar, D. C. M. Wong, H. C. Shen and A. K. Jain, Personal verification using palmprint and hand geometry biometric, In *Proceedings of 4th International Conference on Audio and Video-based Biometric Person Authentication (AVBPA)*, Guildford, 668–678, June 2003.

79. A. Ross, A. K. Jain and J. Reisman, A hybrid fingerprint matcher, *Pattern Recognition*, vol. 36, 1661–1673, July 2003.

80. A. K. Jain, S. Prabhakar and S. Chen, Combining multiple matchers for a high security fingerprint verification system, *Pattern Recognition Letters*, vol. 20, 1371–1379, 1999.

81. X. Lu, Y. Wang and A. K. Jain, Combining classifiers for face recognition, In *Proceedings of IEEE International Conference on Multimedia and Expo (ICME)*, Baltimore, MD, vol. 3, 13–16, July 2003.

82. E. Bigun, J. Bigun, B. Duc and S. Fischer, Expert conciliation for multimodal person authentication systems using Bayesian statistics, In *First International Conference on AVBPA*, Crans-Montana, 291–300, March 1997.

83. L. Hong and A. K. Jain, Integrating faces and fingerprints for personal identification, *IEEE Transactions on Pattern Analysis and Machine Intelligence*, vol. 20, 1295–1307, December 1998.

84. R. W. Frischholz and U. Dieckmann, Bioid: A multimodal biometric identification system, *IEEE Computer*, vol. 33, no. 2, 64–68, 2000.

85. Y. Bulatov, S. Jambawalikar, P. Kumar and S. Sethia, Hand recognition system using geometric classifiers, In *DIMACS Workshop on Computational Geometry, (14–15 November 2002)*, Piscataway, NJ, 14–15, 2002.

86. Hand, Wikipedia [Online]. Available: https://en.wikipedia.org/wiki/Hand [Accessed: 13-September-2016], 2016.

87. C. Ward, M. Tocheri, J. Plavcan, F. Brown and F. Manthi, Early Pleistocene third metacarpal from Kenya and the evolution of modern human-like hand morphology, *Proceedings of the National Academy of Sciences*, vol. 111, no. 1, 121–124, 2013.

88. T. Rhee, Human hand modeling from surface anatomy, In *ACM SIGGRAPH Symposium on Interactive 3D Graphics and Games*, Redwood City, CA, 1–9, 2006.

89. C. L. Taylor, The anatomy and mechanics of the human hand, 22–35, 2016.

90. HealthLine, Hand anatomy, pictures & diagram | Body maps, Healthline.com [Online]. Available: www.healthline.com/human-body-maps/hand [Accessed: 13-September-2016], 2016.

91. P. Varchol and D. Levický, Using of hand geometry in biometric security systems, *Radioengineering*, vol. 16, no. 4, 82–87, December 2007.

92. M. Adan, A. Adan, A. S. Vázquez and R. Torres, Biometric verification/identification based on hands natural layout, *Image and Vision Computing*, vol. 26, 451–465, 2008.

93. M. A. Ferre, J. Fàbregas, M. Faundez, J. B. Alonso and C. Travieso, Hand geometry identification system performance, In *IEEE*, 2009.

94. B. Aghili and H. Sadjedi, Personal identification/verification by using four fingers, In *3rd International Congress on Image and Signal Processing, IEEE*, Yantai, China, 2619–2623, 2010.

95. M. A. Ferrer, A. Morales, C. M. Travieso and J. B. Alonso, Low cost multimodal biometric identification system based on hand geometry, palm and finger print texture, In *41st Annual IEEE International Carnahan Conference on Security Technology*, Yantai, China, 52–58, 2007.

96. W.-C. Wang, W.-S. Chen and S.-W. Shih, Biometric recognition by fusing palmprint and hand-geometry based on morphology, In *IEEE International Conference on Acoustics, Speech and Signal Processing*, Taipei, Taiwan, 2009.

97. A. Kale, A. N. Rajagopalan, N. Cuntoor and V. Kruger, Gait-based recognition of humans using continuous HMMs, In *Proceedings of the Fifth IEEE International Conference on Automatic Face and Gesture Recognition*, Washington, DC, 336–341, 2002.

98. N. V. Boulgouris, Gait recognition using radon transform and linear discriminant analysis, *IEEE Transactions on Image Processing*, vol. 16, no. 3, 731–740, March 2007.

99. H. Liu, Y. Cao and Z. Wang, Automatic gait recognition from a distance, In *Control and Decision Conference (CCDC), 2010*, Chinese, 2777–2782, 26–28 May 2010.

100. S. Sharma, R. Tiwari, A. Shukla and V. Singh, Identification of people using gait biometrics, *International Journal of Machine Learning and Computing*, vol. 1, no. 4, 409–415, October 2011.

101. Claudia Nickel, Accelerometer-based biometric gait recognition for authentication on smartphones, 2012.

102. S. Yoon, K. Bae, K. R. Park and J. Kim. Pan-Tilt-Zoom Based Iris Image Capturing System for Unconstrained User Environments at a Distance. In: S. W. Lee, S. Z. Li (eds) *Advances in Biometrics*. ICB 2007. Lecture Notes in Computer Science, vol 4642. Springer, Berlin, Heidelberg, 653–662, 2007.

2

Advances in Unconstrained Handprint Biometrics

Gaurav Jaswal, Amit Kaul, and Ravinder Nath
National Institute of Technology

CONTENTS

2.1 Background

At present, one of the security methods has been applied to every consumer devices to prevent impostors from accessing the resources and for ensuring human identity and privacy protection. The previously developed customary authentication/identification methods like ID card, tokens, passwords, PIN codes, etc. provide a limited scope of authentication, as all these can be simply duplicated, cracked or stolen (Jain et al. 2004). Biometrics can be used easily as an alternative in place of such traditional authentication mechanisms (Michael et al. 2008). It uses specific set of physiological or behavioral characteristics for reorganization of an individual. Over the last decade, numerous biometric traits have been explored, including ear, fingerprint, palm print, face, gait and iris for reliable security solutions. But, there does not exist any biometric trait, which satisfies all desired properties strictly. Every method has its own specific challenges that do not allow its usages in all application scenarios. Ideally, any human biological feature (Jaswal et al. 2016) cannot be a biometric trait unless it satisfies the characteristics like universality, accessibility, uniqueness, permanence, etc. To design an automated biometric system, certain points need to be kept in mind: user acceptance, controlled environmental conditions, accuracy, computational time, device cost and security (Jaswal et al. 2017a). Depending upon the application, any biometric system is commonly a part of following main stages, i.e., enrollment, verification (1:1 matching) or identification (1:N matching) (Jaswal et al. 2016). However, the unimodal biometric solutions have been shown to be quite vulnerable to spoofing attacks, which involve presenting a fake copy of the genuine user's sample to the biometric sensor. On the other hand, multimodal biometric solutions combine information extracted from two or more than two biometric traits, thereby exhibiting higher robustness against such attacks. In addition to this, one can explore the possibility of fusing more than one biometric sample of same trait or multiple algorithms. This is termed as multi-biometrics. For instance, biometric fusion is classified under four main schemes such as score level, rank level, feature level and decision level (Jaswal et al. 2017a). In particular, multimodal systems are more relevant when the number of enrolled users is very large. Also, they provide better biometric performance and lower imaging simplicity, which are essential for performing recognition on large databases (Jaswal et al. 2017b). The use of multimodal biometric systems, however, has two major drawbacks. Firstly, multiple biometric sensors are required to acquire the biometric samples and higher user cooperation is needed at the time of enrollment (Jain et al. 2004). These drawbacks, however, can be controlled to a certain limit if multiple modalities can be acquired from a single location. This is possible if the biometric characteristics lie close to each other. For example, palm print and finger knuckles can be captured by single ordinary camera with a single shot. Therefore, in order to maximize user cooperation while still enjoying the benefits of multimodal biometrics, we propose to capture the whole frontal hand image such that it covers the palm and finger regions. In fact, the frontal surface of hand includes other patterns too like hand geometry, palm vein, fingerprint and finger vein. But, we extract the biometric characteristics from palm and finger knuckle regions in the image. This reduces the complexity of the image capturing procedure as well as the number of image acquiring devices, than other modalities. Illustrations of a few hand biometric traits are shown in Figure 2.1.

Among hand biometric traits, fingerprint is of outstanding merit. It has been widely studied and is an integral part of the many real-world biometric systems. However, fingerprints are subject to occupational hazard. Also, they require high-resolution sensors

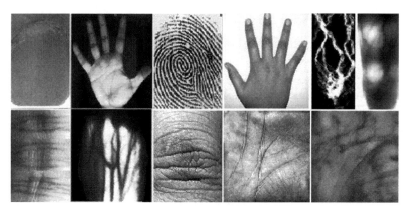

FIGURE 2.1
Hand biometric traits.

TABLE 2.1

Comparison of Hand Biometric Traits

	Sensor		Characteristics		
Trait	Type	Cost	Universality	Uniqueness	Permanence
Palm print	Contact	High	High	High	High
Knuckle print	Contactless	Low	High	High	High
Palm vein	Contactless	High	High	High	High
Fingerprint	Contact	Low	Moderate	High	High
Hand geometry	Contact	High	Moderate	Moderate	Moderate

for better image quality. The fingerprint is of abysmal quality for hand laborer, blacksmith, etc. due to the nature of their work (Jaswal et al. 2016, Zhang et al. 2010a). If a fingerprint-based biometric system has a large number of such users, then its precision is greatly affected. In such scenario, an alternate is to use finger knuckle or palm print, which possesses almost comparable feature as fingerprint while being unaffected by such occupational hazards (Zhang et al. 2003). Table 2.1 provides a brief comparison between fingerprint, and other available biometric traits lie on front side of hand surface. While, the hand geometry features may not be unique for biometric identification as they are found common among people and can change over time (Michael et al. 2008, Jaswal et al. 2016). In the similar way, hand vein patterns require additional NIR light source for image acquisition.

2.1.1 Palm Print and Knuckle Print Biometrics

The palm print and finger knuckle are physiological traits and consist of various line-like texture that has been utilized for personal identification (Zhang et al. 2003).

2.1.1.1 Palm Print

A palm print refers to inner portion of hand that lies in between wrist and base of the fingers. The entire palmer surface contains plenty of lines (majorly three dominant lines, some thin and asymmetrical lines), creases (the ridge and valley structures), geometry, minutiae and singular points (Michael et al. 2008, Jaswal et al. 2018). An example of palm print ROI image is shown in Figure 2.2b. The additional usable merits of palm print are economic or contactless imaging setup, non-intrusive and less sensitive to injury on the

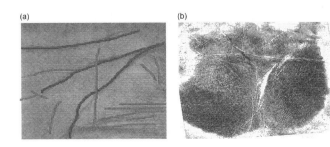

FIGURE 2.2
Anatomical structure: (a) palm print (<100 dpi). (b) Palm print (>400 dpi).

palm surface (Zhang et al. 2003). A palm print image can be categorized into (i) low resolution (<100 dpi) and (ii) high resolution (>400 dpi) as shown in Figure 2.2. The set of features extracted from a high-resolution palm image are appropriate for forensic applications, while images with lower pixel density are fit for commercial usages (Kong et al. 2004).

Challenges of palm print: Variation of illumination robustness and rotation as well as translation and handling the problem of occlusion are the major challenges of palm print recognition.

2.1.1.2 Inner Knuckle Print

The horizontal lines like skin patterns on finger inner surface specifically at Proximal Inter Phalangeal (PIP) joint is called as inner knuckle pattern (IKP) (Jaswal et al. 2016, Xu et al. 2015). These line features can be easily captured using contactless ordinary cameras without much user efforts. A sample IKP image is shown in Figure 2.3b. In addition, the merits like lesser user cooperation, non-intrusiveness and simple imaging setups make inner knuckles to gain increasing interest than others (Jaswal et al. 2017b). In literature, there are other important studies available on knuckle image that make use of flexion creases outside the finger surface as a biometric trait. In particular, the convex lines appearing on the dorsal surface of finger knuckle joints are called the finger knuckle dorsal patterns (FKP) (Kumar and Xu 2016).

Challenges of Inner Knuckle print: The main challenges are the ways to handle the problem of illumination variation, non-rigid distortion as well as rotation and translation. In addition, quality assessment and consistent ROI segmentation of major and minor knuckles are still an open areas of research.

The remaining chapter is structured into several sections. Section 1.2 describes the state-of-the-art works related to finger knuckle and palm print recognition methods. Next section reviews the sensor technology used for finger knuckle and palm print traits. In Section 1.4, proposed multimodal schemes have been discussed. Then, experiment results are described in Section 1.5. The conclusion and prospective directions on this topic will be discussed in the last section of this chapter.

(a)

(b)

FIGURE 2.3
Anatomical structure: (a) finger dorsal knuckle print; (b) finger inner knuckle print.

2.2 Related Works

The skin patterns over palm and finger knuckle surfaces have a lot of discrimination ability in order to be broadly recognized as a biometric identifier. In recent years, researchers have attempted to address the various trait specific challenges. Majority wise it has been observed that region of interest (ROI) detection, feature extraction, dimension reduction and classification play an important role in the design of personal recognition system. The other requirements like the template size, memory storage space, etc. are also essential but it is difficult to attain collectively. In this section, state-of-the-art studies in the areas of palm print and finger knuckle are presented.

Palm print: Palm print-based personal authentication is well implicit and extensively represented in state-of-the-art algorithms over the last 14 years (Zhang et al. 2003). The several well-known algorithms have been reported that were successful in improving the effectiveness and efficiency of palm print recognition systems. The existing work is mainly grouped into five categories: subspace learning, line detection, coding, transform, local descriptor and fusion schemes. In Kong et al. (2004), authors used CompCode by encoding the orientation information of lines with the help of multiple 2D Gabor filters. In addition to this, Palm Orientation Code (POC) (Wu et al. 2005a), Robust Line Orientation Code (RLOC) (Jia et al. 2008), Discriminative Competitive Code (DCC) (Leng and Zhang 2011), Double Orientation Code (DOC) (Fei et al. 2016), Competitive Orientation Map (COM) (Tabejamaat and Mousavi 2017), etc. are well defined in the literature. Later, cancelable coding schemes based on texture features have been suggested. In Connie et al. (2004), authors combined token-generated random data with palm print features. In Huang et al. (2008), authors utilized modified finite Radon transform to detect shape and position of principal lines and major wrinkles. On transform front, Discrete Cosine Transform (DCT) (Kumar and Zhang 2006), multi-resolution wavelet transform (Wu et al. 2005b), dual-tree complex wavelets with SVM (Chen and Xie 2007), Contourlets-based local fractal texture analysis (Pan et al. 2008), etc. were employed in various texture-based approaches. In Badrinath et al. (2012), palm print images are divided into small overlapping square blocks. The 2D block is first converted into two 1D signal by performing averaging operation in horizontal and vertical directions. The phase difference between horizontal and vertical signals is binarized using zero-crossing to obtain the binary vector. Likewise in Saedi and Charkari (2014), authors characterize the frequency contribution of palm texture based on 2D-DOST. As for as local descriptor is concerned, authors extracted histogram using Local Binary Pattern (LBP) (Wang et al. 2006), authors implemented LBP with Gabor response (Shen et al. 2009), authors proposed a joint framework with L2 norm regularization for 3D palm print (Zhang et al. 2015).

In Li and Kim (2017), author presented a unique local descriptor called as Local Micro Structure Tetra Pattern (LMTP) for palm print recognition. In contrast to 2D recognition approaches, 2D and 3D palm print information were fused at the feature level (Cui 2014). Moreover, some studies explored to utilize palm print features in mobile/consumer device scenarios to protect identity theft (Han et al. 2007). In another work (Bhilare et al. 2017), authors attempted to address the problem of presentation attack for palm print and proposed novel presentation attack detection (PAD) approach to discriminate between real and artifacts biometric samples.

Knuckle print: The shape of effective creases and lines on minor and major finger knuckles are observed to be highly stable in individuals in order to be considered as a biometric identifier. Most of the conventional IKP and FKP recognition algorithms can be grouped into following categories: coding, subspace methods, texture analysis methods and other image processing methods (Jaswal et al. 2016, Xu et al. 2015). The first time in 1989, authors exposed about the significance of knuckle texture for forensic applications (Jungbluth 1989). In Joshi et al. (1998), authors proposed the use of inner-side creases of fingers for biometric identification. The matching was carried out using the normalized correlation function. The results of some state-of-the-art IKP local detectors such as LBP (Liu et al. 2013), phase congruency (Zhang et al. 2012), CompCode (Kong and Zhang 2004), magnitude map of CompCode (Zhang et al. 2010b) are well studied in literature. In Zhang et al. (2009), authors proposed a multi-algorithmic approach based on matching score fusion of three subspace methods and achieved an EER of 1.39%. In Zhang et al. (2009), authors focused to use major knuckle patterns near PIP joint and employed 2D Gabor filter to extract orientation information. The similar coding approaches include: improved CompCode&MagCode (Zhang et al. 2011), knuckle code (Kumar and Zhou 2009), Adaptive Steerable Orientation Coding (ASOC) (Li et al. 2012), monogenic code (based on phase and orientation information of knuckle images (Zhang et al. 2010b), etc. In Shariatmadar and Faez (2011), authors computed the Gabor filter response and applied dimension reduction by projecting PCA weights into LDA. In Kong et al. (2014), authors applied two stage classification using Gabor filter and SURF methods and obtained EER of 0.22% and 0.218%, respectively. In Kumar and Wang (2015), the authors proposed to recuperate minutiae feature points over major/minor knuckles and perform matching using a dataset of 120 users. In Kumar (2014), authors made efforts to emphasize minor knuckles that lie in between proximal and the metacarpal phalanx bones (MCP joint) of fingers. In Kumar and Xu (2016), authors presented an approach that extracts the complete textural information on hand dorsal region. In particular, the textural features of palm dorsal and lower minor knuckle were concatenated together. The two challenging datasets were used for validation of results, which confirmed the importance of dorsal texture in biometric purposes.

2.3 Palm Print and Finger Knuckle Print Traditional Sensors

The acquisition of high-resolution (>400 dpi) hand images like fingerprint or palm print has been historically carried out by spreading the fingerprint or palm print with ink and pressing it against a paper (Michael et al. 2008, Jaswal et al. 2016). The paper is then scanned resulting in digital representation. This process is called as offline acquisition and is still used in law enforcement and forensic applications. Nowadays, it is possible to acquire the biometric image by pressing the fingerprint or palm print against the flat surface of electronic sensor. The example images of such type of sensors are given in Figure 2.4. This process is known as online acquisition (Zhang et al. 2003). Such impression based electronic imagings are of four types: thermal, optical, ultrasonic and silicon sensors.

FIGURE 2.4
High-resolution acquisition sensors.

On the other side, digital cameras and document scanners are sufficient for acquiring low-resolution (<100 dpi) hand images in different ways (Zhang et al. 2003). Initially, the palm print acquisition devices were contact-based, i.e., users had to put their hand on a planar scanner using a guiding peripheral and faced a lot of restrictions especially for physically disabled users. As a result, the trend shifted towards the use of contactless, peg-free hand biometrics system. As such type of studies offers free movement of the hand, thus encountered the common problem like the alignment of hand. Recently, studies have progressed towards 3D hand-acquisition devices to overcome such problem (Cui 2014), but the 3D recognizers are expensive, bulky and do slow data acquisitions make them undesirable for real-time applications. In the beginning, limited numbers of studies are reported due to lack of datasets and expert finger knuckle imaging devices (Nigam et al. 2016). Thereafter, peg-based acquisition devices are used, which constitute camera and panel on which user can place his or her finger.

Recently, to alleviate hygiene concerns and to provide a user-friendly environment for finger acquisition devices unlike fingerprint scanners, a platform-free, non-contact or touchless formats have been considered (Cheng and Kumar 2012). All the acquired input images are further preprocessed to extract clear finger or palm contour and remove artifacts because of varying light conditions, guiding pegs, user rings, overlapping cuffs or creases due to too light or too heavy pressing hand or hand rotation.

2.4 Proposed Single Sensor Single Shot Multi-Biometric System

A multimodal biometric recognition system based on single sensor single shot acquisition is proposed. From the input hand image, the ROI's of a palm print and three inner knuckle prints (index, middle and ring) are efficiently extracted. These images are enhanced using G-L fractional differential algorithm and next transformed into illumination invariant representation using LLBP. Two types of features from each modality are computed using UR-SIFT (local) and BLPOC (global) methods. Finally two stage fusion framework has been implemented for palm print and three IKP modalities.

The complete architecture of proposed multimodal biometric system showing preprocessing, feature extraction, matching and fusion is depicted in Figure 2.5. The stepwise detail of each module is given below:

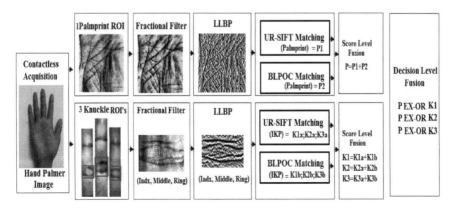

FIGURE 2.5
Block diagram of proposed hand-based multimodal biometric system.

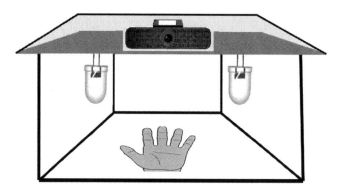

FIGURE 2.6
Contactless acquisition setup.

2.4.1 Image Acquisition

To capture the palm print and inner knuckle print images, a simple imaging device is constructed as shown in Figure 2.6. The device is designed to capture the palm print and inner knuckle print simultaneously using an ordinary camera (Logitech C925e). The camera for the image acquisition is allocated at the upper part of the device. The dimension of the device is $9 \times 10 \times 12\,\text{cm}^3$. The distance between the camera and the target hand position is based on the focus length of the camera. The focus distance for top camera is 6 cm. To remove illumination robustness problem, this system has been tested on yellow, white and blue light. However, we select white light bulb for data acquisition. An important point is that the location of the middle knuckle may vary due to the length of fingers of each individual.

2.4.2 Preprocessing Techniques

Since the low-resolution images are easily corrupted due to sensor-level noise, which lowers the overall performance. Also, the palm region signifies a fairly deformable surface result in uncertain reflections. Thus, it is important to reduce these effects at input level. In this work, three main steps are involved during the preprocessing stage.

2.4.3 ROI Extraction

As part of the pre-processing, the acquired hand image is binarized using gray level thresholding to segment the hand region. This segmented image is further processed to separate the palm print and inner knuckle print ROIs. The state-of-the-art techniques have been used to segment a palm print ROI and three knuckle ROIs (Jaswal et al. 2017b,c) from the acquired hand image. A ROI of size 176×176 for palm print and three ROIs of size 160×80 are cropped for knuckle print.

2.4.3.1 Palm Print ROI Extraction

The detail description of palm print ROI extraction algorithm is as follows:

1. First, resize the input image to 284×384 and contrast adjusts the input image using MATLAB® built-in function (imadjust).

2. Smooth the image using Gaussian low-pass filter of size 15×15 and standard deviation 4.

3. Obtain gradient image using MATLAB in-built function (imgradient) for the Gaussian filtered image.

4. Multiply the last obtained image by a factor 2 and add the resultant with filtered image. This is done to brighten the Gaussian filtered image at the boundary of the foreground.

5. Find otsu threshold for the image obtained in previous step using MATLAB built-in function (multithresh).

6. Obtain the otsu thresholded (BW) image by passing otsu threshold to the MATLAB built-in function (imquantize).

7. Erode or remove small objects and dilate the otsu thresholded image using diamond structural element of size 5 to get fingers separated from each other in case of noise in the image.

8. Fill holes in the binary image and take out all white linked regions excluding the prime one. This eliminates any other small objects in the background and removes background portions that appear white in BW image.

9. Obtain a cropped palm region over hand. Now, locate two points at the underside of middle and ring fingers in the image obtained in step 9. Find centroid and boundary of palm region using MATLAB in-built functions.

10. Find distance from the left end of image to the finger boundary. Take only three least prominent peaks that represent the bottom points of index, middle and ring finger. Take the first and third points.

11. Find out the angle between the line joining the two points and horizontal line. Next, rotate the image to make the line joining the two point horizontal.

12. Find out the largest square inside the palm region that satisfies the following conditions: The top edge of square should be below the two bottom points of finger, and size of square must be larger than the distance between the two bottom points of finger.

13. All four corner points of finger are in the foreground of the rotated black and white image. Thus, eliminate 35 numbers of pixels from all four boundaries of the ROI image to remove dark regions. Finally, resize the ROI obtained in previous step to 176×176. The stepwise output of ROI extraction method is shown in Figure 2.7.

2.4.3.2 Inner Knuckle Print ROI Extraction

A ROI of fixed size (180×90) is cropped from the original hand image for reliable recognition. We exclude the thumb and little finger since the geometrical configuration of thumb and little finger is different and also has very little line information.

To extract the ROI, firstly, segment the middle, index and ring finger region from input hand image. The detail description of IKP ROI extraction algorithm is as follows:

1. Smooth the input gray scale image $I(x, y)$ using low-pass filtering. Apply ostu thresholding on convolved result to get a binarized image.

2. Use Sobel edge detection to obtain the correct hand region. Obtain a rectangular box surrounding the hand region by dilating the binarized image.

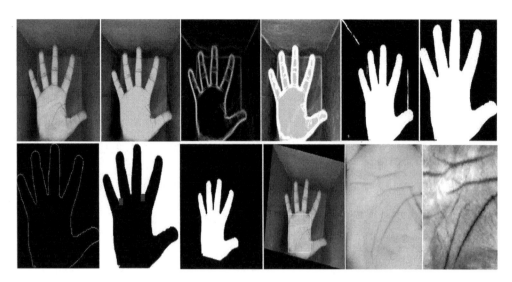

FIGURE 2.7
Palm print ROI extraction.

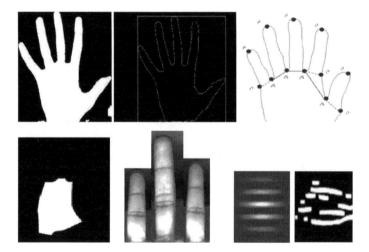

FIGURE 2.8
Inner knuckle print ROI extraction.

3. Locate two valley points (a, b) and (p, q) for middle and index fingers, respectively. Set the mid-points m for (a, b) and n for (p, q), respectively.

4. Estimate a rough location of ROI near PIP joint such that horizontal lines exist between 1/2–3/4 length of middle or index fingers. Detect the horizontal lines using modified Gabor filter (with curvature parameter) which achieves higher filter response for pixels lie over horizontal lines.

Hence, ROI is cropped based on location of horizontal lines, and resize the image empirically to 180 × 60. Finally, we get three types of ROI samples: Index (I), Middle (M) and Ring (R). The stepwise output of ROI extraction is shown in Figure 2.8.

2.4.4 Image Enhancement and Transformation

In this study, 2D discrete Grunwald–Letnikov (G–L) fractional derivative is used to enhance the overall visual effect of palm print and IKP ROIs. For $\alpha > 0$, αth order GL-based FD with respect to x for the duration [a, x] is defined as:

$$D_x^\infty f(x,y) = \lim_{h \to 0} h^{-v} \sum_{m=0}^{n-1} \frac{1}{\tau(-v)} \frac{\pi(m-v)}{\tau(m+1)} f(x - mh) \qquad (2.1)$$

where α is a real number. Similarly, αth order GL-based FD w.r.t. y can also be realized. It is a fractional derivative operator when $\alpha = 0$. It is observed that at negative values it behaves as a low-pass filter while as the order increases the transition band gets constricted and high-pass characteristics become prominent. Table 2.2 represents fractional order differential masks of a 2D signal f(x,y) in x and y direction such as:

Suppose, an ROI image f(x, y) is convolved with a (z × t) size moving window function which applies the mask to every pixel in ROI, then the response to signal $I^\alpha(x, y)$ is given by:

$$I^\infty(x,y) = \sum_{z=-a}^{a} \sum_{t=-b}^{b} w(z,t) f(x+z, y+t) \qquad (2.2)$$

where f(x, y) represents spatial coordinates of a pixel and w(z, t) is the value of the fractional differential mask. Thus, the image gets overall improvement through edges, contours and retains texture information. The enhanced ROI images are shown in Figure 2.9.

TABLE 2.2

Fractional Order Differential Mask in x and y Directions

$\alpha 2 - \alpha/12$	$\alpha 2 - \alpha/3$	$\alpha 2 - \alpha/12$	$\alpha 2 - \alpha/12$	$-\alpha/5$	$-\alpha/6$
$-\alpha/5$	1	$-\alpha/5$	$\alpha 2 - \alpha/3$	1	$-2\alpha/3$
$-\alpha/6$	$-2\alpha/3$	$-\alpha/6$	$\alpha 2 - \alpha/12$	$-\alpha/5$	$-\alpha/6$

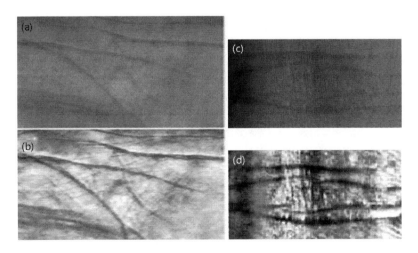

FIGURE 2.9
Image enhancement using fractional mask (a) palmprint original image (b) palmprint enhanced Image (c) inner knuckle original image (d) inner knuckle enhanced image.

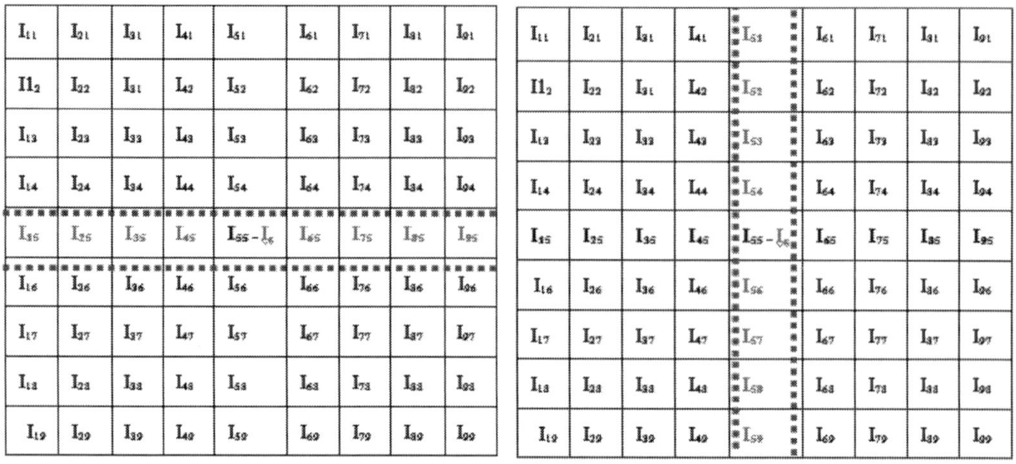

I_{11}	I_{21}	I_{31}	I_{41}	I_{51}	I_{61}	I_{71}	I_{81}	I_{91}
Il_2	I_{22}	I_{31}	I_{42}	I_{52}	I_{62}	I_{72}	I_{82}	I_{92}
I_{13}	I_{23}	I_{33}	I_{43}	I_{53}	I_{63}	I_{73}	I_{83}	I_{93}
I_{14}	I_{24}	I_{34}	I_{44}	I_{54}	I_{64}	I_{74}	I_{84}	I_{94}
I_{35}	I_{25}	I_{35}	I_{45}	$I_{55}-I_{65}$	I_{65}	I_{75}	I_{85}	I_{95}
I_{16}	I_{26}	I_{36}	I_{46}	I_{56}	I_{66}	I_{76}	I_{86}	I_{96}
I_{17}	I_{27}	I_{37}	I_{47}	I_{57}	I_{67}	I_{77}	I_{87}	I_{97}
I_{18}	I_{28}	I_{38}	I_{48}	I_{58}	I_{63}	I_{73}	I_{83}	I_{93}
I_{19}	I_{29}	I_{39}	I_{49}	I_{59}	I_{69}	I_{79}	I_{89}	I_{99}

I_{11}	I_{21}	I_{31}	I_{41}	I_{53}	I_{61}	I_{71}	I_{81}	I_{91}
Il_2	I_{22}	I_{31}	I_{42}	I_{52}	I_{62}	I_{72}	I_{82}	I_{92}
I_{13}	I_{23}	I_{33}	I_{43}	I_{53}	I_{63}	I_{73}	I_{83}	I_{93}
I_{14}	I_{24}	I_{34}	I_{44}	I_{54}	I_{64}	I_{74}	I_{84}	I_{94}
I_{15}	I_{25}	I_{35}	I_{45}	$I_{55}-I_{65}$	I_{65}	I_{75}	I_{85}	I_{95}
I_{16}	I_{26}	I_{36}	I_{46}	I_{56}	I_{66}	I_{76}	I_{86}	I_{96}
I_{17}	I_{27}	I_{37}	I_{47}	I_{67}	I_{67}	I_{77}	I_{87}	I_{97}
I_{13}	I_{23}	I_{33}	I_{43}	I_{53}	I_{63}	I_{73}	I_{83}	I_{93}
I_{12}	I_{29}	I_{39}	I_{49}	I_{59}	I_{69}	I_{79}	I_{39}	I_{99}

FIGURE 2.10
Neighborhood pattern considered for image transformation.

In the example images, the size of the G-L mask is (3×3), but the non-integer order of mask is different for each image. These enhanced images are further processed into illumination invariant representation in the next step.

The objective to implement image encoding scheme is not only to get illumination invariance but as to achieve discriminative edge features. The image encoding schemes like local binary pattern (LBP) has been mostly studied in computer vision problems due to its highly discriminative strength and robustness against illumination (Wang et al. 2006, Shen et al. 2009). LBP mainly recapitulated the neighborhood spatial structure of an image by thresholding it with binary weight and set up the decimal number as a texture donation (Figure 2.10).

Moreover, it consumes less computational cost. As the palm and knuckle regions consist of lines and wrinkles, thus we have employed another variant of LBP, namely LLBP (Local Line Binary Pattern) (Petpon and Srisuk 2009) to efficiently encode the structure of palm and knuckle.

One can see the neighborhood pattern of LLBP along horizontal and vertical direction in both LBP and LLBP are alike but the main differences are as follows:

- The neighborhood shape of LLBP is a straight line, but the neighborhood in LBP is a square shape.
- The binary weight is allocated starting from the left and right neighboring pixel of middle pixel to the end of left and right side.

However, LLBP gets the binary code along with vertical and horizontal direction separately and then compute its magnitude, which mainly describes the change in image gray scale level. The mathematical expression for LLBP (magnitude) is given by:

$$LLBP_m = \sqrt{LLBP_H{}^2 + LLBP_V{}^2} \qquad (2.3)$$

The palm print and IKP images after LLBP transformation are shown in Figure 2.11.

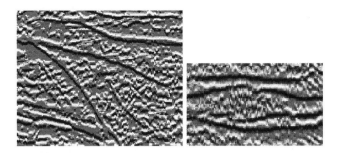

FIGURE 2.11
Example of palm print and IKP processed by LLBP.

2.4.5 Feature Extraction and Matching

Two feature extraction techniques viz UR-SIFT (Sedaghat et al. 2011) and BLPOC (Zhang et al. 2011) are applied to design the system so that they complement to each other for efficient recognition results. In particular, a solitary concatenation algorithm using local (UR-SIFT) and global (BLPOC) features which can overcome all fundamental challenges has been presented.

UR-SIFT: To perform image matching by using the set of local interest points is an important aspect of local texture descriptor-based approaches. Local image features like blobs, points and micro-image regions are unique and facilitate such schemes to better handle varying illumination, translation and scale (Jaswal et al. 2017a). The SIFT algorithm and its variants have been effectively applied in different computer vision and image processing applications. However, standard SIFT causes problems with the multi-source remote sensing images, particularly unable to extract required number of feature points because such images composed information over a large range of variation on frequencies. In this paper, a strongest variant of SIFT known as uniform robust SIFT (UR-SIFT) is employed for image matching (Sedaghat et al. 2011). The UR-SIFT is based on an idea of selecting high-quality SIFT features with a uniform distribution in both the scale and image spaces. It has been observed that UR-SIFT extracts more key points than standard SIFT and perform matching efficiently, as shown in Figure 2.12.

BLPOC: BLPOC is an improvement of POC-based matching that concentrates on the low-frequency components filtering out noisy higher frequencies (Ito et al. 2008). The POC value between two images can be used as similarity score and is obtained using the cross-spread spectrum of the Fourier transform of both images. In brief, BLPOC confines the limits of the inverse DFT of the phase correlation function in the frequency domain. For genuine matching, the BLPOC provides a quick correlation spike than the original POC. Also it is more robust beside the noise and is better at distinguishing between genuine from impostor matching.

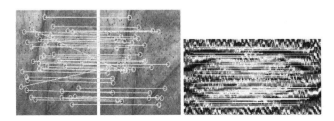

FIGURE 2.12
UR-SIFT matching.

2.4.6 Two-Stage Fusion

In this study, a two-stage fusion strategy for integrating the palm print and IKP modalities is deployed. At first stage, score level fusion (Jain et al. 2004) is implemented to obtain individual score of palm print and IKP. In palm print score level fusion, the matching scores of palm print obtained by two methods, i.e., BLPOC and UR-SIFT are fused. Similarly in IKP score level fusion, the matching scores of BLPOC and UR-SIFT are combined. However, any two scores are first normalized using Z score normalization before fusion. In this way, we have final scores of palm print and IKP. In the second stage, a decision level fusion rule is used to consolidate the scores outputs by the palm print and knuckles that reveal the actual identity of individual.

2.5 Experimental Analysis and Evaluation Parameters

In this study, a total of two databases, namely IIT Delhi Palmprint database (Kumar 2008) (public database) and NITH (In-House) database are used. The sample images are shown in Figure 2.13. These two databases are discussed in detail in the following section.

IIT Delhi Palmprint Database: The image database consists of hand images collected from more than 235 persons. Among them, 150 individuals have at least six image samples per hand, which have been utilized in the tests. The images of the dataset are in BMP format which are acquired using a digital CMOS device.

NITH Hand Database: The image database consists of 800 images, collected 8 images per hand (left and right) of 100 subjects. The database is collected in single session under varying illumination and changing background using ordinary camera. Each image shows major challenge of hand pose variation up to 5°.

Testing Protocol: The algorithms are performed by using MATLAB (R2016a), on i5 processor (2.33 GHz) with 4 GB Ram, Windows 10 operating system (64 bit). The results are determined under equal number of training/test samples, because most of the state-of-the-art systems had adopted this. For verification, a normalized value of the threshold is assumed such that if the threshold is greater than the distance score, a person is rejected otherwise authorized. While in identification, a person is recognized by comparing with entire enrolled users on the basis of extracting features. The performance is evaluated in terms of Equal Error Rate (EER), Correct Recognition Rate (CRR), ROC and CMC performance curves. The description about these parameters is given below:

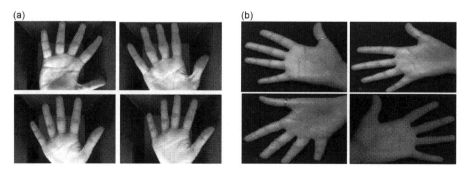

FIGURE 2.13
Sample images: (a) IITD palm print; (b) NITH palm print.

Equal Error Rate (EER): A predetermined threshold at which FAR linearly varies with respect to FRR in ROC plot, then the common value is called as Equal Error Rate. It means, EER is a point of intersection of FAR and FRR curves. The lower value of EER provides better system performance.

Correct Recognition Rate (CRR): The probability of correctly identifying a person from total number of individuals available in dataset. Besides, comparing the training and test images, it also arranges them on the basis of matching scores. This is also termed as rank-1 recognition rate.

Decidability Index (DI): This measures separability between imposter and genuine matching scores.

Computation Time: This measures the computation time to process the each and individual tasks in a biometric system (Table 2.3).

Experiment-1: The objective of the first test is to check the baseline performance of the UR-SIFT and BLPOC algorithms separately over palm print and IKP datasets separately. For palm print, right hand IIT Delhi images are used while three types of IKP ROIs (I, M, R) are extracted using in-house hand database (Figures 2.14 and 2.15).

The following conclusions are made in this test: (i) In case of palm print recognition methods, BLPOC performs overall better than UR-SIFT. BLPOC achieves CRR–99.85%, EER–0.95% and DI–2.60. (ii) The CMC, score distribution and ROC graphs for palm print are shown in Figures 2.14, 2.16 and 2.18, respectively. Score level fusion (palm print) of BLPOC and UR-SIFT outperforms any other methods and achieves CRR–100%, EER–0.42% and DI–2.90. The CMC, score distribution and ROC graphs for inner knuckle print are shown in Figures 2.14, 2.16 and 2.18, respectively. (iii) Among palm print and IKP, the individual performance of palm print recognition is superior. In case of knuckle performance,

TABLE 2.3

Testing Specifications of Datasets (Full Images)

Type of Database	Subjects	Poses (R)	Total	Testing	Training	Genuine Matchings	Impostor Matchings
IIT Delhi (palm print)	235	6	1,410	705	705	2,115	494,910
NITH (IKP)	100	8	800	400	400	1,600	158,400

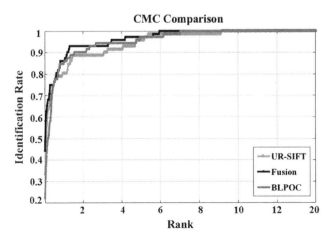

FIGURE 2.14
CMC analysis for palm print.

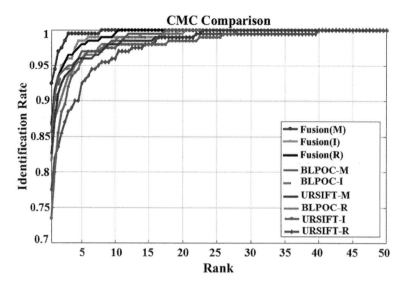

FIGURE 2.15
CMC analysis for IKP.

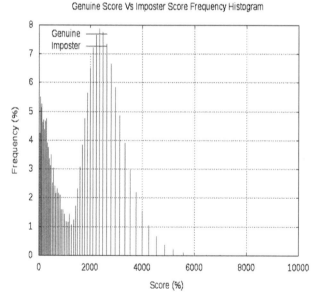

FIGURE 2.16
Genuine vs imposter (palm print).

middle IKP performs better. (iv) Score level fusion (middle IKP) of BLPOC and UR-SIFT achieves CRR–99.24%, EER–0.86% and DI–2.66 (Figures 2.17–2.19).

Experiment-2: In the first experiment, the individual scores of palm print and three IKPs are obtained. Now we have again fused these scores for the development of multimodal biometric system. For this, decision level fusion using simple AND rule is implemented to obtain the final outcome. This will clarify that the given hand belongs to genuine user or imposter. The detail description of overall results is given in Table 2.4. Among all the

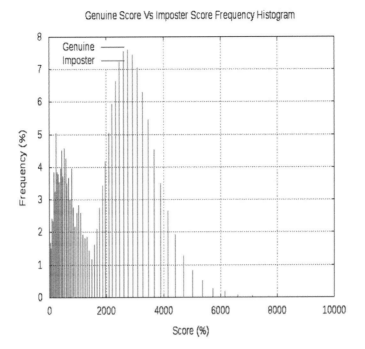

FIGURE 2.17
Genuine vs imposter (IKP).

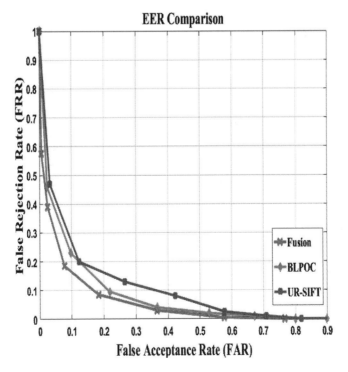

FIGURE 2.18
EER comparison (palm print).

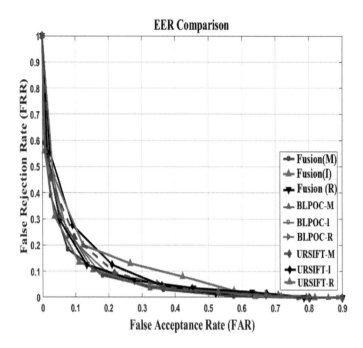

FIGURE 2.19
EER comparison (IKP).

TABLE 2.4

Comparative Analysis with State-of-the-Art Methods

Methods	CRR (%)	EER (%)	DI	Speed (ms)
PalmCode (palm print) (Zhang et al. 2003)	-	GAR-98.4	-	1,100
MSIFT (palm print) (Morales et al. 2011)	-	0.31	-	6,187
Deep-matching (FKP) (Jaswal et al. 2017a)	99.39	0.92	3.30	756.82
MoriCode&MtexCode (FKP) (Gao et al. 2014)	-	1.0481	-	-
Multimodal (palm print, FKP) (Jaswal et al. 2017c)	100	0.68	-	-
Proposed-IITD right palm (BLPOC)	99.85	0.95	2.60	1,356
Proposed-IITD right palm (UR-SIFT)	99.56	1.16	2.48	1,582
Proposed-NITH middle (M) knuckle (BLPOC)	98.35	0.98	1.94	1,108
Proposed-NITH index (I) knuckle (BLPOC)	98.22	1.44	2.02	1,108
Proposed-NITH ring (R) knuckle (BLPOC)	97.18	1.72	2.87	1,108
Proposed-NITH middle (M) knuckle (UR-SIFT)	98.08	1.32	2.44	1,405
Proposed-NITH index (I) knuckle (UR-SIFT)	97.33	1.25	2.58	1,405
Proposed-NITH ring (R) knuckle (UR-SIFT)	96.14	1.38	2.36	1,405
Proposed-palm print fusion (P)	100	0.42	2.90	-
Proposed-middle knuckle fusion (K1)	99.44	0.86	2.66	-
Proposed-index knuckle fusion (K2)	99.12	1.04	2.82	-
Proposed-ring knuckle fusion (K3)	99.02	1.15	2.70	-
Proposed-palm print (P); IKP(K1)	100	0.49	3.64	-
Proposed-palm print (P); IKP(K2)	100	0.92	3.24	-
Proposed-palm print (P); IKP(K3)	100	0.99	3.06	-

multimodal combination, decision level fusion of palm print and middle knuckle print is more superior.

Experiment-3: Apart from this, the performance of proposed methods has been compared with reported state-of-the-art multimodal systems. We have considered two image datasets and chosen equal number of test and training images per person. Table 2.4 presents comparison of four state-of-the-art methods with our proposed work on the basis of CRR, EER, DI and speed parameters. The results clarify that our method shows comparable results with other existing methods. This shows the strength of the proposed architecture in which improvement is done at each module.

2.6 Conclusions

This work explains the working principle of multimodal biometric authentication system that makes use of palm print and inner knuckle features. We have proposed an idea to acquire both of the traits using single sensor imaging device. Initially, palm and knuckle ROIs are extracted and then enhanced using Fractional Filter methods. Further, the images textures of palm and finger knuckle are transformed using LLBP-based encoding schemes that mainly works over pixel intensities and obtain more robust texture representations. Then the local and global features are extracted from each modality using UR-SIFT and BLPOC methods, respectively. Finally, a two-stage fusion strategy is implemented to obtain the final outcome which show high-performance results (CRR–100%, EER–0.49%, DI–3.64). In addition to this, the proposed multimodal algorithm is also compared with some well-known existing systems which reveal the significance of proposed multi-feature fusion strategy. Our unimodal system performance is also better than the reported state-of-the-art algorithms.

References

G. S. Badrinath, K. Tiwari, and P. Gupta. 2012. An efficient palmprint based recognition system using 1d-DCT features. In *International Conference on Intelligent Computing* (Vol. 7389, pp. 594–601). Berlin Heidelberg: Springer.

S. Bhilare, V. Kanhangad, and N. Chaudhari. 2017. A study on vulnerability and presentation attack detection in palm print verification system. *Pattern Analysis and Applications*, Vol. 21(3), 1–14.

W.S. Chen, Y. S. Chiang, and Y. H. Chiu. 2007. Biometric verification by fusing hand geometry and palm print. In *Proceedings of Third International Conference on Intelligent Information Hiding and Multimedia Signal Processing* (Vol. 2, pp. 403–406). Kaohsiung: IEEE.

K. Cheng, and A. Kumar. 2012. Contactless finger knuckle identification using smart phones. In *Proceedings of the International Conference of the Biometrics Special Interest Group* (pp. 1–6). Darmstadt: IEEE.

T. Connie, A. Teoh, M. Goh, D. Ngo. 2004. PalmHashing: A novel approach for dual-factor authentication. *Pattern Analysis and Applications*, Vol. 7(3), 255–268.

J. Cui. 2014. 2D and 3D palm print fusion and recognition using PCA plus TPTSR method. *Neural Computing and Applications*, Vol. 24, 497–502.

L. Fei, Y. Xu, W. Tang, and D. Zhang. 2016. Double-orientation code and nonlinear matching scheme for palm print recognition. *Pattern Recognition*, Vol. 49, 89–101.

G. Gao, J. Yang, J. Qian, and L. Zhang. 2014. Integration of multiple orientation and texture information for finger-knuckle-print verification. *Neurocomputing*, Vol. 135, 180–191.

Y. Han, T. Tan, Z. Sun, and Y. Hao. 2007. Embedded palmprint recognition system on mobile devices. In *Proceedings of International Conference on Biometrics* (pp. 1184–1193). Berlin Heidelberg: Springer.

D.-S. Huang, W. Jia, and D. Zhang. 2008. Palmprint verification based on principal lines. *Pattern Recognition*, Vol. 41(4), 1316–1328.

K. Ito, T. Aoki, H. Nakajima, K. Kobayashi, T. Higuchi. 2008. A palmprint recognition algorithm using phase-only correlation. *IEICE Transaction on Fundamentals*, Vol. 91(4), 1023–1030.

A. K. Jain, A. Ross and S. Prebhakar. 2004. An introduction to biometric recognition. *IEEE Transaction on Circuits and Systems for Video Technology*, Vol. 14(1), 4–20.

G. Jaswal, A. Kaul, and R. Nath. 2016. Knuckle print biometrics and fusion schemes—Overview, challenges, and solutions. *ACM Computing Surveys (CSUR)*, Vol. 49(2), 1–34.

G. Jaswal, A. Nigam, and R. Nath. 2017a. DeepKnuckle: Revealing the human identity. *Multimedia Tools and Applications*, Vol. 76(18), 18955–18984.

G. Jaswal, A. Kaul, and R. Nath. 2017b. IKP based human authentication using multi-feature fusion. In *Proceedings of 4th International Conference on Signal Processing and Integrated Networks* (pp. 154–159). Delhi: IEEE.

G. Jaswal, A. Kaul, and R. Nath. 2017c. Palmprint and finger knuckle based person authentication with random forest via kernel-2DPCA. In *International Conference on Pattern Recognition and Machine Intelligence* (pp. 233–240). Berlin Heidelberg: Springer.

G. Jaswal, A. Kaul, R. Nath and A. Nigam 2018. DeepPalm-A Unified Framework for Personal Human Authentication. In *International Conference on Signal Processing and Communications (SPCOM)* (pp. 322–326). Bangalore: IEEE International Conference on Signal Processing and Communications 2018.

W. Jia, D. Huang, and D. Zhang. 2008. Palmprint verification based on robust line orientation code. *Pattern Recognition*, Vol. 41, 1504–1513.

D. Joshi, Y. V. Rao, S. Kar, V. Kumar, and R. Kumar. 1998. Computer-vision-based approach to personal identification using finger crease pattern. *Pattern Recognition* Vol. 31(1), 15–22.

W. Q. Jungbluth. 1989. Knuckle print identification. *Journal of Forensic Identification*, Vol. 39, 375–380.

A. W. Kong, and D. Zhang. 2004. Competitive coding scheme for palmprint verification. In *Proceedings of 17th International Conference on Pattern Recognition* (pp. 520–523). Cambridge, UK: IEEE.

T. Kong, G. Yang, and L. Yang. 2014. A hierarchical classification method for finger knuckle print recognition. *EURASIP Journal on Advances in Signal Processing*, Vol. 14(1), 1–44.

A. Kumar. 2008. Incorporating cohort information for reliable palmprint authentication. In *Proceedings of Sixth Indian Conference on Computer Vision, Graphics & Image Processing* (pp. 583–590). Bhubaneswar: IEEE.

A. Kumar. 2014. Importance of being unique from finger dorsal patterns: Exploring minor finger knuckle patterns in verifying human identities. *IEEE Transactions on Information Forensics and Security*, Vol. 9(8), 1288–1298.

A. Kumar, and C. Ravikant. 2009. Personal authentication using finger knuckle surface. *IEEE Transactions on Information Forensics and Security*, Vol. 4, 98–109.

A. Kumar, and B. Wang. 2015. Recovering and matching minutiae patterns from finger knuckle images. *Pattern Recognition Letters*, Vol. 68, 361–367.

A. Kumar, and Z. Xu. 2016. Personal identification using minor knuckle patterns from palm dorsal surface. *IEEE Transactions on Information Forensics and Security*, Vol. 11(10), 2338–2348.

A. Kumar, and D. Zhang. 2006. Personal recognition using hand shape and texture. *IEEE Transactions on Image Processing*, Vol. 15, 2454–2461.

A. Kumar, and Y. Zhou. 2009. Human identification using knuckle codes. In *Proceedings of 3rd International Conference on Biometrics, Theory and Applications* (pp. 147–152). Washington DC: IEEE.

L. Leng, and J. Zhang. 2011. Dual-key-binding cancelable palmprint cryptosystem for palmprint protection and information security. *Journal of Network and Computer Applications*, Vol. 34(6), 1979–1989.

G. Li, and J. Kim. 2017. Palmprint recognition with local micro-structure tetra pattern. *Pattern Recognition*, Vol. 61, 29–46.

Z. Li, K. Wang, and W. Zuo. 2012. Finger-knuckle-print recognition using local orientation feature based on steerable filter. In *Emerging Intelligent Computing Technology and Applications* (pp. 224–230). Berlin Heidelberg: Springer.

M. Liu, Y. Tian, and Y. Ma. 2013. Inner-knuckle-print recognition based on improved LBP. In *Proceedings of the International Conference on Information Technology and Software Engineering* (Vol. 212, pp. 623–630). Berlin Heidelberg: Springer.

G. K. Michael, T. O. Connie, and A. B. J. Teoh. 2008. Touch-less palm print biometrics: Novel design and implementation. *Image and Vision Computing*, Vol. 26(12), 1551–1560.

A. Morales, M. A. Ferrer, and A. Kumar. 2011. Towards contactless palmprint authentication. *IET Computer Vision*, Vol. 5(6), 407–416.

A. Nigam, K. Tiwari, and P. Gupta. 2016. Multiple texture information fusion for finger-knuckle-print authentication system. *Neurocomputing*, Vol. 188, 190–205.

X. Pan, Q. Ruan, and Y. Wang. 2008. Palmprint recognition using contourlets based local fractal dimensions. In *Proceedings of 9th International Conference on Signal Processing* (pp. 2108–2111). Beijing: IEEE.

A. Petpon, and S. Srisuk, 2009 September. Face recognition with local line binary pattern. In *IEEE Fifth International Conference on Image and Graphics* (pp. 533–539). Shanxi: IEEE.

S. Saedi, and N. M. Charkari. 2014. Palmprint authentication based on discrete orthonormal S-Transform. *Applied Soft Computing*, Vol. 21, 341–351.

A. Sedaghat, M. Mokhtarzade, and H. Ebadi. 2011. Uniform robust scale-invariant feature matching for optical remote sensing images. *IEEE Transactions on Geoscience and Remote Sensing*, Vol. 49(11), 4516–4527.

Z. S. Shariatmadar, and K. Faez. 2011. Novel approach for finger knuckle print recognition based on Gabor feature fusion. In *Proceedings of 4th International Congress on Image and Signal Processing* (Vol. 3, pp. 1480–1484). Shanghai: IEEE.

L. Shen, Z. Ji, L. Zhang, and Z. Guo. 2009. Applying LBP operator to Gabor response for palmprint identification. In *Proceedings of International Conference on Information Engineering and Computer Science* (pp. 1–3). Wuhan: IEEE.

M. Tabejamaat, and A. Mousavi. 2017. Concavity-orientation coding for palmprint recognition. *Multimedia Tools and Applications*, Vol. 76(7), 9387–9403.

The IIT Delhi. Touch less palm print database, version 1.0: http://web.iitd.ac.in/~ajaykr/Database_Palm.htm.

X. Wang, H. Gong, H. Zhang, B. Li, and Z. Zhuang. 2006. Palmprint identification using boosting local binary pattern. In *Proceedings of 18th International Conference on Pattern Recognition* (Vol. 3, pp. 503–506). Hong Kong: IEEE.

X. Wu, K. Wang, and D. Zhang 2005a. Palmprint authentication based on orientation code matching. In *Proceedings of International Conference on Audio and Video Based Biometric Person Authentication* (pp. 555–562). Berlin Heidelberg: Springer.

X. Q. Wu, K. Q. Wang, and D. Zhang. 2005b. Wavelet energy feature extraction and matching for palmprint recognition. *Journal of Computer Science & Technology*, Vol. 20, 411–418.

X. Xu, Q. Jin, L. Zhou, J. Qin, T. T. Wong, and G. Han. 2015. Illumination-invariant and deformation-tolerant inner knuckle print recognition using portable devices. *Sensors*, Vol. 15(2), 4326–4352.

D. Zhang, W. K. Kong, J. You, and M. Wong. 2003. On-line palmprint identification. *IEEE Transactions on Pattern Analysis and Machine Intelligence*, Vol. 25(9), 1041–1050.

L. Zhang, S. Ying, L. Hongyu, and L. Jianwei. 2015. 3D palmprint identification using block-wise features and collaborative representation. *IEEE Transactions on Pattern Analysis and Machine Intelligence*, Vol. 37(8), 1730–1736.

L. Zhang, L. Zhang, and D. Zhang. 2009. Finger-knuckle-print: A new biometric identifier. In *Proceedings of 16th International Conference on Image Processing* (pp. 1981–1984). Cairo: IEEE.

L. Zhang, L. Zhang, and D. Zhang. 2010a. Monogeniccode: A novel fast feature coding algorithm with applications to finger-knuckle-print recognition. In *International Workshop on Emerging Techniques and Challenges for Hand-Based Biometrics* (pp. 1–4). Istanbul: IEEE.

L. Zhang, L. Zhang, D. Zhang, and Z. Guo. 2012. Phase congruency induced local features for fingerknuckle-print recognition. *Pattern Recognition*, Vol. 45, 2522–2531.

L. Zhang, L. Zhang, D. Zhang, and H. Zhu. 2010b. Online finger-knuckle-print verification for personal authentication. *Pattern Recognition*, Vol. 43(7), 2560–2571.

L. Zhang, L. Zhang, D. Zhang, and H. Zhu. 2011. Ensemble of local and global information for finger-knuckle print recognition. *Pattern Recognition*, 1990–1998.

3

Voiceprint-Based Biometric Template Identifications

Akella Amarendra Babu

St. Martin's Engineering College

Sridevi Tumula and Yellasiri Ramadevi

Chaitanya Bharati Institute of Technology

CONTENTS

3.1 Introduction

Research on Voiceprint Biometric Speaker Identification (VBSI) technology has attracted the attention of the speech researchers during the past six decades or so, due to its increasing utility in our day-to-day life (Huang et al., 2014). During 1980s, VBSI problem was formulated as Bayes statistical inference (Stern and Morgan, 2012), which triggered paradigm shift in the development of VBSI technology. Today, most of the state-of-the-art VBSI systems are based on statistical methods. Doubling the computing power for a given cost in every 12–18 months (Mattys and Scharenborg, 2014) combined with shrinking costs of memory, enabled the VBSI researchers to perform the experiments practically in real time. There is a surge in VBSI performance during the past three decades due to the availability of abundant speech corpora (Baker et al., 2009).

Humans have the inherent ability to learn new pronunciations and new words through conversations with others whereas VBSI systems are developed using labeled speech corpus and tested with unlabeled "Everyday Speech". It is expensive and time consuming to label "Everyday Speech". Therefore, it is impossible to label "Everyday Speech" corpus for real time applications. The above predicament steered the researchers to ponder over the possibility of incorporating the data-driven unsupervised dynamic adaptation capability in VBSI systems mimicking the human adaptation process (Zhou, 2013).

The VBSI systems lack the unsupervised learning and adaptation capability. The humans are data-driven whereas the VBSI systems are trained using supervised methods. The humans learn and adapt using everyday speech whereas the VBSI systems need transcribed speech utterances. Therefore, the supervised adaptation is not suitable for real time systems implementation.

Pronunciation differs from person to person. Speakers can be identified by their pronunciation. Pronunciation-based speaker identification and authentication systems may be developed using phonetic distance measurements.

The chapter is organized into seven sections with related work that is covered in next section. Section 3.3 gives the computation and compilation of Phoneme Substitution Cost Matrix (PSCM). Section 3.4 describes the Dynamic Phone Warping (DPW) algorithm and its illustration. Section 3.5 deals with procedure to estimate critical distance parameter. Section 3.6 describes the design of pronunciation-based voiceprint biometric identification model. Section 3.7 gives the conclusions of this chapter.

3.2 Related Work

3.2.1 Unsupervised Learning

The difference between the supervised learning and unsupervised learning depends on class availability of data samples which are used for parameter estimation. Let's denote the pair (x, ω) as a sample data, where x is the observed data and ω is the class from which the data x comes. In supervised learning, ω, the class information of the sample data x, is given. In contrast, data is incomplete in unsupervised learning where the class information ω is missing. Vector quantization is one of the unsupervised learning techniques (Ma et al., 2013).

3.2.2 Pronunciation Modeling

The summary of pronunciation modeling is as follows.

- Knowledge-based methods use pronunciation dictionaries which are derived from the phonological linguistic rules like G2P and P2P convertors. The performance of these methods degrades while dealing with the conversational speech which is characterized by high pronunciation variability (Hahn et al., 2012).

- Data-driven methods combine G2P convertors with the original speech utterance and produce pronunciation surface forms. The static data-driven methods implement the pronunciation variability with such representations like Finite State Transducer (FST) and use speech utterances to train parameters. These methods are not suitable for real world conversational speech applications, as they need labeled speech corpus which is expensive and time consuming.

- The current research is focusing on dynamic data-driven methods which use speech as input to build pronunciation models (Lu et al., 2013). These methods are called unsupervised data-driven adaptation methods.

3.2.3 Phonetic Distance Measurements

Phonetic distance between a pair of words is used in speech research for different purposes by various researchers.

- Ben Hixon et al. (2011) used a modified Needleman–Wunsch sequence alignment algorithm to calculate the phonetic distance between a pair of phoneme sequences. The performance of three G2P methods is compared using above phonetic distance measurements.

- Martijn Wieling et al. (2011) used Levenshtein distance to measure the phonetic distance between a pair of dialect variations. Bark scale is used to measure acoustic distances between vowels to get better human perception.

- Michael Pucher et al. (2007) investigated correlation between word confusion matrix and phonetic distance measurements. These distance measurements are used for developing grammars and dialogues by dialogue system developers.

- Pietquin and Dutoit (2006) used modified ratio model to measure the substitution cost of a pair phonemes and used the DP algorithm to measure the edit distance between two phoneme sequences. The acoustic distance is used to dynamically estimate the ASR performance by correlating the acoustic perplexity versus Word Error Rate (WER).
- Sriram et al. (2004) used phonetic distance measures for matching query to database records in cross lingual search applications.

3.3 Phoneme Substitution Cost Matrix

The study to measure the sound distance between two words starts with computation of distance between a pair of speech sounds. Linguistically distinct speech sounds are called phonemes.

The distance between the two phonemes gives the cost of substituting one phoneme by the other. The substitution costs of all pairs of phonemes compiled in a matrix form are termed as the Phoneme Substitution Cost Matrix (PSCM).

3.3.1 Classification of Phonemes Based on Articulatory Features

Classification of Standard English phonemes into sound classes that consists of 11 vowel sounds, 4 diphthongs, 4 semi-vowels, 4 nasal sounds, 6 stops, 8 fricatives, 2 affricates and 1 whisper (Rabiner et al., 2010).

3.3.2 Phonetic Distance

The phonetic distance between a pair of phonemes is measured by using the articulatory feature sets of two phonemes. JC (Pa, Pb) the Jaccard coefficient of similarity between the two phonemes Pa and Pb is given by

$$JC\,(Pa, Pb) = (Fa \cap Fb)/(Fa \cup Fb) \tag{3.1}$$

where phoneme Pa is generated using a set of m features, Fa and phoneme Pb is generated using a set of n features Fb.

One minus the Jaccard coefficient is called the Jaccard distance JD.

$$JD\,(Pa, Pb) = 1 - JC\,(Pa, Pb) \tag{3.2}$$

The cost of insertion or deletion is computed as half of the average substitution cost. It is termed as one Indel.

3.3.3 Phonetic Distance Computations

Results of computation of distances between various pairs of phonemes are summarized in this section. The phonetic distances from the front vowel /i/ to other phonemes are summarized in Table 3.1.

TABLE 3.1

Phonetic Distance between Front-Vowel /i/ and Other Phonemes

S. No	Phoneme	Jaccard (Phonetic) Distance
1	Front vowel	0.182
2	Back vowel	0.412
3	Diphthong	0.714
4	Semi-vowel	0.75
5	Nasal	0.75
6	Fricative	0.765

TABLE 3.2

Phonetic Distance between Fricative /z/ and Other Phonemes

S. No.	Phoneme	Jaccard (Phonetic) Distance
1	Fricative	0.10
2	Stops	0.681
3	Nasal	0.667
4	Diphthong	0.789
5	Semi-vowel	0.809
6	Vowel	0.765

The phonetic distances from the fricative /z/ to other phonemes are summarized in Table 3.2.

Results shown in Tables 3.1 and 3.2 reveal the following.

- The phonetic distance between two vowels is less than the phonetic distance between a vowel and a consonant.
- The phonetic distance between a pair of front vowels is less than the distance between a front vowel and a back vowel.
- The phonetic distance between a fricative and semi-vowels, diphthongs, other consonants and vowels is more than the distance between a pair of fricatives.

The phonetic distance between a pair of phonemes is taken as the cost of substitution edit operation. The average cost of substitution operation computed from PSCM is 0.36 and the cost of one Indel is 0.18.

3.4 Dynamic Phone Warping (DPW)

The phonetic distance between various pairs of phonemes is computed and compiled in the form of PSCM in the previous section. In this section, the PSCM is used to compute the phonetic distance between a pair of pronunciations.

There are various algorithms in the field of software engineering which use dynamic programming for solving complex problems. One such algorithm is the

Needleman–Wunsch (NW) sequence alignment algorithm. The modified NW algorithm is used to compute the phonetic distance between a pair of phoneme sequences. It is termed as Dynamic Phone Warping.

3.4.1 DPW Algorithm

DPW algorithm uses dynamic programming for global alignment. The DPW algorithm is mentioned below.

Algorithm 3.1: DPW Algorithm

Input: Two phoneme sequences (SeqA and SeqB)
Output: Normalized phonetic distance between SeqA & SeqB
Begin
Step 1: Initialization
Declare a matrix M with m rows and n columns.
First column and the first row initialization.
  for i = 1 to m

$$M(i, 1) = i * Indel \tag{3.3}$$

  for j = 1 to n

$$M(1, j) = j * Indel \tag{3.4}$$

Step 2: Fill up the Matrix
Using

$$M(i, j) = (\min(M_{i-1, j-1} + C(\varphi_i, \varphi_j), (M_{i-1, j}) + C, (M_i, j-1) + C) \tag{3.5}$$

where $C(\varphi_i, \varphi_j)$ is the cost of replacing phoneme i with phoneme j.

$$C(\varphi_i, \varphi_j) = 0 \quad if \; \varphi_i \equiv \varphi_j \tag{3.6}$$

$$= Cost \; of \; replacing \; \varphi_i \; with \; \varphi_j \tag{3.7}$$

Distance between the two sequences = Value at the bottom right hand corner entry.

$$D(m, n) = M(m, n) \tag{3.8}$$

Step 3: Calculate Normalized Phonetic Distance Dn

$$Dn = D(m, n) / \max(m, n) \tag{3.9}$$

End

The DPW algorithm estimates the minimum distance between a pair of phoneme sequences.

TABLE 3.3

Summary of DPW Results

Test Case No.	Description	Analysis Word	Comparison Word	Normalized Phonetic Distance
1	A word with same pronunciation compared with itself	ACTIVISTS AE K T AH V AH S T S	ACTIVISTS AE K T AH V AH S T S	0
2	A pair of pronunciations of the same word	ACTIVISTS AE K T AH V AH S T S	ACTIVISTS AE K T IH V IH S T S	0.08
3.	A pair of different words	ACTIVISTS AE K T AH V AH S T S	ANNUITY AH N UW IH T IY	0.26

3.4.2 Experimentation Details, Results and Analysis

Data sets are selected from CMU pronunciation dictionary. Three test case results are summarized in Table 3.3. Test case 1 results show the DPW results between the phoneme sequences of the same word, "ACTIVISTS". There is zero phonetic distance between the two sequences. The DPW results with two different pronunciations are given in test case 2. The absolute distance is 0.72. The normalized phonetic distance is 0.08.

The phonetic distance between two different words is more than the phonetic distance between a pair of pronunciations. The result is given in test case no. 3.

An important hypothesis emerged from analysis of above test case results. The phonetic distance between a pair of pronunciation variations of a word is significantly less than the phonetic distance between a pair of words. Therefore, it is possible to distinguish pronunciation variations from different words based on phonetic distances.

3.5 Critical Distance Criteria

A classifier is designed to categorize the input utterance into either a pronunciation variation or an Out of Vocabulary (OOV) word in this section. The classifier is termed as Accent-Word Classifier (AWC). The critical distance value is empirically estimated to threshold the classification. A new algorithm to estimate the critical distance termed as Critical Distance Estimation (CDE) algorithm is presented.

3.5.1 Definitions

The parameters used in the estimation of critical distance are defined below.

- **Critical Distance (Dc):** Dc is used to distinguish between an OOV word and a pronunciation variant.
- **Insertion and Deletion Costs:** Instead of taking the same value for insertion operation (Ix) and deletion operation (Dx) costs, different values are assumed using parameter m in the following formula

$$Ix = m * Dx \quad 0.2 < m < 2.0 \tag{3.10}$$

- **Estimation of Critical Distance:** The parameter γ is used in estimation of Dc using the formula

$$Dc = Indel * \gamma \quad 0 \leq \gamma \leq 1.0 \tag{3.11}$$

- **Substitution Costs:** The parameter δ is used to vary the value of the substitution cost. Cv (φi, φj), the cost of exchanging two vowels is calculated using

$$Cv\,(\varphi i, \varphi j) = \delta * Cc\,(\varphi i, \varphi j)\; 0.6 < \delta < 1 \tag{3.12}$$

where Cc (φi, φj) is the substitution cost.

3.5.2 Critical Distance Estimation (CDE) Algorithm

The algorithm to estimate the critical distance is given below.

Algorithm 3.2: CDE Algorithm

Input: Input file1 (SeqA) and Input file2 (SeqB)
Output: Value of Dc
Begin
1. Set the value of γ with $0 \leq \gamma \leq 1.0$ using the formula (5.2).
2. Set the value of δ between 1.0 and 0.6 using the formula (5.3).
3. Select SeqA from file1 and SeqB from file2.
4. Phonetic distance is calculated and normalized, Dn.
5. Repeat step 3 and 4 for all SeqA with all SeqB.
6. Count the number of errors in the resolution as per the following criteria:
   a. Set True-Positives = 0; False-Positives = 0;
   b. If SeqA and SeqB are the same word pronunciation variations and Dn <= Dc, then increment True-Positives.
   c. If SeqA and SeqB are different word pronunciations and Dn <= Dc, then increment False-Positives.
7. Compute precision = True-Positives/(True-Positives + False-Positives).
8. Select the value of Dn as Dc, corresponding to maximum precision.
9. Repeat steps 1–8 varying the values of γ and δ.
Select the values of γ and δ at which the Dc value gives the highest precision.
End

3.5.3 Experimentation

The details of experimentation for estimation of Dc are given below. Data sets are drawn from two data bases; CMU pronunciation dictionary and TIMIT speech data corpus. The pronunciation phoneme sequences of all words with multiple pronunciations are divided into nine groups of 1,000 words each. The base form pronunciations of the words are listed in input file1 and pronunciation variants of these words are listed in input file2.

Each pronunciation base form listed in input_file1 is compared with its pronunciation variant listed in input_file2. Also, pronunciation of each word in input_file1 is compared

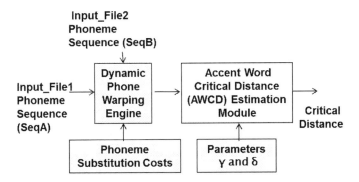

FIGURE 3.1
Critical distance estimation.

with the pronunciation of a different word in input_file2. Thus, there are 1,000 pronunciation variant pairs and 1,000 pairs with different words. Therefore, there are nine data sets
with 2,000 pairs in each data set. A ninefold cross validation scheme is adopted, where
eight data sets are used in training phase while the ninth data set is used for testing. In the
same way, a total of nine data sets are prepared with the transcribed pronunciations taken
from TIMIT speech corpus.

The block diagram of the experimental setup for the estimation of critical distance is
given in Figure 3.1.

Two sequences of phonemes are input to the DPW engine. The distance between them
is calculated by the DPW engine and is normalized. The critical distance is estimated by
varying the values of γ and δ parameters.

3.5.4 Results and Analysis

Effect of using different insertion and deletion operations costs is experimented. The
experimental results showed that the performance of AWC remained unchanged when
value of m is varied from 0.8 to 1.3. However, there is performance degradation when value
of m goes below or above the aforesaid range. Therefore, the costs of insertion and deletion
operations are taken as equal.

3.5.4.1 Estimation of Parameter γ

The results of five data sets each from CMUDICT and TIMIT are summarized in Table 3.4.

In Table 3.4, the precision and accuracy metrics are highest at γ = 0.35 for data sets drawn
from CMUDICT. In case of data sets drawn from TIMIT, the precision metric is highest at
γ = 0.30 and accuracy metric is highest at γ = 0.40.

3.5.4.2 Estimation of Parameter δ

The results show that the performance improved as the value of δ decreased from 1.0
to 0.6. The performance improvement is graphically shown in Figure 3.2. Pronunciation
Classification Error Rate (PCER) percentage has reached the minimum value at δ equal to
0.8 and is flat for 0.7 and 0.6. The results of further experiments show minimum error at δ
equal to 0.6.

TABLE 3.4

Performance of the Accent Classifier

Data Set #	Source of Data Set	Highest Precision Point		Accuracy		Sensitivity at Highest Precision	Accuracy at Highest Precision
		Operating Point (y)	Precision Value	Operating Point (y)	Value		
1	CMUDICT	0.35	1.00	0.35	0.998	0.903	0.998
2	CMUDICT	0.35	1.00	0.50	0.996	0.839	0.995
3	CMUDICT	0.35	1.00	0.35	0.996	0.839	0.996
4	CMUDICT	0.30	1.00	0.40	0.997	0.774	0.994
5	CMUDICT	0.35	0.96	0.50	0.995	0.839	0.995
6	TIMIT	0.30	0.95	0.35	0.987	0.594	0.986
7	TIMIT	0.30	0.94	0.40	0.982	0.469	0.982
8	TIMIT	0.30	0.83	0.40	0.982	0.313	0.977
9	TIMIT	0.30	0.94	0.40	0.985	0.531	0.984
10	TIMIT	0.30	0.95	0.40	0.985	0.591	0.986

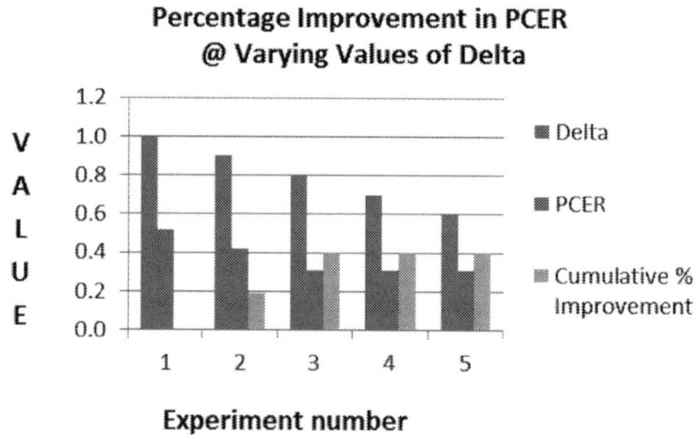

FIGURE 3.2
Graphical view results of PCER varying values of δ.

3.6 Biometric Template for Speaker Identification

3.6.1 Typical Speaker Recognition System

Block diagram of a typical speaker recognition system is given in Figure 3.3.

During the training phase, speaker models are built for all the authorized speakers using speaker-dependent training data. Feature extraction module takes the input speech from the authorized speakers and extracts the feature vectors which are used to generate the speaker models. The speaker models are recognized in the testing phase (Gaikwad and Ahmed, 2014).

The feature extraction module uses spectral-based feature sets such as mel-cepstrum. In speaker identification, the target speaker with the smallest distance from the test speaker is identified.

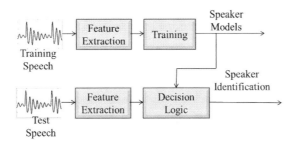

FIGURE 3.3
Typical speaker recognition system.

3.6.2 Speaker Recognition Models

Following speaker recognition models are discussed:

1. Minimum-distance classifier Model.
2. Acoustic Model
3. GMM Model
4. Proposed Model

3.6.2.1 Minimum-Distance Classifier Model

In the minimum-distance classifier model, the speaker recognition is based on the distance between the averages of the feature vectors over multiple analysis frames between training and testing data. The main drawback of the minimum-distance classifier is that it does not distinguish between acoustic speech classes such as quasi-periodic, noise like and impulse like sounds. One way to segment speech signal in terms of speech classes is through speech recognition at phoneme level or word level.

3.6.2.2 Acoustic Model

One way to obtain the phoneme level classification is by using Hidden Markov Model (Jelinek, 1998). Another way to get acoustic classes is through vector quantization (VQ) using k-nearest neighbor clustering algorithm. In VQ approach, a single class is picked up for each feature vector selected in testing phase.

3.6.2.3 GMM Model

Probabilistic models such as Gaussian Mixture Models (GMM) using a multi-dimensional probability distribution functions (pdf) of feature vectors will provide variability in speech production (Campbell, 1997).

3.6.2.4 Hybrid HMM/VQ-Based Model

Hybrid HMM/VQ-based speaker recognition model uses voiceprints of speaker-dependent pronunciations. The input speech signal is segmented using an acoustic model like HMM and the resulting phoneme sequences are clustered in VQ codebook using

Multi-Layer Code Book Architecture		
Layer 1 - Front End	Recently and Frequently Used	Speaker Models
Layer 2 -	Recently Used	
Layer 3 -	Frequently Used	
Layer 4 - Back End	Total Lexicon	

FIGURE 3.4
Enhanced MLCB.

phonetic distance measurements. The code words in the VQ codebook are mapped to their respective speaker models. A speaker identified using the KL divergence among various speaker models.

3.6.3 Pronunciation Voiceprint-Based Speaker Recognition Model

The front end processing includes feature extraction and generation of phoneme sequences using an acoustic model. During training phase, speaker models are built up and in testing phase, the test data is compared with the speaker models in the database and the target speaker is identified or verified.

3.6.3.1 Preparation of Speaker Models

The speaker models are built up using the adaptation model that the design of the MLCB memory is enhanced and the pronunciations in the memory are indexed to speaker models. Each pronunciation variant is mapped to the corresponding speaker. Probability of each pronunciation for the corresponding speaker class is computed. The speaker models are built up using the probability distribution function pertaining to that speaker. Similarly, the speaker models for all speakers are prepared and kept in the database during the training phase. During testing phase, the speaker model of the test speaker is compared with the speaker models using KL distance and speaker is identified or verified. The design of the MLCB memory is given in Figure 3.4.

The MLCB memory is organized into four layers. All the speakers enrolled so far are kept at the back end of its memory in layer 4. The recently and frequently referred speakers are stored in the front end layer 1. If the speaker is not found in the front end, the middle layers are searched which contain the recently used and frequently referred speakers in layers 2 and 3, respectively.

3.6.3.2 Speaker Identification

The Kullback–Leibler divergence is the relative entropy between two probability distributions on a random variable. It is a measure of the distance between them. Given two probability distributions $p(x)$ and $q(x)$ over a discrete random variable X, the relative entropy given by $D(p||q)$ is defined as follows:

$$D(p \| q) = \sum_{x \in X} p(x) \ln \frac{p(x)}{q(x)} \tag{3.13}$$

Each speaker model contains a set of words with their pronunciations mapped to it. The pronunciation differs from person to person and therefore the pronunciation mapping for various speaker models differs. The distance between a pair of speaker models is computed using Kullback–Leibler (KL) divergence (Distance) method (Quatieri, 2012). The KL divergence of speaker B from speaker A is given by

$$D_{KL}(A \| B) = \sum_{w \in W} A(w) \ln \frac{A(w)}{B(w)} \tag{3.14}$$

where A(w) is the discrete probability of pronunciation of speaker A for word w and B(w) is that of speaker B measured over random variable W.

If the word w has n instances of pronunciations and pronunciation p of speaker A occurs m times, then the probability of finding the P in W is given by

$$A(w) = m/n \tag{3.15}$$

Let the test speaker be A and the target speakers be B, C, … , S, then the target speaker with minimum KL divergence from test speaker is identified as the recognized speaker.

3.6.4 Speaker Recognition Algorithm

The speaker recognition algorithm covers preparation of speaker models, computation of KL divergence between two speaker models and identification of the speaker. The algorithm is given below.

Algorithm 3.3: Speaker Identification Algorithm

Input: Analysis Phoneme Sequence mapped to the respective speakers.
Output: Speaker identified

1. Begin
Step 1: Initialization
// Read phoneme sequences from input and memory files
2. long Dmin <= 1.0 //Initialize Dmin to 1;
3. Open input_file, MLCB_file, database_file;
4. String[] seqA <= Analysis_phoneme_Sequence;
5. ArrayList < String[] > seqB <= MLCB_phoneme_seq;
6. ArrayList < String[] > seqC <= database_phoneme_seq;
Step 2: Search
// Compare Analysis sequence with phoneme sequences in memory
7. String[] word_hypothesis <= null;
8. while (SeqB is not null);
9.   phonetic_distance D <= compare (seqA, seqB);
10.   If (D < Dmin) {
11.   Dmin <= D;
12.   word_hypothesis <= seqB;
13.   } end while

Step 3: Preparation of speaker models
// Index seqA to the respective speaker and recalculate the PDF of the respective speaker.
14.   switch (Dmin) {
15.   case (0): output_word <= word_hypothesis;
16.   case (<= D_C): output_word <= word_hypothesis;
17.   addRecordToMemory ();
18.   index_Pronunciation_to_speaker();
19.   ComputePDF()
20.   case (> D_C) : Repeat step 2 with seqA and seqC
21.   addRecordToMemory ();
22.   index_OOV_to_speaker();
23.   ComputePDF()
24.   } end switch
Step 4: Speaker Identification
25.   input A(w);   // input probability distribution of test speaker A
26.   input B(w);   // input probability distribution of target speaker B
27. $D_{KL}(A \| B) = \sum_{w \in W} A(w) \ln \dfrac{A(w)}{B(w)}$ // find relative entropy
28.   Repeat steps 25 and 26 for all target speakers
29.   Select the target with minimum KL divergence
30.   **end** of algorithm

3.6.5 Experimentation Details

3.6.5.1 Data Sets Source and Sample Size

Data sets are taken from TIMIT and NTIMIT databases. The TIMIT speech database recorded the speech samples from 630 speakers both males and females in quiet environment.

NTIMIT database consists of same utterances as TIMIT but transmitted through the actual telephone channel which limits the bandwidth to 3.3 kHz and has additive noise. TIMIT speech corpus consists of connected word speech utterances and is popularly used in speech recognition research. Distribution of the speakers is shown in Table 3.5.

The training data set consists of utterances from 462 speakers and the test data set consists of utterances from 168 speakers. Each speaker read ten utterances.

A total of five data sets are prepared with the transcribed pronunciations taken from TIMIT speech corpus. Each data set consists of utterances from 31 speakers spoken by both male and female distributed over all the eight dialect regions. Four data sets are drawn from the training speech corpus, while the fifth data set is taken from test speech corpus.

TABLE 3.5

Distribution of Speakers in Training and Test Sets

Set	#Male	#Female	Total
Training	326	136	462
Test	112	56	168
Total	438	192	630

3.6.5.2 Experimental Setup

While the pronunciation phoneme sequences are listed in input_file, the words along with the pronunciations are listed in database_file. The MLCB_File is kept empty.

3.6.5.3 Experimentation Methodology

3.6.5.3.1 Preparation of Input Files

The data set of 4,000 words is divided into 15 input files with size in the ascending order. For example, the first input file is with 25 words and the second input file is with 80 words, the third input file is with 109 words, the fourth input file is with 184 words and so on.

3.6.5.3.2 Adaptation Algorithm

In training phase, the adaptation algorithm and speaker model are mapped to the new words and their pronunciations, and in testing phase, the models are compared.

3.6.5.3.3 Cross Validation

Fivefold cross validation is adopted to ensure that there is isolation and cross verification between the training and test sets. The data sets 1–4 are drawn from the training speech corpus, while the fifth data set is taken from test speech corpus.

3.6.5.3.4 Experimentation

The speaker models are prepared using eight sentences spoken by the speaker (80%) and tested with two sentences (20%). The results are recorded.

3.6.6 Results and Discussion

3.6.6.1 Speaker Models

The speaker models are built using the probability distribution of the pronunciations which are mapped to the respective speakers. Table 3.6 gives the discrete probabilities for nine speakers and it is graphically shown in Figure 3.5.

TABLE 3.6

PDF for Speaker Models

Speaker	\multicolumn{11}{c}{Word Number}
	1	2	3	4	5	6	7	8	9	10	11
MCPM0	0.89	0.33	0.11	0.78	0.11	0.11	0.22	0.67	0.11	0.89	0.56
FAKS0	0.89	0.22	0.22	0.78	0.11	0.33	0.67	0.67	0.11	0.89	0.56
FDAC1	0.89	0.22	0.11	0.11	0.11	0.11	0.67	0.67	0.11	0.89	0.22
FDAW0	0.11	0.22	0.11	0.78	0.33	0.11	0.67	0.11	0.11	0.11	0.11
FDML0	0.89	0.22	0.22	0.78	0.33	0.11	0.67	0.67	0.22	0.89	0.56
MDAC0	0.89	0.11	0.44	0.11	0.33	0.11	0.11	0.22	0.22	0.89	0.56
MEDR0	0.89	0.33	0.44	0.78	0.33	0.33	0.67	0.67	0.33	0.89	0.22
FAJW0	0.89	0.33	0.44	0.78	0.33	0.33	0.67	0.22	0.33	0.89	0.56
FCMM0	0.89	0.11	0.44	0.78	0.33	0.11	0.67	0.67	0.33	0.89	0.11

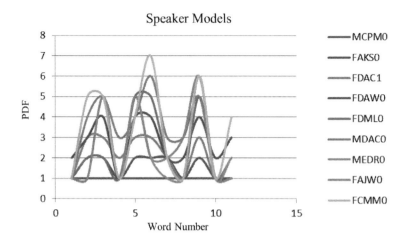

FIGURE 3.5
Speaker models.

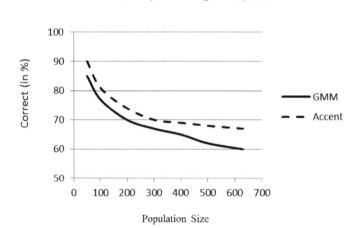

FIGURE 3.6
Performance of speaker identification systems.

3.6.6.2 *Speaker Identification*

Experiments are conducted for speaker identification using various population sizes using NTIMIT database. The results are recorded with varying population sizes and are given in Figure 3.6.

3.7 Conclusions

Phonetic distance-based data-driven approaches for unsupervised adaptation will find its applications in several fields of the speech processing. Dynamic phone warping algorithm can be used in any language, provided the set of phonemes of that language is used. The phonetic distance measurements are independent of the language used. Techniques used in this chapter may be used for any language. The phoneme set for target language should be used in place of phoneme set of English language. Languages with sparse transcribed speech corpus can use this pronunciation adaptation approach for building up the pronunciation dictionary. The performance of the phonetic distance-based speaker identification model is better than the other models.

References

J. M. Baker, L. Deng, S. Khudanpur, C.-H. Lee, J. Glass, N. Morgan, (2009), Historical developments and future directions speech recognition and understanding, *IEEE Signal Processing Magazine*, vol 26 (4) 78–85.

P. Campbell, (September 1997), Speaker recognition: A tutorial, *Proceedings of IEEE*, vol 85 (no. 9) 1437–1462.

U. Gaikwad, M. Ahmed, (2014), State of the art in speaker recognition systems, *International Journal of Research in Engineering and Scientific Applications*, vol 1 1.

S. Hahn, P. Vozila, M. Bisani, (2012), Comparison of grapheme-to-phoneme methods on large pronunciation dictionaries and LVCSR tasks, *IEEE Proceedings of INTERSPEECH 2012*. Lyon.

B. Hixon, E. Schneider, S. L. Epstein, (2011), Phonemic similarity metrics to compare pronunciation methods, *INTERSPEECH 2011*. Florence.

X. Huang, J. Baker, R. Reddy, (2014), A historical perspective of speech recognition, *Communications of the ACM*, vol 57 (1) 94–103.

F. Jelinek, (1998), *Statistical Methods for Speech Recognition*. The MIT Press, Cambridge, MA.

L. Lu, A. Ghoshal, S. Renals, (2013), Acoustic data-driven pronunciation lexicon for large vocabulary speech recognition, *IEEE Workshop on Automatic Speech Recognition and Understanding (ASRU), 2013*.

Z. Ma, A. Leijon, W. B. Kleijn, (2013), Vector quantization of LSF parameters with a mixture of Dirichlet distributions, *IEEE Transactions on Audio, Speech, and Language Processing*, ISSN 1558-7916, vol 21 (9) 1777–1790.

S. L. Mattys, O. Scharenborg, (2014), Phoneme categorization and discrimination in younger and older adults: A comparative analysis of perceptual, lexical, and attentional factors, *Psychology and Aging, American Psychological Association*, vol 29 (1) 150–162.

O. Pietquin, T. Dutoit, (2006), A probabilistic framework for dialog simulation and optimal strategy learning, *IEEE Transactions on Audio, Speech and Language Processing*, vol 14 (2) 589–599.

M. Pucher, A. Türk, J. Ajmera, N. Fecher, (2007), Phonetic distance measures for speech recognition vocabulary and grammar optimisation, *3rd Congress of the Alps Adria Acoustics Association 27–28 September 2007*, Graz.

T. F. Quatieri, (2012), *Discrete Time Signal Processing*. Pearson, Upper Saddle River, NJ.

L. Rabiner, B. Juang, B. Yegnanarayana, (2010), *Fundamentals of Speech Recognition*, second ed. Prentice Hall, Englewood Cliffs, NJ.

S. Sriram, et al., (2004), Phonetic distance based cross lingual search, Report from HP Labs India and BITS Pilani. Hyderabad.

R. M. Stern, N. Morgan, (November 2012), Hearing is believing biologically-inspired methods for robust automatic speech recognition, *IEEE Signal Processing Magazine* 34–43.

M. Wieling, E. Margaretha, J. Nerbonne, (2011), Inducing phonetic distances from dialect variation, *Computational Linguistics in the Netherlands Journal*, vol 1 109–118.

B. Zhou, (2013), Statistical machine translation for speech: A perspective on structures, learning, and decoding, *Proceedings of the IEEE*, vol 101 (5) 1180–1202.

4

Behavioral Biometrics: A Prognostic Measure for Activity Recognition

Ankit A. Bhurane and Robin Singh Bhadoria

Indian Institute of Information Technology

CONTENTS

4.1 Introduction to Behavioral Biometrics

Biometrics has emerged as one of the most essential technologies in recent years. Biometrics refers to a set of various techniques used for characterization of an individual's identification or verification or both. Note that identification refers to one-to-many (inter-class) matching whereas the verification refers to one-to-one (intra-class) matching. Behavioral biometrics deal with the capturing and training of behavior of the user. Such behavior can be readily obtained from user interaction with various devices. AADHAR (a unique identification number) (UIDAI, 2018) is one of the examples of the biometric data management system in practice that has been well accepted and linked to a majority of government organizations in India. However, the possible leakage of the database raises several concerns about the security of the user data. Behavioral biometrics may act as a protection cover to

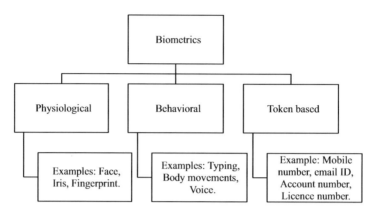

FIGURE 4.1
Classification of biometrics.

avoid such leakage and strengthen the security. This chapter intends to give an overview and useful resources related to some popular behavioral biometrics. A broad classification tree of biometrics is shown in Figure 4.1.

The behavioral class of biometric techniques offers several advantages as follows.

1. **Silent data acquisition:** The user data can be quietly captured on-the-fly during the interaction with the objects (e.g. personal computer), without any special preparation by the user.
2. **Worry-free verification:** Unlike others, behavioral biometrics does not require any password(s) or PINs to be remembered for identity verification.
3. **Theft-free**: As every human being is unique, if chosen properly, the features can be very difficult to mimic. This makes the system more secure.
4. **Dynamicity:** The data itself gets adaptively tuned to the user interactions keeping the acquired data up-to-date. This is in contrast with the physiological biometrics that is inherently static.

Depending upon the device that the user interacts, the behavioral biometrics can be classified into different categories. Some commonly used devices and corresponding data are listed below.

1. Computer keyboard: The typing style varies from person to person. The keyboard button press details (keystroke dynamics) (Monrose and Rubin, 2000; Ardamax, 2018) can be recorded as a feature.
2. Microphone: A mic can be used to record "voice"-related features of a user.
3. Handwriting or signature style: A user may have specific style of writing text or signature. The pen strokes can thus determine individual identity. Such signals can be recorded from devices like touch, pressure-sensitive pen tablets, or scanned documents.
4. Computer software and internet: An individual may use some software more often than the rest. Thus, software usage statistics can also act as a feature. The usage of internet data depends upon an individual's interest. The website browsing statistics features can, therefore, act as one of the features for identity verification.

5. Medical instruments: Various medical instruments used to record biomedical signals like electrocardiogram, electroencephalogram, electromyogram, etc. can be used to record user specific biological features.

6. Camera: An individual can be characterized by his/her body movements. These movements of body parts can act as features.

We present the details of these biometric techniques in the subsequent sections.

4.2 Keystroke Dynamics

A generic framework of keystroke biometric learning and authorization is shown in Figures 4.2 and 4.3. The role of each block is discussed as follows.

4.2.1 User Enrollment

As a very first step, a user has to get enrolled to form a specific profile. The input to the system could be fixed or variable text or phrases. The text data may be simple short text or a long text that can be customized in order to test the typing skills of the user. The system may request the user to enter the desired text multiple times until a required precision is obtained. Such enrollment is called as static enrollment, unlike the dynamic where the user gets enrolled on-the-fly while working on the computer.

The data recorded by the system is subjected to the next block in order to extract the characteristic features. Various feature extraction techniques are given in the next section.

4.2.2 Feature Extraction

The most widely used features extracted from computer keyboard are:

1. Key code: This is a unique code associated with the keys.

2. Duration of key press (Bergadano, Gunetti, and Picardi, 2002): This is the amount of time for which a user holds a particular key while typing. Figure 4.4 shows an example time stamp features for the types word "CRC".

FIGURE 4.2
A generic flow of keystroke data registration.

FIGURE 4.3
A generic flow of keystroke data evaluation.

Key transition time

Typed key	C	C	R	R	C	C
	↓	↑	↓	↑	↓	↑
Key press and release time	T_{0p}	T_{0r}	T_{1p}	T_{1r}	T_{2p}	T_{2r}

T_{01} spans from the second C to R; T_{12} spans from R to the second-last C.

FIGURE 4.4
An example of n-graph and corresponding attributes.

3. Key transition time: This is the time required by a user to switch between the keys.

4. n-graphs (Revett, 2008): A combination of two or more successive typed letters are called as n-graphs. Most commonly used features are the first- and second-order statistics, probability distribution function, and entropy of the digraphs.

5. Keystroke sound (Joseph Roth, 2015): The sound generated from key typing can also act as a discriminative feature for biometric authentication.

6. Hand shape features (Joseph Roth, 2014): The hand shape features while typing can be recorded by initially segmenting the foreground, followed by hand detection, separation, and shape context histograms.

7. Authentication time: This is the time taken by a user (speed) to successfully log on to the system.

8. The Number of attempts: The total number of attempts made by a user to successfully log on to the system.

4.2.3 Classification of Keystroke Dynamics

The features are normalized and preprocessed before classification. The methodology used for classification includes:

1. Simple distance metrics like p-norm, Mahalanobis, City block, Minkowski, Chebychev, Cosine, Hamming, Spearman, Jaccard, and so on.

2. Hidden Markov Model (HMM).

3. Unsupervised data clustering algorithms like k-means.

4. Supervised algorithms like Support Vector Machines (SVM).

5. Fuzzy logic (FL) and artificial neural networks (ANN), rough sets, etc. It has been found that the classification accuracy increases with the increase in the text length used for user profile creation.

A detailed survey of keystroke biometrics can be found in Teh, Teoh, and Yue (2013).

4.2.4 Public Database for Keystroke Dynamics

This section lists some of the popular publicly available databases for keystroke dynamics. Table 4.1 gives the details of the database with the number of subjects used for creating the database and features extraction technique used in each of these databases.

TABLE 4.1

A List of Publicly Available Database for Keystroke Dynamics

S. No.	Authors/Database Name	# Subjects	Features Extracted
1	Kevin S. Killourhy and Roy A. Maxion (Maxion, 2018)	51	Key press, release, and hold time
2	Aythami Morales Moreno (Morales, 2018)	300	Scan code, key press, and release time
3	Mohamad El-Abed, Mostafa Dafer and Ramzi El Khayat (Mohamad El-Abed, 2018)	51	Key press and release time, multiple key press and release, correction factor
4	BeiHang Keystroke Dynamics Database (Yilin Li, 2018)	117	Press and release time
5	BioChaves Working Group (Group, 2018)	48	ASCII code and keydown elapsed time
6	How-We-Type Dataset (Anna Maria Feit, 2018)	52	Key symbol, key press, and transition time, eye gaze, hand, and finger alternation
7	Visual and Acoustic Typing behaviour database (Joseph Roth, 2018)	56	Hand shape descriptors

4.2.5 Commercial Applications

A few commercial application based on keystroke dynamics are given below.

1. KeyTrac (2018): This is an initiative by University of Regensburg to provide high-performance keystroke dynamic-based biometric solution. The application records the characteristics of the typist on-the-fly. The application provides both password and text solutions. The webpage-based service asks user to enter the predefined text and extracts features in the background during the enrollment. The application uses key press and release timings. During enrollment, if the number of typos exceeds certain threshold value, the user is asked to re-enroll. During authentication, the user is asked to enter a text to validate the dynamics and generate a match score in percentage.

2. TypingDNA (2018): This application records the typing pattern during the user account creation. The account creation demands user's email ID, password, and phone number and in turn, generates the *apikey* and *apisecret* (codes specific to the user). The authentication can be done from any unknown location. The application makes use of the "Postman", an addon to create and test the application programming interface (API) requests from the user device. The API offers both the same text and any text pattern for enrollment and authentication.

4.2.6 Limitations of Keystroke Biometrics

1. Lack of consistency: As the authentication in keystroke-based biometric is dependent mainly on the rhythm of the user, the consistency may get hampered by the moods of the user.

2. Dynamic features: The rhythm of the keystrokes changes with time. A static system would, therefore, find it difficult to determine the authenticity of a person if the rhythm gets changed beyond the expected margin. A regular update of the enrolled data becomes necessary to avoid such situations.

4.3 Speaker Recognition

A person can be characterized by his/her voice. A system that recognizes a person based on the voice is called as a speaker recognition system. This is in contrast with the voice recognition system that differentiates between different voice signals. The example of voice recognition is Google input voice typing that recognizes words, not the speaker. The speaker recognition system differentiates between different people, whereas the voice recognition system determines the words spoken by a speaker, irrespective of the speaker.

4.3.1 User Enrollment

The voice signals are typically recorded using built-in microphone available on laptop or phone. A generic framework of the speaker recognition system is shown in Figure 4.5.

Similar to keystroke dynamics-based system, the speaker recognition system can be of fixed and variable text type. In fixed text type, the same text is used for user enrollment and authentication, whereas in variable text type, the user is asked to read the randomly generated text. Both of the systems have their own advantages and disadvantages. The fixed text systems are shown to be more efficient as compared to the variable text system. However, some crucial problems like false authentication using voice mimicking, low permanence of the extracted features pose several challenges in building an efficient speaker recognition system.

4.3.2 Voice Feature Extraction

Popularly used features for speaker identification and/or verification are as follows.

1. Figure 4.6 targets to recover the input excitation x[n] by minimizing the mean of the squared difference between the approximation of the input voice and the actual voice signal, i.e. $\min \Sigma \left(y[n] - \hat{y}[n] \right)^2$.

2. Mel frequency cepstral coefficient (MFCC): The concept of MFCC was introduced in Davis and Mermelstein (1980) in an attempt to represent monosyllable word representation. MFCC are the coefficients obtained as follows.

FIGURE 4.5
A generic framework of speaker recognition system.

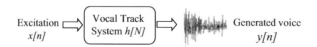

FIGURE 4.6
Calculating linear predictive coefficients of voice signal.

- The input voice signal is subjected to the windowed Fourier transform.
- Calculate the energy spectrum of the windowed signal using triangular filter banks. The power spectrum is calculated on the pitch scale (melody/*mel* scale).
- Calculate the logarithm of the energies.
- Calculate the discrete cosine transform and retain M coefficients.

The algorithm is based on the pattern in which the sound is perceived by humans.

3. Toeplitz matrix minimal eigenvalues algorithm (Saeed and Nammous, 2007): In this method, the spectral estimate of the speech based on linear predictive coefficients (LPC) is calculated in order to get an envelope of the spectral. This is followed by the formation of Toeplitz matrix using the coordinates of the samples of the spectral envelope. The minimum eigen values of the obtained Toepliz matrix then acts as a feature vector that can be further given for distance matching or classification.

4. Discrete wavelet transform (DWT): Multiresolution-based approaches using the wavelet transform for robust feature extraction. One such attempt is presented in Hsieh, Lai, and Wang (2002).

For speaker recognition, various approaches have been investigated in the literature including artificial neural networks (ANN) based on probabilistic neural network, radial basis function (RBF), deep learning, HMM, independent component analysis, wavelet transform.

4.3.3 Speaker Recognition Public Databases

A few lists of the publicly available database for speaker recognition is as follows.

- TIMIT database (Rémy, 2018): The database consists of 630 speakers with their recordings and transcripts.
- CHAIN speaker database (Informatics, 2018): The database is provided by UCD school of Computer Science and Informatics for a maximum of 36 speakers.
- IIT Multivariability Guwahati Database (S.R.M. Prasanna, 2018): The database was recorded in four phases of 100–200 subjects. The database is recorded in different controlled and uncontrolled environments.
- YOHO database (Joseph Campbell, 2018): The database consists of the recordings from 138 speakers in a practical working environment. The database is provided by International Telephone & Telegraph (ITT) defense communication.
- Australian National Database of Spoken Language (ANDOSL) (School of Electrical Engineering Sydney University, 2018): The database is provided by Australian National University. The database consists of speech signals recorded by assigning different speaking tasks to 129 speakers in a restricted environment.
- Switchboard-1 (Godfrey and Holliman, 1993): This database is a collection of two-way telephonic conversation of more than 500 speakers.

Limitations of voice-based behavioral biometrics:
The main challenge of voice-based biometric is to deal with the dynamicity of the voice that changes with time, mental and physical situations. Further, the classifier finds it difficult to distinguish between the authentic voice and mimicked voice. To deal with such

cases, prosody (suprasegmental) speech features (Mary and Yegnanarayana, 2008) that take into account the auditory and acoustic impression of the speaker have been found to improve the overall performance of the system (Jin and Yoo, 2010).

4.4 Handwriting and Signature Recognition

Handwriting of the person can be used in the characterization of an individual's identity. Handwriting can be acquired by scanning the document written previously by a user (static) or by allowing the user to write the document on-the-fly (dynamic). For dynamic acquisition, a pressure-sensitive and angle-sensitive pen tablet can be used. The dynamic acquisition of the handwriting provides several advantages over the static, as it records additional information such as writing angle, pressure levels, negligible preprocessing, etc.

The static recording of the handwriting data invites many preprocessing overheads including enhancements, cropping, alignment, background removal, segmentation, etc. Further, if an acquired data is signature, the information is low. Thus, making the recognition a difficult task. A snapshot of scanned handwritten notes with the bounding boxes of the connected regions are shown in Figure 4.7.

4.4.1 Handwritten Text and Signature Public Databases

The public databases are available in various formats like isolated characters, digits, words, and sentences. Some of the publicly available databases are:

- International Unipen Foundation database (UNIPEN, 2018): The dataset is available for testing both online and offline user authentication algorithms.

- MNIST: This is the database of handwritten digits in the form of normalized images. The database has been tested by several methods. The comparison of efficiencies obtained by various approaches can be found in LeCun, Cortes, and Burges (2018).

- CASIA handwriting database (National Laboratory of Pattern Recognition (NLPR)|Institute of Automation, 2018): This database is provided by Chinese Academy of Science in both Chinese and English languages. The database consists of handwritten sentences recorded from 187 Chinese and 134 English users, respectively. This database is recorded online and has information about spatial coordinates, pressure, and pen angles.

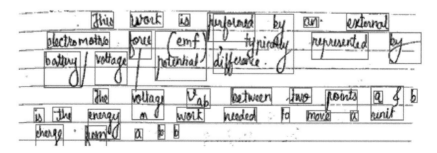

FIGURE 4.7
An example of handwritten text recorded offline. The bounding boxes show the connected regions.

- IAM database (INF, 2018): The database consists of handwritten text recorded from about 657 users from different categories of publications.
- CEDAR database (CEDAR, 2018): This database was recorded in multiple phases and consists of text from various categories of English usage written by about 200 users.
- University of Las Palmad de Gran Canaria (ULPGC) database: The database consists of online and offline original and synthetic signatures. The link also provides various algorithm toolboxes and performance comparison of various classifiers.
- Signature Verification Competition database (SVC, 2014): This database consists of five users with genuine and forged signatures and was used for international signature verification competition.

The complete list of datasets provided by International Association for Pattern Recognition (IAPR) is available at IAPR-TC11 (2018).

4.4.2 Feature Extraction

This section gives various feature extraction techniques used for offline and online handwriting recognition.

a. Offline handwriting
    The features used for offline handwritten text includes the following.
- Aspect ratio: The ratio of width and height of the signature.
- Thickness of the line segments.
- Shape and trajectory of handwriting strokes: This corresponds to the changes in the angles. In order to determine the angles, the signature image can be divided into small regions and appropriate angles or similar shape features are calculated starting from initial point. An example of an image to be processed for calculating the trajectory is shown in Figure 4.8.
- Continuity of the strokes.
- Density and spacing of the pen strokes.
- Transform-based features: Different transform-based features like Hough, Radon, and Wavelet are proven to be effective in representing the handwritten text.

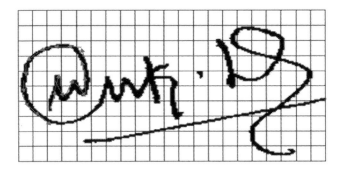

FIGURE 4.8
An example of a signature recorded offline segmented into small regions. The segmented regions are matched with the database shape segments. The image is annotated with initial angles of the segments that can act as a descriptor.

b. Online handwriting

   Online recording of handwritten text allows to capturing various system-level features that can be readily captured during the enrollment. These features include statistics of:

   • Current pen tip location: these are the spatial coordinates (x, y).

   • Velocity of the strokes: the instantaneous velocity of the pen strokes.

   • Angle of the strokes: the instantaneous alignment/holding angle of the pen.

   • Pressure levels of the pen strokes: the real-time pressure magnitudes of the pen during recording.

   • Pen up and down instances: the continuity of the pen strokes.

   • Maximas and minimas of the strokes: this takes into account the spatial bounds of the pen strokes.

For classification, some popular techniques include template matching, dynamic time warping (DTW), transform-based similarity matching, ANN, SVM, and HMM. A comparative analysis of correlation, DTW, and tree matching is given in Parizeau and Plamondon (1990).

Limitations of handwriting-based behavioral biometrics

1. Dynamicity of the data: The handwriting of a person varies with time and the database needs to be timely updated for efficient performance.

2. Non-uniqueness: In a large pool of database, signature style can be similar, making it difficult for recognition.

3. High chance of forgery: The percentage of forgery is high in handwriting recognition, especially if the data is acquired offline.

4.5 Computer-/Mobile Application-Based Behavioral Biometrics

The usage of any computer platform for a mobile device or a laptop varies from one user to others. Some people may use specific event-related software more frequently than generic apps. For example, an office employee may use the accounting software, document editing tools, or email applications more often as compared to a gamer who would make frequent use of graphics software. Same is true with the browsing or shopping pattern. A techie is more likely to shop for gadgets or electronic devices or related stuff as compared to groceries and cosmetic shopping. The search and sorting of items on shopping websites can help in creating a shopping profile of a user. An example of the browsing history of a biometric researcher and a gamer is shown in Figures 4.9 and 4.10.

An application that records all such computer activities would silently capture the characteristics of the owner. Such software records all the computer activities like browsing, typing, and also records the screenshots and camera samples from the computer to keep track of the user.

Based on the activities of the user, a hit count of the events is created to form a feature vector. For the matching purpose, the profile of the user is matched with the saved features and a similarity score is generated that helps in the authorization.

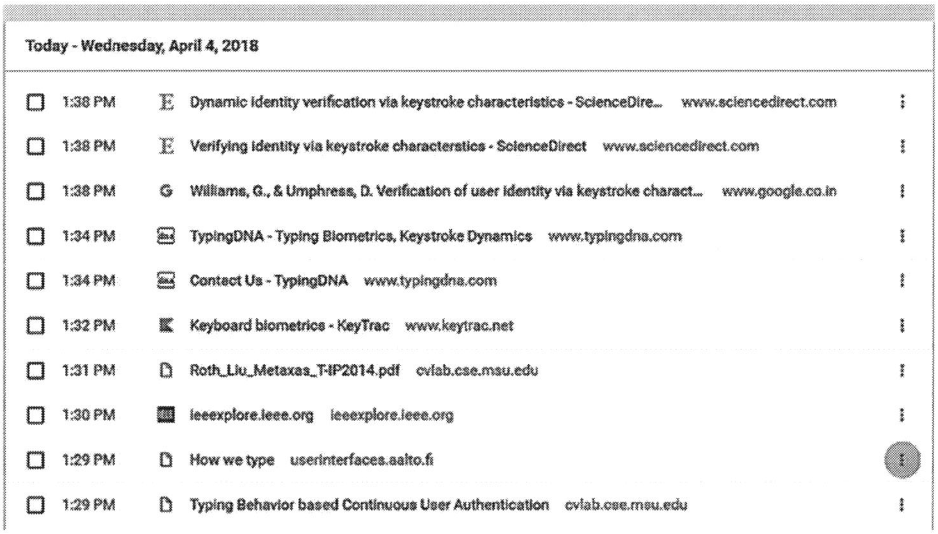

Today - Wednesday, April 4, 2018			
☐ 1:38 PM	E	Dynamic identity verification via keystroke characteristics - ScienceDire... www.sciencedirect.com	⋮
☐ 1:38 PM	E	Verifying identity via keystroke characteristics - ScienceDirect www.sciencedirect.com	⋮
☐ 1:38 PM	G	Williams, G., & Umphress, D. Verification of user identity via keystroke charact... www.google.co.in	⋮
☐ 1:34 PM	▨	TypingDNA - Typing Biometrics, Keystroke Dynamics www.typingdna.com	⋮
☐ 1:34 PM	▨	Contact Us - TypingDNA www.typingdna.com	⋮
☐ 1:32 PM	▨	Keyboard biometrics - KeyTrac www.keytrac.net	⋮
☐ 1:31 PM	▯	Roth_Liu_Metaxas_T-IP2014.pdf cvlab.cse.msu.edu	⋮
☐ 1:30 PM	▨	ieeexplore.ieee.org ieeexplore.ieee.org	⋮
☐ 1:29 PM	▯	How we type userinterfaces.aalto.fi	⊙
☐ 1:29 PM	▯	Typing Behavior based Continuous User Authentication cvlab.cse.msu.edu	⋮

FIGURE 4.9
An example of browsing history of a biometric researcher.

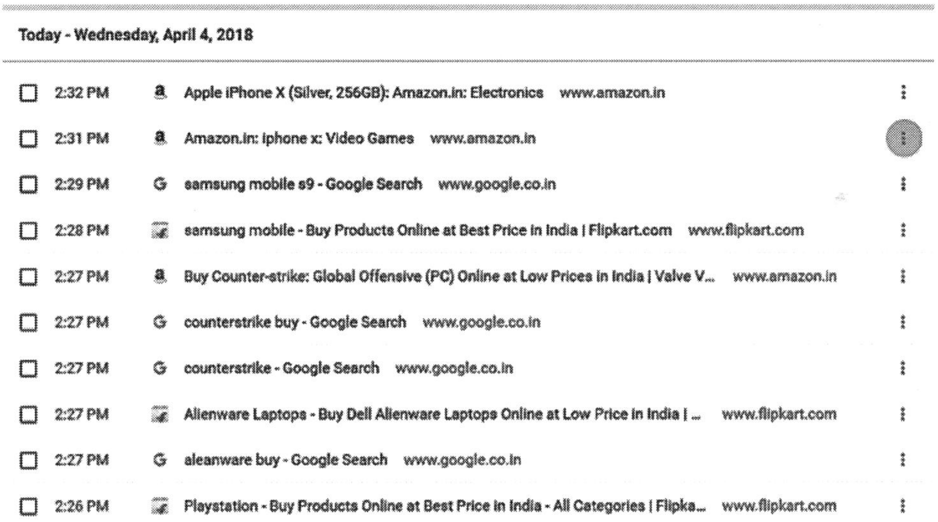

Today - Wednesday, April 4, 2018			
☐ 2:32 PM	a	Apple iPhone X (Silver, 256GB): Amazon.in: Electronics www.amazon.in	⋮
☐ 2:31 PM	a	Amazon.in: iphone x: Video Games www.amazon.in	⊙
☐ 2:29 PM	G	samsung mobile s9 - Google Search www.google.co.in	⋮
☐ 2:28 PM	▨	samsung mobile - Buy Products Online at Best Price in India	Flipkart.com www.flipkart.com	⋮
☐ 2:27 PM	a	Buy Counter-strike: Global Offensive (PC) Online at Low Prices in India	Valve V... www.amazon.in	⋮
☐ 2:27 PM	G	counterstrike buy - Google Search www.google.co.in	⋮
☐ 2:27 PM	G	counterstrike - Google Search www.google.co.in	⋮
☐ 2:27 PM	▨	Alienware Laptops - Buy Dell Alienware Laptops Online at Low Price in India	... www.flipkart.com	⋮
☐ 2:27 PM	G	aleanware buy - Google Search www.google.co.in	⋮
☐ 2:26 PM	▨	Playstation - Buy Products Online at Best Price in India - All Categories	Flipka... www.flipkart.com	⋮

FIGURE 4.10
An example of browsing history of a gamer.

4.6 Biomedical Signals as Biometrics

Various biomedical signals, like electrocardiogram (ECG) and electroencephalogram (EEG), can characterize an identity of the user. A sample ECG of and EEG signals are shown in Figure 4.11. ECG signals are recorded by placing electrodes on the chest and limbs and measuring the electrical activity of the heart. A typical ECG waveform takes a specific PQRST pattern. The prominent time stamps P, Q, R, S, T, corresponding amplitudes, and the intervals between the successive waveforms can act as a feature vector specific to an

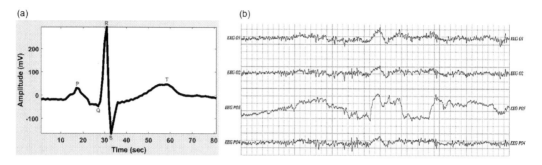

FIGURE 4.11
An example of PQRST interval of ECG waveform (a) and EEG signals from the PhysioBank database (b) (Goldberger, 2000).

individual. The non-fiducial features that are independent of any time stamps are also available in the literature. A detailed survey of various fiducial and non-fiducial features-based ECG biometric recognition is available in Odinaka et al. (2012). In Nemirko and Lugovaya (2005), the authors have used principal component analysis (PCA) and linear discriminant analysis (LDA) for biometric classification and have obtained the classification accuracy of as high as 96%. A survey of ECG databases for biometric systems can be found in Merone, Soda, Sansone, and San (2016).

Unlike ECG, EEG waveform does not have any fixed pattern. The signals can be recorded while asking the users to perform certain tasks like imagining the movement of limbs. In the case of EEG, time-domain features like information content, correlation, or transform domain-based features like spectral statistics, or multiresolution features can be used to form the feature vector. The EEG features have been claimed to be significantly permanent (Emanuele Maiorana, 2015). In this work, the authors have analyzed the autoregressive (AR) reflection coefficients against power spectral density and frequency coherence in the θ-β band of the EEG signals. In EEG, the strength and duration of event-related potentials (ERPs), signals associated with the responses of the brain can also act as a good measure for individual characterization. A review of EEG-based biometric recognition including time, frequency, and wavelet domain approaches is given in Abo-Zahhad, Ahmed, and Abbas (2015).

The EEG or ECG signals are specific to a person and are impossible to exactly mimic the responses. Thus the biometrics based on such signals serves as a powerful protective layer for authorization. The limitations of such biomedical signal-based biometrics would be the requirement of specialized devices for capturing the signals.

4.7 Conclusions

In this chapter, we have given an overview of popular behavioral biometrics. The keystroke dynamics is recommended to be used as auxiliary protection to the existing authentication instead of treating it as standalone biometric. A detailed evaluation of the performance can be found in work by Khalid Saeed with Marcin Adamski (2017). In the case of long text verification, the efficiency and time required for evaluation increases with an increase in the number of users. Further, it should be noted that the database needs to be regularly updated due to the dynamicity of the rhythm of the typists. However, the ease of data capture and processing makes the keystroke dynamics as one of the top choices in behavioral biometrics.

The speaker recognition also offers simple acquisition setup and is found to be effective for speaker identification and recognition purposes. For improved performance, the speaker recognition can be used in association with speech recognition. Also, the system needs to be resistant to the voice mimic attempts for authorization. Similarly, the offline handwriting recognition faces the issues of higher false acceptance due to high possibility of copying of patterns. The efficiency is better in case of online data acquisition and recognition in which additional behavioral features are recorded. For biomedical signal acquisition, specialized devices are required making it less user friendly. However, the biomedical data can be secret and unique to a person making it difficult to copy and reproduce. Thus the biomedical-based behavioral dynamics can provide higher security and better characterization of an individual.

Apart from the behavioral biometrics discussed in this chapter, other biometrics like mouse dynamics, gait (nature of human walking), and face are also investigated in the literature. To make best of these biometrics, a multimodal biometrics using a combination of two or more behavioral biometrics can improve the performance of the overall system. A survey of multimodal biometrics can be found in Wang and Liu (2011). Based on the type and level of fusion, the multimodal biometrics can be classified into different types. For example, the weighted average or non-linear weighting of the feature vectors at feature extraction or classification levels can be treated as a fusion technique. Overall, the behavioral biometrics provide promising platform that can help boosting the security of the existing biometric authentication system.

References

Abo-Zahhad, M., Ahmed, S. M., & Abbas, S. N. (2015). State-of-the-art methods and future perspectives for personal recognition based on electroencephalogram signals. *IET Biometrics*, 4 179–190.

Anna Maria Feit, D. W. (2018, 3 4). How we type: Movement strategies and performance in everyday typing. Retrieved from Aato University: http://userinterfaces.aalto.fi/how-we-type/.

Ardamax, S. (2018, 4 4). Ardamax software. Retrieved from Ardamax Keylogger 4.8: www.ardamax.com/keylogger/.

Bergadano, F., Gunetti, D., & Picardi, C. (2002). User authentication through keystroke dynamics. *ACM Transactions on Information and System Security*, 5 367–397.

CEDAR. (2018, 3 4). Handwriting recognition language models. Retrieved from Center of Excellence for Document Analysis and Recognition (CEDAR): www.cedar.buffalo.edu/handwriting/HRdatabase.html.

Davis, S., & Mermelstein, P. (1980). Comparison of parametric representations for monosyllabic word recognition in continuously spoken sentences. *IEEE Transactions on Acoustics, Speech, and Signal Processing*, 28 357–366.

Emanuele Maiorana, D. L. (2015). On the permanence of EEG signals for biometric recognition. *IEEE Transactions on Information Forensics and Security*, 11 163–175.

Godfrey, J., & Holliman, E. (1993). Switchboard-1 release 2. Retrieved from Linguistic Data Consortium University of Pennsylvania: https://catalog.ldc.upenn.edu/ldc97s62.

Goldberger, A. L. -K. (2000, 6 13). PhysioBank, PhysioToolkit, and PhysioNet: Components of a new research resource for complex physiologic signals. Retrieved from PhysioNet: https://physionet.org/physiobank/database/.

Group, B. W. (2018, 3 4). UFS - Universidade Federal de Sergipe. Retrieved from: www.biochaves. com/en/download.htm.

Hsieh, C. -T., Lai, E., & Wang, Y. -C. (2002). Robust speech features based on wavelet transform with application to speaker identification. *IEE Proceedings - Vision, Image and Signal Processing*, 149 108–114.

IAPR-TC11. (2018, 4 4). IAPR-TC11: Reading systems. Retrieved from International Association for Pattern Recognition (IAPR) Technical Committee Number 11: www.iapr-tc11.org/mediawiki/.

INF, F. R. (2018, 3 4). IAM handwriting database. Retrieved from University of Bern: www.fki.inf. unibe.ch/databases/iam-handwriting-database.

Informatics, U. S. (2018, 3 4). CHAINS: Characterizing individual speakers. Retrieved from: http:// chains.ucd.ie/.

Jin, M., & Yoo, C. D. (2010). Speaker verifcation and identification. In L. Wang, & X. Geng (eds) *Behavioral Biometrics for Human Identifcation: Intelligent Applications* (pp. 264–289). Hershey, PA: Medical Information Science Reference, IGI Global.

Joseph Campbell, A. H. (2018, 3 4). YOHO speaker verification. Retrieved from Linguistic Data Consortium University of Pennsylvania: https://catalog.ldc.upenn.edu/ldc94s16.

Joseph Roth, X. L. (2014). On continuous user authentication via typing. *IEEE Transactions on Image Processing*, 23 4611–4624.

Joseph Roth, X. L. (2015). Investigating the discriminative power of keystroke sound. *IEEE Transactions on Information Forensics and Security*, 10 333–345.

Joseph Roth, X. L. (2018, 3 4). Computer vision lab. Retrieved from Typing Behavior based Continuous User Authentication: http://cvlab.cse.msu.edu/project-typing-behavior.html.

KeyTrac. (2018, 3 4). Keyboard biometrics made simple for you. Retrieved from KeyTrac: www.key-trac.net.

Khalid Saeed with Marcin Adamski, T. B. (2017). *New Directions in Behavioral Biometrics*. New York: CRC Press.

LeCun, Y., Cortes, C., & Burges, C. J. (2018, 4 4). The MNIST database of handwritten digits. Retrieved from The MNIST Database: http://yann.lecun.com/exdb/mnist/.

Mary, L., & Yegnanarayana, B. (2008). Extraction and representation of prosodic features for language and speaker recognition. *Speech Communication*, 50 782–796.

Maxion, K. K. (2018, 3 4). Keystroke dynamics - Benchmark data set. Retrieved from: www.cs.cmu. edu/~keystroke.

Merone, M., Soda, P., Sansone, M., & San, C. (2016). ECG databases for biometric systems: A systematic review. *Expert Systems With Applications*, 67 1–40.

Mohamad El-Abed, M. D. (2018, 3 4). RHU keystroke dynamics benchmark dataset. Retrieved from: www.coolestech.com/rhu-keystroke/.

Monrose, F., & Rubin, A. D. (2000). Keystroke dynamics as a biometric for authentication. *Future Generation Computer Systems*, 16 351–359.

Morales, A. (2018, 3 4). Keystroke biometrics ongoing competition (KBOC). Retrieved from: https://sites.google.com/site/btas16kboc/home.

National Laboratory of Pattern Recognition (NLPR)|Institute of Automation, C. A. (2018, 3 4). CASIA handwriting database. Retrieved from Biometric Ideal Test: http://biometrics.idealtest.org/ dbDetailForUser.do?id=10.

Nemirko, A. P., & Lugovaya, T. S. (2005). Biometric human identification based on electrocardiogram. *Proceedings of XII-th Russian Conference on Mathematical Methods of Pattern Recognition* (pp. 387–390). Moscow: MAKS Press.

Odinaka, I., Lai, P. -H., Kaplan, A. D., O'Sullivan, J. A., Sirevaag, E. J., & Rohrbaugh, J. W. (2012). ECG biometric recognition: A comparative analysis. *IEEE Transactions on Information Forensics and Security*, 7 1812–1824.

Parizeau, M., & Plamondon, R. (1990). A comparative analysis of regional correlation, dynamic time warping, and skeletal tree matching for signature verification. *IEEE Transactions on Pattern Analysis and Machine Intelligence*, 12 710–717.

Rémy, P. (2018, 3 4). The DARPA TIMIT acoustic-phonetic continuous speech corpus. Retrieved from: https://github.com/philipperemy/timit.

Revett, K. (2008). *Behavioral Biometrics*. Singapore: John Wiley & Sons Ltd.

S.R.M. Prasanna. (2018, 3 4). Electro medical and speech technology (EMST) laboratory. Retrieved from: www.iitg.ac.in/eee/emstlab/index.php.

Saeed, K., & Nammous, M. K. (2007). A speech-and-speaker identification system: Feature extraction, description, and classification of speech-signal image. *IEEE Transactions on Industrial Electronics*, 54 887–897.

School of Electrical Engineering Sydney University. (2018, 3 4). Australian National Database of Spoken Language (ANDOSL). Retrieved from: http://andosl.rsise.anu.edu.au/andosl/.

SVC, 2. (2014, 4 4). Signature verification competition 2004. Retrieved from: www.cse.ust.hk/svc2004/download.html.

Teh, P. S., Teoh, A. B., & Yue, S. (2013). A survey of keystroke dynamics biometrics. *The Scientifc World Journal*, 2013 1–24.

TypingDNA. (2018, 3 4). Retrieved from: www.typingdna.com

UIDAI. (2018, 4 3). Unique Identification Authority of India|Government of India. Retrieved from UIDAI: https://uidai.gov.in.

UNIPEN. (2018, 4 4). International Unipen Foundation-iUF. Retrieved from Data and Benchmarks for Handwriting Recognition: www.unipen.org/products.html.

Wang, Y., & Liu, Z. (2011). A survey on multimodal biometrics. In G. Lee (eds) *Advances in Automation and Robotics*, Vol. 2. Lecture Notes in Electrical Engineering (pp. 387–396). Berlin, Heidelberg: Springer.

Yilin Li, B. Z. (2018, 3 4). The BeiHang keystroke dynamics database. Retrieved from: http://mpl.buaa.edu.cn/detail1.htm.

5

Finger Biometric Recognition with Feature Selection

Asish Bera

Haldia Institute of Technology

Debotosh Bhattacharjee and Mita Nasipuri

Jadavpur University

CONTENTS

5.1 Introduction

Biometrics has been inevitable in diverse application areas of pattern recognition and machine learning for the last five decades (Jain, Nandakumar, and Ross, 2016). Its inherent benefits surmount its applicability in secure human authentication compared to the common password- and token-based verification methods. Several obtrusive physiological (e.g., fingerprint, face, hand geometry, hand vein, retina, etc.) and behavioral (e.g., gait, keystroke, signature, etc.) modalities have been explored to find a reliable biometric solution (Jain, Ross, and Prabhakar, 2004). Numerous biometric traits are deployed in the government (national ID card), financial (banking transaction), and industrial (attendance maintenance) sectors for secure authentication. Moreover, biometrics is a viable tool for individualization in real-time systems, surveillance, forensic investigation, internet of things (IoT), and other online-based services. However, a single trait is not always adequate to render satisfactory accuracy with a more significant population especially, thousands, or even more. In such cases, multimodal biometrics is an alternative solution (Ross, Nandakumar, and Jain, 2006). Several multibiometric and fusion-based approaches have been implemented for performance and security enhancement. Astoundingly, billions of devices have been connected through the IoT with biometric verification facilities (Borgia, 2014). It has been predicted that more than 20 billion devices will be connected through the IoT in 2020 (Hung, 2017). Thus, biometric is essentially an invulnerable tool for human authentication in this contemporary society (Dutağaci, Sankur, and Yörük, 2008).

Personal verification using hand biometric properties also renders a robust solution that was pioneered with the earliest patents of Miller (U.S. Patent No. 3576538, 1971) and Ernst (U.S. Patent No. 3576537, 1971). Then, after almost three decades, the prospects of hand geometry have been investigated rapidly, with the state-of-the-art method proposed by Reillo, Avila, and Macros (2000). The image acquisition framework of that system is based on guiding pegs. After the hand image preprocessing, geometric measurements of the hand are computed, and salient features are selected for experiments. Further, the hand biometric approaches are presented with hand images acquired in a contact-free manner, i.e., without the assistance of any rigid peg. According to the study, most of the unconstrained hand biometric approaches have been developed after the year 2000. A compendious study was summarized by Duta (2009). Most of the novel methods emphasize on pose flexibility in a contactless environment. Moreover, several unconstrained and contact-free proprietary hand databases have been created in different nations to facilitate the researchers for devising novel techniques. In this regard, other than 2-D hand images (Yörük, Konukoğlu, Sankur, and Darbon, 2006; Kumar, 2008), the databases of 3-D (Kanhangad, Kumar, and Zhang, 2011), infrared (Wang and Chung, 2012), hyperspectral (Ferrer, Morales, and Diaz, 2014), and synthetic (Morales, Ferrer, Cappelli, Maltoni, Fierrez, and Ortega-Garcia, 2015) hand images have been tested by the researchers. Hand geometry is also related to an elementary forensic investigation (Bera, Bhattacharjee, and Nasipuri, 2014).

Every trait carries its pros and cons which signifies the specificity in multiple real-world applications. Hand biometrics also bears specific intrinsic characteristics and provides nonintrusive identity verification solutions reliably. The principal advantages and drawbacks of hand biometrics are summarized in Table 5.1. However, regardless of several limitations, the usefulness of this biometric modality is attractive and deployed ubiquitously for automatic attendance management in the nuclear plant, identity verification

TABLE 5.1

The Pros and Cons of Hand Biometrics

Advantages	Disadvantages
a. Low cost of the sensors, such as a traditional scanner or camera.	a. Interface contact between the hand and device may raise hygienic issues.
b. Size of feature template is small, typically only a few bytes.	b. Intraclass posture variation is significantly more than interclass.
c. User-friendliness due to simple operations for verification.	c. Low uniqueness and universality for identification.
d. High user acceptability and less invasiveness.	d. Larger device surface prevents its deployment in smaller hand-held devices, like smartphone.

for international border control, and others. Due to its inherent limits, hand biometrics is mainly suitable for verification purposes rather than the identification of a larger population. The fundamental weakness pertinent to posture variation cannot be eradicated accurately because of the intricate anatomic foundation of the thumb. Hence, the difficulty is avoided just by removing the thumb from feature computation (Kang and Wu, 2014; Hu, Jia, Zhang, Gui, and Song, 2012). In some approaches, only the four fingers are segmented while the thumb and palm are neglected before feature extraction. This modality is known as the four-finger biometrics that provides equivalent or even better accuracy compared to the five-finger-based methods.

A traditional biometric system is divided into four key modules, namely, the image acquisition (or sensor) module, feature extraction (and selection) module, template matching module, and decision-making module. Prior to feature computation, a robust preprocessing method is followed for finger segmentation. Then, either the geometric measurements or the shape-based descriptors from the fingers are computed (Dutağaci et al., 2008; Luque-Baena, Elizondo, López-Rubio, Palomo, and Watson, 2013). The pixel-level information is transformed into a descriptor, such as the Fourier descriptors (FDs) (Kang and Wu, 2014), shape context (SC) (Hu et al., 2012), wavelet descriptors (WDs) (Sharma, Dubey, Singh, Saxena, and Singh, 2015), scale-invariant feature transform (SIFT) (Charfi, Trichili, Alimi, and Solaiman, 2014), and others. Both the two types of features, that is, geometric and shape based) are equally imperative to represent an identity uniquely. Notably, a conflation of both types of features can also endow satisfactory results. The defined features (i.e., either geometric or shape based) are not capable enough to contribute adequately in performance evaluation. Hence, only the significant features should be chosen using a suitable feature selection technique in the training phase (Guyon and Elisseeff, 2003; Lashkia and Anthony, 2004). The aim is not only to compute a diverse set of salient features but also to select a minimal subset of discriminative features. The insignificant features are discarded which implies the dimensionality reduction of the input feature set. A pictorial representation of the finger biometric system with the precise tasks during the training and testing phases is shown in Figure 5.1.

In the context of hand biometrics, different schemes for feature selection are implemented.

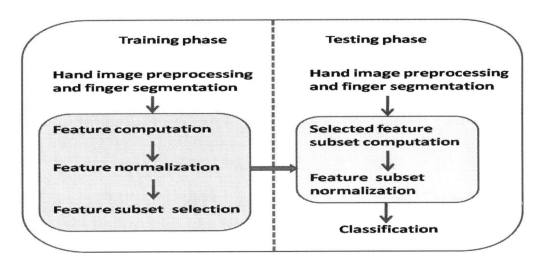

FIGURE 5.1
A finger biometric approach with the feature selection strategy.

5.1.1 Objectives

It can be averred that other than a simple unimodal hand biometric approach, mainly, three different types of schemes are adopted in the literature for improving the robustness and accuracy of a hand biometric system:

a. *Multimodal scheme:* Two or more modalities, such as the hand and face, are employed (Tsalakanidou, Malassiotis, and Strintzis, 2007). The feature extraction methods differ according to the modalities applied. The matching scores of different systems or traits are combined at the score-level or decision-level fusion.

b. *Monomodal with a fusion strategy:* A suitable fusion technique is applied at any stage of implementation, e.g., matching score-level (Kang and Wu, 2014), decision-level fusion using Dempster–Shafer theory (Bera, Bhattacharjee, and Nasipuri, 2015), etc. A diverse set of feature extraction methods on the underlying trait has also been emphasized in other works, described later.

c. *Unimodal with a feature selection technique:* Only relevant features (e.g., hand geometric) are selected, and extraneous features are discarded to reduce feature dimensionality using any apposite metric, such as mutual information (MI), correlation, etc. (Luque-Baena et al., 2013).

These schemes have been experimented over various feature sets of fingers or hand shape in different existing works. The finger biometric features emphasize the local characteristics of fingers. It has been stated earlier that a sizeable intraclass posture variation of the thumb affects the accuracy while included during experimentation (Bera and Bhattacharjee, 2017). Herein, a finger-geometry recognition approach with feature selection algorithms is described. The principal objectives of this chapter are summarized as follows.

i. An exhaustive vignette on the state-of-the-art methods of finger biometric authentication is presented with an emphasis on the feature selection techniques.

ii. Finger geometric feature extraction and the selection of global and local relevant features that optimize the feature space and improve the performance are described.

iii. An in-depth experimental analysis is illustrated for performance evaluation with the weighted and selected feature subspaces defined over the Bosphorus database.

Section 5.2 presents a study on the related hand and finger state-of-the-art biometric approaches. Preprocessing methodology and feature computation of finger biometric modality are described in Section 5.3. Feature selection algorithms are described in Section 5.4. The experimental conception and results are analyzed in Section 5.5. A final remark is drawn in Section 5.6.

5.2 Related Works

Hand biometrics has been excogitated in several possible research directions. Based on the finger biometric feature extraction module, existing approaches can be classified into four categories, shown in Figure 5.2. A summary of those methods is briefed here.

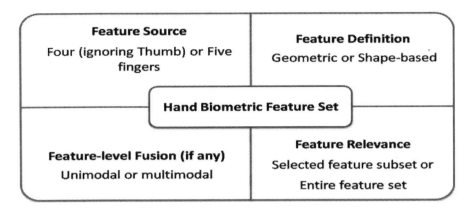

FIGURE 5.2
Different approaches on the hand biometric feature set.

Feature resource: R raw hand images are collected using a scanner or a camera from the users at different imaging sessions. After the preprocessing, a uniform representation is followed for all the images, and the hand is segmented. This process is also known as hand normalization (Yörük et al., 2006). Next, a set of specific features are computed from the whole hand and/or particular to the fingers. In the case of finger biometry, the fingers are segmented and separated from the actual hand. Next, the features are computed from the normalized fingers. The anatomic intricacy of the thumb precludes its feature computation in some cases that implicates the significance of four-finger biometrics.

Feature definition: A feature set is computed based on the geometric measurements (e.g., area, length, width, etc.) of various aspects and/or shape descriptors using different spatial transformation techniques, such as the Fourier transform, wavelet transform, and others (Dutağaci et al., 2008; Luque-Baena et al., 2013). The dimension of a geometric feature set (usually, within a hundred) is somewhat smaller than shape-based features (generally, few hundred or more). Thus, dimensionality reduction method (such as principal component analysis) is usually followed, and only the essential attributes are chosen while the rest are discarded. Notably, geometric feature extraction and feature matching tasks are slightly more straightforward than the silhouette-based features due to the smaller feature space regarding the computational complexity. The main advantage of a shape descriptor is the invariance property under one or more affine transformation(s), such as translation, rotation, scaling, and others (Charfi et al., 2014). Some common features are summarized in Table 5.2.

Feature relevance: The uniqueness of a feature determines its significance in the feature set. Relevant features are selected to define the identity while insignificant and noisy features are discarded. Feature relevance exhibits the aptness for dimensionality reduction too. The relevant features are chosen, and an optimal subset is formulated using a suitable metric, such as the MI. The number of useful features is a general condition for subset formulation and stopping criterion. The relevance computation is explained in Section 5.4. A few approaches on the hand and finger biometrics having the benefits of feature selection are summarized in Table 5.3.

Feature-level fusion: Another paradigm is to combine different feature spaces. Homogeneous feature sets (e.g., two different hand geometric feature vectors of the same subject) are conflated with the mean values of corresponding features (Bera et al., 2015). Compatible

TABLE 5.2

Finger Biometric Feature Definition

Geometric Features (Luque-Baena et al., 2013; Bera and Bhattacharjee, 2017)

Area (A): the number of boundary pixels of the hand (or finger).

Length: the Euclidean distance between the finger-tip pixel and mid-pixel of the finger-baseline.

Widths: the breadths of a finger at different latitudes along its length.

Extent: the ratio of the area and minimum bounding rectangle area (MBRA), given as A/MBRA

Perimeter (P): the cumulative distance between each pair of adjacent contour pixels.

Major-axis length and minor-axis length: the major-axis and minor-axis lengths of the fitted ellipse over every finger. The biaxial lengths are invariant to rotation and translation.

Solidity: the ratio of area and convex hull area (CA), given as A/CA

Circularity: defined as $4\pi A/P^2$

Compactness: measured as P^2/A

Interfinger angle: defined concerning the pivotal finger axes or located vital points.

Equivalent diameter: computed as $(4 \times \text{Area}/\pi)^{0.5}$

Distance from Centroid (DC): the distance from the centroid (x_c, y_c) of a finger to a contour pixel (x, y), defined as $[(x - x_c)^2 + (y - y_c)^2]^{0.5}$

Shape-Based Features

FD: magnitudes of the points after the discrete Fourier transformation (Kang and Wu, 2014).

WD: the coefficients after the wavelet transformation with multiresolution analysis (Sharma et al., 2015).

SC: defined with n pixels representing a shape. For any pixel, a histogram of the relative coordinates of remaining $(n - 1)$ pixels is computed. The histogram describes the SC of that pixel, and the bins of the histogram are defined in a log-polar space (Hu et al., 2012).

SIFT: the local maxima and minima key points using Gaussian linear transformation at various scale-spaces. Noisy and contour pixels are removed (Charfi et al., 2014).

Hu moment invariant (HMI): defined with the second and third normalized central moments; consisting of six complete orthogonal invariants and one skew orthogonal invariant based on algebraic invariants, and representing the statistical properties (Luque-Baena et al., 2013).

Zernike moment invariant (ZMI): based on a set of complex polynomials that form a complete orthogonal set over the interior of a unit circle and defined as the projection of the image on these orthogonal basis functions (Choraś and Choraś, 2006).

Angular radial transform (ART): based on the polar coordinate system in which the sinusoidal basis functions are defined on a unit disc (Yörük, Dutağaci, and Sankur, 2006).

Invariance property	DC	FD	WD	SC	HMI	ZMI	ART	SIFT
Translation	√	√	√	√	√	√	√	√
Rotation		√	√	√	√	√	√	√
Scaling		√	√	√	√	√	√	√
Noise							√	√
Illumination								√
Distortion								√

feature sets are more comfortable to manipulate and formulate a uniform feature template (Ross et al., 2006). Alternatively, heterogeneous feature templates (e.g., fingerprint minutiae and hand geometric features) are concatenated after normalizing them into a standard feature space. Combining two different vectors into one also increases the overall dimension of the feature vector. Another solution is to match the heterogeneous feature vectors separately and fuse the matching scores to render the final score or to make a decision about the claimed identity (Ross and Jain, 2003; Jain, Nandakumar, and Ross, 2005; Hanmandlu, Grover, Madasu, and Vasirkala, 2010). A vignette about the various levels of fusion techniques in the context of hand biometrics is illustrated in Figure 5.3. Mainly, the sensor-level (Kanhangad, Kumar, and Zhang, 2011), feature-level (Dutağaci, Sankur, and Yörük, 2008),

TABLE 5.3

Hand Biometric Approaches with Feature Selection

S. No.	Author	Feature Extraction	Feature Selection Method	Experimentation
1	Reillo et al. (2000)	In total, 31 geometric features are measured from hand. It includes the widths and heights of fingers, deviations, and interfinger angles.	*Class variability ratio:* relevance of the features is estimated by the proportion of the interclass variability and intraclass variability. The class-variability is determined using the standard deviation and the mean of feature and number of classes. A higher value ratio implies a more relevant feature.	Database: 20 persons. 25 discriminative features are selected. In total, 97% classification accuracy and much lesser than 10% EERs have been achieved using Gaussian mixture model (GMM).
2	Kumar and Zhang (2005)	In total, 23 geometric features are computed from hand. It includes the perimeter, solidity, extent, palm length and width, finger length and widths, etc.	*Correlation-based feature selection (CFS):* it is a classifier-independent algorithm, and the *Pearson* correlation metric is used. The CFS has effectively reduced the number of features with maintaining similar performances.	Database: 100 subjects. In total, 15 relevant features are selected. Identification accuracy of 87.8% has been achieved using the logistic model tree (LMT) classifier.
3	Luque-Baena et al. (2013)	Total of 403 features are computed from each finger and hand. It defines the finger length, compactness, widths, rectangularity, extent, solidity, FDs, HMI, and others.	*Genetic algorithm (GA) with Mutual information (GA-MI):* GA is applied for feature selection and MI is employed in the fitness function to find out the correlation between a pair of features and to eliminate the redundancy among features. Total of 100 different executions of GA have been tested for relevant feature selection.	Databases: GPDS (144 subjects), IITD (137 subjects), and CASIA (100 subjects). Selects about 50 features, identification precision about 97% has been achieved with the latter two databases using GA-LDA and EERs of 4%–5%.
4	Bera et al. (2017)	Total of 30 geometric features per finger are defined. It includes the area, solidity, extent, major and minor axes lengths of an ellipse fitted over the finger, 10 CDs, widths at 10 different positions of the finger, etc.	*Rank-based forward-backward (RFoBa):* A rank is assigned to every feature according to its relevance. Based on the rank, the forward selection is followed first. Then, the backward elimination is applied successively to formulate two optimal subsets of selected features.	Database: Bosphorus (638, subjects). With the 4 fingers and 12 selected features per finger of the right hand, identification success of 96.56% using the random forest classifier and EER of 7.8% have been achieved.

score-level (Hanmandlu, Grover, Madasu, and Vasirkala, 2010), rank-level (Kumar and Shekhar, 2011), and decision-level (Hanmandlu, Kumar, Madasu, and Yarlagadda, 2008) fusion techniques are shown in Figure 5.3.

In conjunction with hand and finger geometry, some other hand-based modalities are fused at various levels of design, such as the fingerprint, palmprint, hand vein, finger

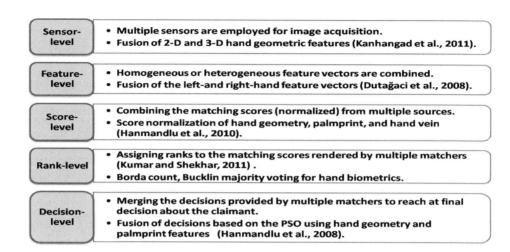

FIGURE 5.3
Different implementation levels of fusion-based hand biometric approaches.

knuckle print, etc. (Peng, Li, Abd El-Latif, and Niu, 2015). In addition, a non-hand-based trait, like the facial pattern, is fused by Tsalakanidou et al. (2007).

Furthermore, other modules (such as the sensor module) have been improved for a robust solution regarding the accuracy and computational time. However, the approaches directed toward feature selection techniques need more attention than the traditional all-feature usability approach. Hence, this chapter also focuses on this objective.

5.3 Preprocessing Methodology of Finger Geometry

The preprocessing of hand images necessitates a few successive steps to segment the hand from a darker background. In the literature, different techniques have been devised for hand segmentation. For example, the k-means clustering algorithm has been used for hand segmentation in Kang and Wu (2014). After segmenting the hand's region of interest (ROI), the fingers are isolated and separated from each other in the case of finger biometry. Bera and Bhattacharjee (2017); and Bera, Bhattacharjee, and Nasipuri (2017) briefly describe the finger segmentation method.

5.3.1 Elementary Preprocessing of Hand Image

The preprocessing tasks described in different available works follow similitude operations. First, a color hand image (L_C) is converted into an equivalent grayscale image (L_G). Noise at the sensor level may be introduced due to environmental factors, such as dirt, inconsistent lighting condition, etc. The noise associated with an input image is removed by a suitable filter, such as a median or Gaussian filter. Next, the grayscale image is converted into a binary image (L_B) using the Otsu's (1979) method for thresholding. A modification of Otsu's approach has also been applied in Kang and Wu (2014). The hand contour is obtained from the binary image using an edge detection algorithm, such as the *Sobel's*

(Luque-Baena et al., 2013) or *Canny's* edge detector (Canny, 1986). The hand contour comprises several components due to any worn hand accessory, such as a wristwatch, bracelet, ring(s), etc. A ring artifact removal method has been followed to overcome the occlusion due to hand accessories in Yörük et al. (2006). Morphological opening corrections are inevitable to recuperate the left-out contour pixels and smooth the silhouette as a connected component of interest. Next, the essential contour component is selected while other insignificant parts (if any) are neglected for further operations.

In a contact-free image acquisition system, a consistent orientation of every hand silhouette is necessary for hand normalization. For rotating the hand contour, the ellipse fitting method is followed in some works such as Sharma et al. (2015). To achieve a uniform orientation, an ellipse is fitted around the ROI so that the major axis passes through the centroid of ROI and coincides with the Cartesian y-axis. The centroid (x_c, y_c) is defined based on the moments, $m_{i,j}$, of the binary ROI.

$$\left(x_c, y_c\right) = \left(m_{1,0}/m_{0,0}, m_{0,1}/m_{0,0}\right) \tag{5.1}$$

The angle of rotation (θ) is estimated as:

$$\theta = 0.5 \times \tan^{-1}\left(\frac{2m_{1,1}}{m_{2,0} - m_{0,2}}\right) \tag{5.2}$$

After the rotation of the image containing ROI, the resulting image is denoted as L_R. These simple processing steps are common in several state-of-the-art techniques of hand biometrics. However, finger isolation can be achieved by different methods.

An approach for finger separation from the hand ROI is followed in Bera and Bhattacharjee (2017) and Bera, Bhattacharjee, and Nasipuri (2017). In their approach, only simple arithmetic and Boolean operations are followed for finger segmentation while some other methods used the reference points for finger ROI extraction. The critical steps of this method are described below and ideated in Figure 5.4. Image L_G is rotated with the equal θ degree, and resulting in image L_θ. Images L_θ and L_R are used for succeeding operations.

5.3.2 Finger Segmentation

The segmentation of fingers from the entire hand shape is another crucial task. The tip-valley points are located for this purpose. In Kang and Wu (2014), the finger-valley points and necessary reference points are used for segmenting the four fingers. The radial distance map of the hand contour is a familiar method for localization of the tip-valley points (Yörük et al., 2006). Contrarily, the localization of reference points is not required for finger isolation in Bera and Bhattacharjee (2017). In this approach, the primary hand contour component is disunited into the left- and right-profile (RP) at finger level. Consecutive steps of operations have been followed to isolate the finger profiles from the entire hand contour component.

In the first step, grayscale intensity variation between every pair of neighbor pixels of L_θ is computed by subtraction at the pixel level that follows a similitude operation of computing the first-order horizontal gradient magnitude.

$$L_H = \eta \times \left[L_\theta\left(p, q\right) - L_\theta\left(p, q + 1\right)\right] \text{ where } \eta = 1/2^h; \quad h = 1, 2, 3, \dots \tag{5.3}$$

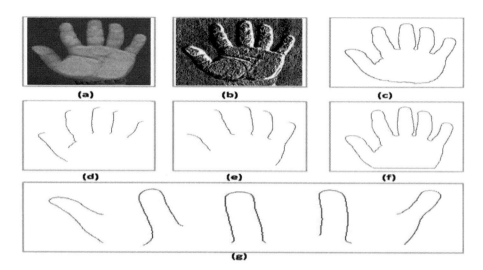

FIGURE 5.4
Significant outcomes of the preprocessing steps. (a) grayscale hand image, (b) grayscale image transformation, (c) most vital component of the hand contour, (d) left sides of the finger contour profiles, (e) right sides of the finger contour profiles, (f) wrist removal and normalized hand shape, (g) segmented fingers from the left to right direction: thumb, index, middle, ring, and little finger.

Where, p defines the altitude index, and q is the breadth index of L_0. Next, L_H is converted into a binary image (L_{BW}) according to the intensity profile of pixels in L_H. The variations of L_H formation at a different scale have been explained in Bera, Bhattacharjee, and Nasipuri (2017) for determining the constant value (here, h = 1, i.e., $\eta = 0.5$ in equation 5.3). The non-zero intensities of L_H are converted into white (1 value) pixels while the remnant pixels are regarded as black (0 value). Image L_H contains white pixels mainly on the right side of each finger due to the intensity changeover from foreground to background.

In the next step, binary image subtraction is applied between L_R and L_{BW} that results in the left profile (LP) of every finger shape.

$$L_{LP} = L_R - L_{BW} \qquad (5.4)$$

Binary image L_{LP} represents the isolated contour segments on the left side of all fingers. Morphological corrections are followed such as closing and bridging with a suitable structure element of 3×3 to eliminate an undesirable effect of noise or intensity alternation.

In the third step, Boolean *xor* operation is applied to retrieve the remnant contour segments on the other side, namely, the RPs of the fingers.

$$L_{RP} = L_R \oplus L_{LP} \qquad (5.5)$$

Image L_{LP} contains the minor part of the profile segment of the little finger, and L_{RP} comprises the lower portion of the contour fragment of the thumb. In the last step, these two lower regions of the respective profile segments are discarded for uniformity in shape representation by using a straight reference line which can be determined in various ways, such as a certain distance from the centroid of the hand. This wrist removal is significant to avert errors during feature computation (Yörük et al., 2006). The conflation of the corresponding left- and right-profiles represents a single and segregated finger.

The entire process from an input image to finger isolation is known as finger normalization. In a four-finger-based approach, after finger segregation, the thumb is excluded explicitly before feature extraction.

5.3.3 Geometric Feature Set Computation

The computation of salient features is another crucial task that may also be inevitably united with some noisy data. The noisy features degrade the performance during the testing on the actual feature space. In hand biometrics, the geometric measurements and shape-based descriptors are extracted as features. Herein, a set of only 13 easily computable geometric features are extracted from each of the normalized four fingers (the number of the feature is mentioned within brackets). The feature set includes:

- Area (1),
- Solidity (1),
- Equivalent diameter (1),
- Major-axis and minor-axis lengths of the ellipse fitted on the finger (2),
- Width per phalanx of a finger (3), and
- Distance from the centroid of the finger to five equidistant contour pixels (5).

These geometric features are defined in Table 5.2. Now, feature normalization is another issue when the magnitudes of the features are in various scalar ranges. It causes a feature fe_i with a higher value to lead over another fe_j with a lower value. It may also cause other features to turn irrelevant during the experiment. So, all the features are normalized to [0, 1] using the *min-max* rule,

$$f_{i,j} = \left(fe_{i,j} - \min\left(fe_i\right)\right) \big/ \left(\max\left(fe_i\right) - \min\left(fe_i\right)\right) \qquad (5.6)$$

where the $fe_{i,j}$ represents the ith feature of the jth sample, and $\max(fe_i)$ and $\min(fe_i)$ represent the maximum and minimum values of the ith feature, respectively. The normalized features are denoted as $f_{i,j}$ or simply f_i in the set $\mathcal{F} = \{f_i\}_{i=1}^{13}$ which is used for subset selection next.

5.4 Feature Selection Algorithms

The essential features are selected mainly using the filter and wrapper methods (Guyon and Elisseeff, 2003; Kohavi and John, 1997). The embedded and hybrid are two other derived methods. The filter method is classifier independent because it selects the features irrespective of any classification model. Mainly, the correlation and information theory metrics are applied to find out the relevance criterion of a feature (Peng, Long, and Ding, 2005; Estévez, Tesmer, Perez, and Zurada, 2009). Based on the order of relevance, a rank is assigned to every feature during selection. On the contrary, the wrapper method uses a classifier to predict the significance regarding its classification correctness. However, its inherent drawback is that the feature selection model may not work on other datasets using some different classifiers. Moreover, the wrapper method tends to lead the overfitting issue, i.e., it may not work well on the test feature sets, and the performance may degrade. The wrapper method is also computationally costly compared to the filter method. Thus, the target is to minimize the error rates while tested on the actual data. The embedded process follows the similitude operation of the wrapper method while the hybrid approach applies both the filter and wrapper methods. The feature selection method is formulated next.

Consider $\mathcal{F}$ is an n-dimensional feature vector where the ith feature is denoted by $f_i \in \mathcal{F}$. According to the uniqueness, a f_i is classified namely as relevant (f_{rel}), irrelevant (f_{irl}), and redundant (f_{red}) (Lashkia and Anthony, 2004). The relevance of f_i is determined in the training phase using a predictive distance metric such as the Euclidean. After the selection process, a subset of selected features, denoted by $S = \{f_i\}_{i=1}^{p}$, is obtained, where p < n. The principal objective of a feature selection algorithm is to determine the right combination of relevant features from the input set $\mathcal{F}$ to attain the minimum classification error. Thus, a subset S with the minimum cardinality must be formulated that maximizes classification accuracy (ψ). Now, related terminologies are defined from the perspective of forward selection and backward elimination.

Relevant and irrelevant feature: A feature f_i is said to be relevant if its inclusion in the subset improves the classification accuracy during the additive forward selection, and if removal of f_i degrades the accuracy during backward elimination. Otherwise, f_i is said to be irrelevant.

Redundant feature: A feature f_i is said to be unnecessary if its inclusion or elimination does not affect the classification accuracy significantly. It means that a f_i that does not improve (while included) or reduce (while discarded) the evaluation criterion and is known to be a redundant feature. Sometimes, the removal of an unnecessary feature can enhance the accuracy.

From the perspective of forward selection, the relevance can be determined as:

$$f_i \rightarrow \begin{cases} f_{rdt} & \text{if } \psi\left(S \cup \{f_i\}\right) - \psi(S) \approx \varepsilon \\ f_{rel} & \text{if } \psi\left(S \cup \{f_i\}\right) - \psi(S) \geq \Delta \\ f_{irl} & \text{otherwise} \end{cases} \qquad (5.7)$$

Alternatively, from the standpoint of backward elimination, significance can be computed as:

$$f_i \rightarrow \begin{cases} f_{rdt} & \text{if } \psi\left(S \setminus \{f_i\}\right) - \psi(S) \approx \varepsilon \\ f_{rel} & \text{if } \psi\left(S \setminus \{f_i\}\right) - \psi(S) \leq \Delta \\ f_{irl} & \text{otherwise} \end{cases} \qquad (5.8)$$

Where, ε and Δ represent the permissible values to determine the redundancy and relevance criteria, respectively. The relevant, irrelevant, and redundant features are denoted by f_{rel}, f_{irl}, and f_{rdt}, respectively. Generally, $\Delta > \varepsilon$ and $\varepsilon \approx 0$ are considered. It (i.e., $\Delta > \varepsilon$) implies that the relevance criterion also validates the checking for redundancy condition. Alternatively, instead of finding two threshold values (ε and Δ), a single limit Δ can be used with a nonzero value that maintains $0 < \Delta < 1$.

According to the present context of four-finger biometrics, a feature can also be classified into two types, namely, global and local. The aim is to eliminate a redundant feature that reduces the cardinality of subset S by considering these two features. However, the researchers may have a different opinion about the classification of global and local features in the context of hand biometrics; here, the definitions are given below.

Global feature: A feature is defined as global when it is attributed with the characteristic computed from the entire hand or from each finger. For example, the area of a normalized hand can be considered as a universal feature. Also, the length of every finger collectively can be termed as a global feature because the same feature is attributed with the same characteristic for the four fingers at the finger level. Thus, a single feature at the hand level or a set of the same features per finger at the finger level is termed as a global feature.

Local feature: A feature is considered local when it is valid for only one or more particular finger(s) rather than all of the fingers. In this case, every feature is considered independently instead of a group of features. For example, the area of a particular finger is local to that finger.

In the case of finger biometrics where only isolated fingers are considered, a global feature is defined as:

$$f_i^{gbl} \rightarrow \left\{ f_i^{Index}, f_i^{Middle}, f_i^{Ring}, f_i^{Little} \right\} \tag{5.9}$$

In the present context, regarding the cardinality, a particular feature can also be defined as global or local while the defined four fingers are considered.

$$f_{i \rightarrow} \begin{cases} f_i^{gbl} & \text{if } |f_i| = 4 \\ f_i^{lcl} & \text{if } |f_i| \leq 3 \end{cases} \tag{5.10}$$

A global feature (f_{gbl}) is called relevant during forward selection if:

$$\psi\left(S \cup \left\{f_i^{gbl}\right\}\right) - \psi(S) \geq \Delta = \psi\left(S \cup \left\{f_i^{Index}, f_i^{Middle}, f_i^{Ring}, f_i^{Little}\right\}\right) - \psi(S) \geq \Delta \tag{5.11}$$

Example: Let $\mathcal{F}$ is an input set of 12 features ($|\mathcal{F}| = 12$), representing the major-axis length (ma), area (ar), and solidity (sl) of each finger. Its representation at local to independent finger level is given as:

$$\mathcal{F} = \left\{ f_{ma}^{Index}, f_{ma}^{Middle}, f_{ma}^{Ring}, f_{ma}^{Little}, f_{ar}^{Index}, \ldots, f_{ar}^{Little}, f_{sl}^{Index}, \ldots, f_{sl}^{Little} \right\} \tag{5.12}$$

According to the order, from the index- to little-finger, numeric labeling is followed as:

$$\mathcal{F} = \left\{ f_{ma}^1, f_{ma}^2, f_{ma}^3, f_{ma}^4, f_{ar}^5, \ldots, f_{ar}^8, f_{sl}^9, \ldots, f_{sl}^{12} \right\}$$

$$= \left\{ \left\{ f_{ma}^{1-4} \right\}, \left\{ f_{ar}^{5-8} \right\}, \left\{ f_{sl}^{9-12} \right\} \right\} \tag{5.13}$$

$$= \left\{ \left\{ f_{ma}^{gbl_1} \right\}, \left\{ f_{ar}^{gbl_2} \right\}, \left\{ f_{sl}^{gbl_3} \right\} \right\}$$

The above formation of $\mathcal{F}$ in equation 5.13 is delineated as the global representation of the features. The reason behind considering local features is to validate a feature explicitly at a finger level which is relevant than a group of a similar feature.

Now, suppose the areas of middle $\left(f_{ar}^{Middle}\right)$ and index $\left(f_{ar}^{Index}\right)$ fingers are relevant while the same features of the other fingers are irrelevant or redundant. Hence, considering the areas of only these two fingers is more appropriate than selecting the areas of the four fingers. Therefore, these two precise local features reduce the cardinality of S by removing the same attribute of the other fingers. Now, a question may arise—why not consider only the local features? The main reason is that the local feature extraction (e.g., SIFT descriptors) and selection requires a high computational cost for a large feature space. It may not also yield an optimal solution in every situation. On the other side, the global features are tested as a group which reduces the computational time regarding a number of required iterations by a factor of four (i.e., the number of fingers or the cardinality of the group of features). Thus, there is a tradeoff between the global and local feature selection regarding the computational complexity and optimality in subset selection regarding the accuracy and cardinality of the subspace.

Algorithm 5.1: Forward Selection

Input: *Feature set:* $\mathcal{F} = \{f_i\}_{i=1}^n$
Output: Selected feature subset: $S_{Fo} = \{f_i\}_{i=1}^m$
   1. Initialize: the number of selected features $m \leftarrow 0$;
      permissible threshold accuracy to determine the relevant feature Δ; and
      selected feature subset, $S_{Fo} \leftarrow \{\Phi\}$.
   2. Select the most pertinent feature as the first feature f_1 from $\mathcal{F}$
      $S_{Fo} \leftarrow f_1$
      $\mathcal{F} \leftarrow \mathcal{F} \backslash \{f_1\}$
      $m \leftarrow 1$
   3. **for** $i \leftarrow 2, 3, \dots, n$
         if $\psi(S_{Fo} \cup f_i) - \psi(S_{Fo}) \geq \Delta$
            $S_{Fo} \leftarrow S_{Fo} \cup \{f_i\}$;
            $F \leftarrow F \backslash \{f_i\}$
            $i \leftarrow i + 1$;
            $m \leftarrow m + 1$;
         end if
      end for
   4. return subset $S_{Fo} = \{f_i\}_{i=1}^m$

Algorithm 5.2: Backward Elimination

Input: Feature set: $S_{Fo} = \{f_i\}_{i=1}^m$
Output: Selected feature subset: $S_{FoBa} = \{f_i\}_{i=1}^p$
   1. Initialize: selected feature subset $S_{FoBa} \leftarrow S_{Fo}$
      the number of selected features $p \leftarrow m$;
      permissible accuracy to remove the irrelevant feature, Δ
   2. **for** $i \leftarrow 1, 2, \dots, m$
         if $\varphi(S_{FoBa} \backslash f_i) - \varphi(S_{FoBa}) \geq \Delta$
            $S_{FoBa} \leftarrow S_{FoBa} \backslash \{f_i\}$
            $i \leftarrow i + 1$;
            $p \leftarrow p - 1$;

end if
end for
3. return subset $S_{FoBa} = \{f_i\}_{i=1}^{P}$
   The time complexity depends on the cardinality of the input set $\mathcal{F}$ and selected feature subset S. It can be estimated as $O(|\mathcal{F}|\cdot|S|)$ for both of the algorithms.

In the literature, several fundamental feature selection algorithms are devised. A comparative study has been provided by Jain, Duin, and Mao (2000). Here, the forward-selection and backward-elimination algorithms are described. A successive combination of both algorithms is formulated by Zhang (2011) to overcome the limitations of both methods, and the algorithm is named as the *adaptive forward-backward (FoBa)* algorithm. In this greedy FoBa algorithm, forward-selection (*Algorithm 1*) is followed to find out an initial subset of relevant features. Subsequently, backward-elimination (*Algorithm 2*) is followed on the ensued subset to reduce further the cardinality and to improve the accuracy. Finally, an optimal subset S of selected features is formulated based on which experiments are conducted on the test data.

A variation of the FoBa algorithm is presented by Bera and Bhattacharjee (2017), known as the *rank-based forward-backward (R-FoBa)* feature selection in the perspective of finger biometrics. In the R-FoBa algorithm, a rank is assigned to every global feature according to the relevance, evaluated by a k-Nearest Neighbor (k-NN) classifier during the learning phase. According to the aptness, the most significant feature is delineated as the most significant (rank-1) feature. Based on the ranking order (i.e., from the highest to the lowest order of relevance), all the global features one at a time are tested for subset formulation. The R-FoBa is validated by the tenfold cross-validation using the *Pearson* correlation metric (Guyon and Elisseeff, 2003). The *out-of-bag* classification error has also been estimated based on the selected subset by the bag of random classification trees (Breiman, 1996, 2001). In this method, the relevance is determined by a simple distance-based k-NN algorithm during training. Finally, testing on the real data is conducted by the wk-NN and random forest **(RF)** classifiers (Figure 5.5).

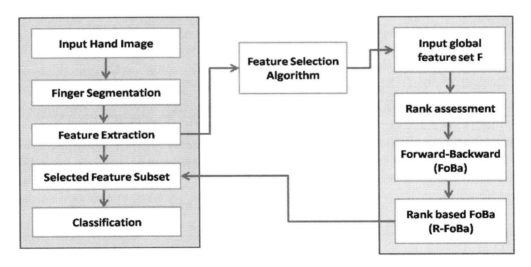

FIGURE 5.5
Finger biometric system with the feature selection scheme.

Another approach is the local feature selection using the same basis of FoBa, with an important target of further decrement of subset cardinality possibly with an improvement of accuracy. In this case, all the features are independently assessed for its inclusion or removal rather than a group of the same attribute. This method is named as the local feature selection. The formulation of the local subset is highlighted in the next section.

5.5 Experimentation

First, a brief description of the hand database and classifiers are presented. Next, the analyses of the various experiments are performed.

5.5.1 Database Description

The Bosphorus hand database has been created for research purposes at Boğaziçi University, Turkey (Dutağaci et al., 2008). This database represents the contact-free color hand images of the individuals who are the students and staff members of different universities of Turkey and France. The age variation of the subjects is between 20 and 50 years. This database contains three images per hand which have been acquired by a scanner (HP Scanjet) with 383×526 pixels at 45-dpi, at three different imaging sessions. The time lapse is between 2 weeks and 3 years, and the average time lapse is 1 year. The intraclass pose variation is sufficient. This publicly and freely available dataset is preferred because of its larger population space.

5.5.2 Classifier Description

The k-NN is a widely experimented supervised algorithm, pertinent for pattern recognition and machine learning tasks. It is defined with a simple algorithmic model based on a specific distance metric with lower time complexity. The number of neighbors is scalable, and the distance between the neighboring feature vectors can be measured using several proximity metrics such as the Euclidean, Mahalanobis, Minkowski, etc. The class label of a query sample feature vector is determined based on the majority voting of the proximity measured from the trained data samples.

(a) **Weighted k-Nearest Neighbor (wk-NN)**

The weight assignment to every feature is a variation of the k-NN algorithm to prioritize a feature over another one (Gou, Du, Zhang, and Xiong, 2012; Bera, Bhattacharjee, and Nasipuri, 2018). The weight can be determined according to the relevance of a feature using a criterion or using a suitable classification algorithm. Mainly, the distance- or correlation-based metric is followed for weight assignment. The weights are estimated based on the learning observations at the training phase and normalized within the [0, 1] range. Here, the assigned weights are based on the Euclidean distance (d) of a simple k-NN classifier with k = 3 during the training (Bera et al., 2018). The weight w_i of a feature f_i is defined as

$$w_i = \frac{d(f_i)}{\displaystyle\sum_{f_i \in F} d(f_i)} \qquad (5.14)$$

where $d(f_i)$ represents the accuracy of f_i evaluated regarding the distance metric. The distance d between training (Tr) and testing (Ts) vector is calculated as:

$$d = \sqrt{\sum_{i=1}^{n} (Tr_i - Ts_i)^2} \qquad (5.15)$$

and $\sum_{i=1}^{n} w_i = 1$. The trained and claimed test features are denoted by Tr_i and Ts_i, respectively. Based on the weight assignment, the distance metric is defined with the w_i values as:

$$d' = d \cdot w_i = \sqrt{\sum_{i=1}^{n} (Tr_i - Ts_i)^2 \cdot w_i} \qquad (5.16)$$

Now, the wk-NN classifier is used for testing. Thresholding on the weight (th_w) is used to judge whether a feature should be included in $\mathcal{F}$ or not, i.e., $f_i \in F$ if $w_i \geq th_w$.

(b) **Random Forest**

This classification model was introduced by Breiman (1996, 2001). The RF is based on a collection of decision trees with a high precision of prediction. It is apposite for a more extensive feature space which may also represent noisy data. Every tree of the RF predicts a decision on the test dataset independently. The tree bagging method is used for classification. An unknown feature vector is tested for which a predicted score is determined based on the trained ensemble dataset. The predicted score represents the weighted average of matching probabilities determined by each classification tree of the RF. The accuracy can be improved with more ensemble trees.

5.5.3 Experimental Description

The experiments for the identification and verification are carried out for the training and testing phases which include the disjoint sets of subjects with the ratio of 2:3. During the training, 200 subjects are chosen randomly while the testing is carried out with another disjoint set of 300 subjects. Altering the subjects for training and testing with maintaining the disjoint property, several experimental sets have been formulated, and the average results are reported. Mainly, different experimental models for identification have been tested to validate the present scheme.

5.5.3.1 Identification

The experimental scenarios are based on the feature selection methods, and the effectiveness of every finger is also assessed singly. Before conducting the significant experiments, the following initial observations are made on the actual testing set having an initial cardinality of 52 features. The k-NN classifier evaluates the accuracies in this method, given in Table 5.4. It is evident that feature normalization and weight assignment are more significant to conduct the experiments rather than using the actual values of the features.

During the training, the weight is computed for each feature and assigned accordingly. An average value is computed from the weights assigned to all features. The average value is regarded as a threshold $(th_w = 3.98)$ that is used to eliminate unnecessary features. If the accuracy of an independent feature is more significant than the threshold, then the feature

TABLE 5.4

Identification Accuracies (%) of the Testing Subjects

Classifier	Subjects	Actual Feature Set	Feature Normalization	Weight Assignment
k-NN	300	93	94.67	95.67

TABLE 5.5

Finger-Level Identification Accuracies (%) with Weight Assignment

Phase	Subjects	Classifier	Index-Finger	Middle-Finger	Ring-Finger	Little-Finger
Training	200	wk-NN	58.5	61	67.5	60.5
		RF	75.5	76	82	81
Testing	300	wk-NN	53	64.34	65.67	58
		RF	68.67	80.34	84	69

TABLE 5.6

Weighted Feature-Level Assessment for Rank Assignment during Training

Assessment	f_1	f_2	f_3	f_4	f_5	f_6	f_7	f_8	f_9	f_{10}	f_{11}	f_{12}	f_{13}
wk-NN	44	47	22	22	19.5	25	26.5	33.5	36	20.5	29	21	24.5
Rank	2	1	10	9	13	7	6	4	3	12	5	11	8

is chosen, otherwise delineated as an unnecessary one. This method formulates a subset of essential features representing only 19 features, rendering 91% identification accuracy by the k-NN classifier. However, this simple weighted threshold is not regarded as an appropriate method for feature selection.

Next, the effectiveness of every finger (excluding the thumb) before feature selection is evaluated by the wk-NN and RF classifiers separately is given in Table 5.5. A rank is assigned to every feature during the training to illustrate the rank-based feature selection, i.e., R-FoBa presented in Bera and Bhattacharjee (2017), according to the highest to the lowest order of classification accuracy rendered by the wk-NN classification scheme, presented in Table 5.6.

Afterward, the feature selection algorithms are applied to the input set $\mathcal{F}$. For this purpose, set F is normalized using the *min-max* rule and weights are assigned accordingly. The features are selected using the wk-NN classifier during the training while the testing is conducted with the wk-NN and RF classifiers. Two feature selection methods, namely the global and local subsets, are formulated based on the trained subject space. Also, the forward-selection is carried out by choosing the features in two ways. First, the features are arbitrarily selected and tested whether the feature should be included in the subset or discarded otherwise. This scheme of randomly feature selection is regarded here as the FoBa which does not involve any prior rank assessment to avoid the pre-computation of rank. To determine the relevant feature, $\Delta = 0.003$ and for the redundant feature elimination $\varepsilon = 0$ have been followed. During the training, the outcomes of the feature selection methods based on the 200 subjects are provided in Table 5.7. It is evident that the R-FoBa performs better than the FoBa method concerning accuracy. On the other side, the dimensionality reduction of the set $\mathcal{F}$ is remarkable. It is evident that the local selection method outperforms the global selection regarding dimensionality optimization, i.e., the cardinality of the optimal subset. Next, using the same strategies adopted for feature selection, the test feature set is assessed, and the results are given in Table 5.8.

TABLE 5.7

Identification Accuracies (%) Using Feature Selection Algorithms during Training

	Global Feature Selection				Local Feature Selection			
Classifier	Cardinality	FoBa	Cardinality	R-FoBa	Cardinality	FoBa	Cardinality	R-FoBa
wk-NN	36	93.5	40	94.5	24	94.34	25	95

TABLE 5.8

Identification Accuracies (%) Using Feature Selection Algorithms during Testing

	Global Feature Selection				Local Feature Selection			
Classifier	Cardinality	FoBa	Cardinality	R-FoBa	Cardinality	FoBa	Cardinality	R-FoBa
wk-NN	36	95	40	96.34	24	96	25	97
RF		97		97.67		96.67		98.67

5.5.3.2 Feature Overfittting

Feature *overfitting* is a crucial issue regarding feature selection. Here, the tenfold cross-validation error and the *out-of-bag* (*oob*) error are estimated based on the selected feature subsets. The tenfold cross-validation errors estimated by the wk-NN using the correlation metric for the global and local R-FoBa are shown in Figure 5.6. The comparison between

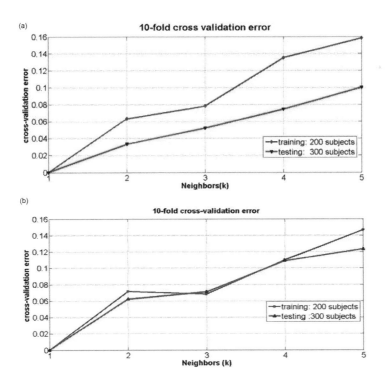

FIGURE 5.6

Ten-fold cross-validation errors by wk-NN using correlation. (a) Global R-FoBa selection, (b) local R-FoBa selection.

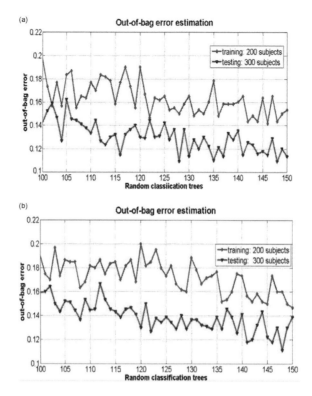

FIGURE 5.7
Out-of-bag error estimated by the ensemble of classification trees. (a) Global R-FoBa selection, (b) local R-FoBa selection.

the selected feature subsets of the training and testing cases are highlighted. For the testing case, altogether 900 (300 subjects × 3 samples per subject) feature vectors are partitioned, out of which 90% samples are trained and 10% samples are used for testing at each fold of validation test. The average validation error of the tenfolds is reported here for each neighbor which has been varied from $k = 1$ to $k = 5$. Regarding all the cases of the neighbors, i.e., $k = 1$ to $k = 5$, the average cross-validation error for the training samples is 0.079, and for the testing samples, the error is 0.073. Hence, the feature-level correlation is more significant for the test samples in this present scenario.

Alternatively, the *oob* errors estimated by the ensemble of random classification trees using the same local feature subsets are ideated in Figure 5.7. For the testing data (900 feature vectors and utmost 150 classification trees), a total of 900 × 150 observations are made. The average mean squared error is computed for the *oob* feature vectors. In this error estimation technique, the error rates of the testing samples are lesser than the training samples too. As the number of trees grows, the *oob* error tends to decrease usually. The average and minimum *oob* errors are given in Table 5.9.

The contribution of each selected feature for both of the subsets are estimated by the wk-NN classifier and is pictorially ideated in Figure 5.8. The local subset contains 25 features (marked with a vertical line in Figure 5.8) which finally offer 97% accuracy. Likewise, the global subset represents 40 features which collectively provide 96.34% correctness. It is evident from Figure 5.8 that the improvement of accuracies using the global subset,

TABLE 5.9

Out-of-Bag (oob) Error Estimation by the RF on the Ensued Subsets

	Global R-FoBa (40 Features)		Local R-FoBa (25 Features)	
Out-of-Bag Error	Average	Minimum	Average	Minimum
Training (200 subjects)	0.163	0.141	0.173	0.146
Testing (300 subjects)	0.130	0.108	0.139	0.111

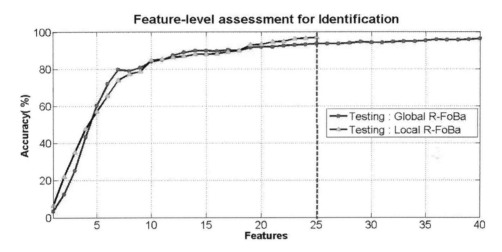

FIGURE 5.8
Feature-level accuracy assessment by the wk-NN classifier during testing.

specifically after the 25th feature is not satisfactory. In this case, a little compromise with accuracy can be preferred to signify the objective of dimensionality optimization. It may also reduce the computation time for the template matching during verification.

In another scheme, the accuracy of each finger is assessed independently. The selected features are used for finger-specific performance evaluation, and the accuracies (the number of corresponding features is mentioned in parenthesis) are provided in Tables 5.10 and 5.11, respectively. However, the accuracies are not enhanced compared to the results provided in Table 5.5. The main reason is a smaller feature dimension of any particular finger. It is evident that for a few number of features, the performance cannot be improved in every situation, which is clear from the results in Tables 5.10 and 5.11. The global selection comparatively performs better than the local selection for finger-level evaluation due to the reason of subset cardinality. However, the trade-off between the subset cardinality and accuracy is notable in this concern. The cardinality of global selection is uniform, and it is 10 features per finger. Contrarily, the cardinalities of finger-level subsets for the local collection are not uniformly distributed. It can be averred that the middle finger is the most discriminative compared to other fingers, regarding the local selection. On the other side, during the global selection, the ring finger is the most significant. Hence, either of the methods cannot be delineated as the more significant as the other. Both of the methods are optimistic based on the evaluation scheme and the trade-off between the optimization of the feature space and accuracy.

TABLE 5.10

Global Finger-Level Identification Accuracies (%) Using R-FoBa during Testing

Classifier	Index (10)	Middle (10)	Ring (10)	Little (10)
wk-NN	45	62.67	63	58.67
RF	67.34	79	82.34	72

TABLE 5.11

Local Finger-Level Identification Accuracies (%) Using R-FoBa during Testing

Classifier	Index (6)	Middle (8)	Ring (6)	Little (5)
wk-NN	36.67	51.34	46	31.67
RF	49.34	70.34	57.34	46.34

5.5.3.3 Verification

Verification is regarded as the one-to-one template matching scheme for the performance evaluation of a biometric system. The assessment metrics regarding the verification are defined in Table 5.12. In the verification, a query feature template is compared to his/her stored templates on an explicit distance threshold. The distances between a claimer and the enrolled feature matrix are calculated. If the differences are within the threshold, then a person is accepted as genuine, otherwise rejected as an imposter. The threshold (T) is defined as

$$T_{e,g} = \sum_{q=1}^{n} \left(\sqrt{\left(\alpha_{e,q} - \beta_{g,q}\right)^2} * w_q \right) \Big/ \text{mean}\left(\alpha_q\right) \tag{5.17}$$

where $\alpha_{e,q}$ denotes the qth feature of the α_e valid user, $\beta_{g,q}$ means the qth test feature of a claimant β_g, and n represents the total number of selected features. The mean value of qth feature is calculated from the training dataset. The metrics and assessment parameters essential for verification have been illustrated with respect to the 300 subjects used for testing, given in Table 5.12. Yet again, the local R-FoBa provides better results for verification experiments regarding the equal error rate (EER). The receiver operating characteristic curves are depicted in Figure 5.9.

TABLE 5.12

Verification Metrics, Parameters, and Equal Error Rates during Testing

Metric	Evaluation Parameter Value	EER
Genuine Acceptance Rate (GAR): True Acceptance/Genuine Comparison	Training set = 2×300 Test set = 300	Global R-FoBa: 0.066
False Acceptance Rate (FAR): False Acceptance/Imposter Comparison	Genuine comparison = 600 Imposter comparison = 179,400	
False Reject Rate (FRR): 1-GAR	Total comparison = 600×300	Local R-FoBa: 0.046
Equal Error Rate (EER): Value at which the FAR and FRR are same.		

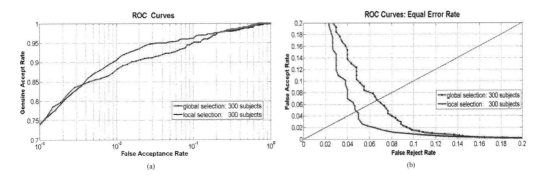

FIGURE 5.9
The ROC Curves. (a) The FAR vs. GAR represented in logarithmic scale for X-axis, (b) the FRR vs. FAR to estimate the EER through the diagonal line.

In summary, the local selection performs better than the global choice. It has also been compared with the R-FoBa that the local R-FoBa provides 98.67% accuracy for the 300 subjects using the RF classifier with 25 features. Contrarily, the global R-FoBa delivers 97% accuracy with 36 features (nine selected features per finger of the left hand) in Bera and Bhattacharjee (2017). Moreover, regarding the accuracy, the present approach is also competitive with the method in Dutağaci et al. (2008) which provides 98.22% identification accuracy for 200 subjects with 28 geometric features. The EERs are also competitive for the cases with a similar setup. Hence, from the perspective of the accuracy and subset cardinality, the local R-FoBa offers improved identification performances using the similitude experimental constraints.

5.6 Conclusion

A finger geometry recognition approach with the selected feature subset is presented in this chapter. The prime concern is to formulate a suitable feature selection method that can render satisfactory accuracy with a smaller number of salient features. The novelty relies on the formulation of an optimal subset of features using the same notion of FoBa. The computational simplicity regarding the geometric feature extraction and selection implicates its suitability in the real-world application for identity verification that mainly requires lesser time. However, the scopes for further improvement pertinent to finger biometrics should be emphasized significantly. Primarily, the preprocessing tasks, salient feature set definition, and most importantly, selection of a right combination of a small number of features should be focused. Some open challenges need immediate attention regarding an accurate preprocessing phase for finger segmentation, nail-effect and clothing occlusion eradication, and others. Even the experimental constraints can be formulated with a large number of people with sufficient posture variations. This trait can also be deployed in the mobile devices, in IoT and other online-based applications within a limited time for verification. Optimistically, it would be expected to uplift the achievements of finger biometrics by exploring these areas of advanced researches with a societal significance.

Acknowledgment

The authors would like to thank Prof. B. Sankur of Boğaziçi University for providing the database of hand image used in this chapter.

References

Bera, A., & Bhattacharjee, D. (2017). Human identification using selected features from finger geometric profiles. *IEEE Transactions on System, Man, and Cybernetics: Systems*, doi:10.1109/TSMC.2017.2744669.

Bera, A., Bhattacharjee, D., & Nasipuri, M. (2014). Hand biometrics in digital forensics. In *Computational Intelligence in Digital Forensics: Forensic Investigation and Applications*, Studies in Computational Intelligence, Muda, A. K., Choo, Y. H., Abraham, A., & Srihari, S. N. (Eds.). *555*, pp. 145–163. Springer, Cham.

Bera, A., Bhattacharjee, D., & Nasipuri, M. (2015). Fusion based hand geometry recognition using Dempster-Shafer theory. *International Journal of Pattern Recognition and Artificial Intelligence, 29*(5), 1556005/1–24.

Bera, A., Bhattacharjee, D., & Nasipuri, M. (2017). Finger contour profile based hand biometric recognition. *Multimedia Tools and Applications, 76*(20), 21451–21479.

Bera, A., Bhattacharjee, D., & Nasipuri, M. (2018). Pose-invariant hand geometry for human identification using feature weighted k-NN classifier. *Information Technology and Applied Mathematics, Advances in Intelligent Systems and Computing, 699*, 115–129.

Borgia, E. (2014). The internet of things vision: Key features, applications, and open issues. *Computer Communications, 54*, 1–31.

Breiman, L. (1996). Bagging predictors. *Machine Learning, 24*, 123–140.

Breiman, L. (2001). Random forests. *Machine Learning, 45*(1), 5–32.

Canny, J. (1986). A computational approach to edge detection. *IEEE Transactions on Pattern Analysis and Machine Intelligence, 8*(6), 679–698.

Charfi, N., Trichili, H., Alimi, A. M., & Solaiman, B. (2014). Novel hand biometric system using invariant descriptors. In *Proceedings of IEEE International Conference on Soft Computing and Pattern Recognition*, pp. 261–266. Tunis.

Choraś, R. S., & Choraś, M. (2006). Hand shape geometry and palmprint features for the personal identification. In *Proceedings of IEEE of 6th International Conference on Intelligent Systems Design and Applications*, pp. 1085–1090. Jinan.

Duta, N. (2009). A survey of biometric technology based on hand shape. *Pattern Recognition, 42*(11), 2797–2806.

Dutağaci, H., Sankur, B., & Yörük, E. (2008). A comparative analysis of global hand appearance-based person recognition. *Journal of Electronic Imaging, 17*(1), 011018/1–19.

Ernst, R. H. (1971). Hand ID system. U.S. Patent No. 3576537. Washington, DC: U.S. Patent and Trademark Office.

Estévez, P. A., Tesmer, M., Perez, C. A., & Zurada, J. M. (2009). Normalized mutual information feature selection. *IEEE Transactions on Neural Networks, 20*(2), 189–201.

Ferrer, M. A., Morales, A., & Diaz, A. (2014). An approach to SWIR hyper spectral hand biometrics. *Information Sciences, 268*, 3–19.

Gou, J., Du, L., Zhang, Y., & Xiong, T. (2012). A new distance-weighted k-nearest neighbor classifier. *Journal of Information & Computational Science, 9*(6), 1429–1436.

Guyon, I., & Elisseeff, A. (2003). An introduction to variable and feature selection. *Journal of Machine Learning Research, 3*, 1157–1182.

Hanmandlu, M., Grover, J., Madasu, V., & Vasirkala, S. (2010). Score level fusion of hand based biometrics using T-norms. In *Proceedings of IEEE*, pp. 70–76. Waltham, MA.

Hanmandlu, M., Kumar, A., Madasu, V. K., & Yarlagadda, P. (2008). Fusion of hand based biometrics using particle swarm optimization. In *Proceedings of IEEE 5th International Conference on Information Technology: New Generations*, pp. 783–788. Las Vegas, NV.

Hu, R. X., Jia, W., Zhang, D., Gui, J., & Song, L. T. (2012). Hand shape recognition based on coherent distance shape contexts. *Pattern Recognition, 45*, 3348–3359.

Hung, M. (Ed.). (2017). Leading the IoT. Gartner. Retrieved 15 May 2018, from www.gartner.com/imagesrv/books/iot/iotEbook_digital.pdf.

Jain, A. K., Duin, R. P. W., & Mao, J. (2000). Statistical pattern recognition: A review. *IEEE Transactions on Pattern Analysis and Machine Intelligence, 22*(1), 4–37.

Jain, A. K., Nandakumar, K., & Ross, A. (2005). Score normalization in multimodal biometric systems. *Pattern Recognition, 38*, 2270–2285.

Jain, A. K., Nandakumar, K., & Ross, A. (2016). 50 years of biometric research: Accomplishments, challenges, and opportunities. *Pattern Recognition Letters, 79*, 80–105.

Jain, A. K., Ross, A., & Prabhakar, S. (2004). An introduction to biometric recognition. *IEEE Transactions on Circuits and Systems for Video Technology, 14*(1), 4–20.

Kang, W., & Wu, Q. (2014). Pose-invariant hand shape recognition based on finger geometry. *IEEE Transactions on Systems, Man, and Cybernetics: Systems, 44*(11), 1510–1521.

Kanhangad, V., Kumar, A., & Zhang, D. (2011). A unified framework for contactless hand verification. *IEEE Transactions on Information Forensics and Security, 6*(3), 1014–1027.

Kohavi, R., & John, G. H. (1997). Wrappers for feature subset selection. *Artificial Intelligence, 97*, 273–324.

Kumar, A. (2008). Incorporating cohort information for reliable palmprint authentication. In *Proceedings of 6th Indian Conference on Computer Vision, Graphics Image Processing (ICVGIP)*, pp. 583–590. Bhubaneshwar.

Kumar, A., & Shekhar, S. (2011). Personal identification using multibiometrics rank level fusion. *IEEE Transactions on System, Man, and Cybernetics, Part C: Applications and Reviews, 41*(5), 743–752.

Kumar, A., & Zhang, D. (2005). Biometric recognition using feature selection and combination. In *Proceedings of 5th International Conference on Audio- and Video-Based Biometric Person Authentication, AVBPA'05*, pp. 813–822. Rye Brook, NY.

Lashkia, G. V., & Anthony, L. (2004). Relevant, irredundant feature selection and noisy example elimination. *IEEE Transactions on System, Man, and Cybernetics: Cybernetics, 34*(2), 888–897.

Luque-Baena, R. M., Elizondo, D., López-Rubio, E., Palomo, E. J., & Watson, T. (2013). Assessment of geometric features for individual identification and verification in biometric hand systems. *Expert Systems with Applications, 40*(9), 3580–3594.

Miller, R. P. (1971). Finger dimension comparison identification system. U.S. Patent No. 3576538. Washington, DC: U.S. Patent and Trademark Office.

Morales, A., Ferrer, M. A., Cappelli, R., Maltoni, D., Fierrez, J., & Ortega-Garcia, J. (2015). Synthesis of large scale hand-shape databases for biometric applications. *Pattern Recognition Letters, 68*(1), 183–189.

Otsu, N. (1979). A threshold selection method from gray-level histograms. *IEEE Transactions on System, Man, and Cybernetics, 9*(1), 62–66.

Peng, H., Long, F., & Ding, C., (2005). Feature selection based on mutual information: Criteria of max-dependency, max-relevance, and min-redundancy. *IEEE Transactions on Pattern Analysis and Machine Intelligence, 27*(8), 1226–1238.

Peng, J., Li, Q., Abd El-Latif, A. A., & Niu, X. (2015). Linear discriminant multi-set canonical correlations analysis (LDMCCA): An efficient approach for feature fusion of finger biometrics. *Multimedia Tools and Applications, 74*(13), 4469–4486.

Reillo, R. S., Avila, C. S., & Macros, A. G. (2000). Biometric identification through hand geometry measurements. *IEEE Transactions on Pattern Analysis and Machine Intelligence, 22*(10), 1168–1171.

Ross, A., & Jain, A. K. (2003). Information fusion in biometrics. *Pattern Recognition Letters, 24*, 2115–2125.

Ross, A., Nandakumar, K., & Jain, A. K. (2006). *Handbook of Multibiometrics.* Springer, New York.

Sharma, S., Dubey, S. R., Singh, S. K., Saxena, R., & Singh, R. K. (2015). Identity verification using shape and geometry of human hands. *Expert Systems with Applications, 42,* 821–842.

Tsalakanidou, F., Malassiotis, S., & Strintzis, M. G. (2007). A 3D face and hand biometric system for robust user-friendly authentication. *Pattern Recognition Letters, 28,* 2238–49.

Wang, M. H., & Chung, Y. K. (2012). Applications of thermal image and extension theory to biometric personal recognition. *Expert Systems with Applications, 39,* 7132–7137.

Yörük, E., Dutağaci, H., & Sankur, B. (2006). Hand biometrics. *Image and Vision Computing, 24,* 483–497.

Yörük, E., Konukoğlu, E., Sankur, B., & Darbon, J. (2006). Shape-based hand recognition. *IEEE Transactions on Image Processing, 15*(7), 1803–1815.

Zhang, T. (2011). Adaptive forward-backward greedy algorithm for learning sparse representations. *IEEE Transactions on Information Theory, 57*(7), 4689–4708.

Part II

The Biometric Computing – Algorithms & Methodologies

6

Iris Recognition Systems in a Non-Cooperative Environment

Mireya Saraí García Vázquez
Instituto Politécnico Nacional–CITEDI

Eduardo Garea Llano
Advanced Technologies Application Center–CENATAV

Juan Miguel Colores Vargas
Universidad Autónomade Baja California–ECITEC

Alejandro Álvaro Ramírez Acosta
MIRAL R&D&I

CONTENTS

6.1 Introduction

The etymology of the word biometrics is composed of the term "bio" which comes from the Greek word "life" and the term "metric", which comes from the Greek word "measure"; these terms form the concept: measure of life. In recent years, the term "biometrics" refers to the use of methods and algorithms for automated recognition of a person's identity; these methods and algorithms make use of the physical and morphological characteristics of the person. History shows that there have been mechanisms and systems to recognize people by their physical and morphological characteristics. In Babylon (500 B.C.), commercial transactions were registered on fresh clay tablets signed with the fingerprints. In 1870 in France, Alphonse Bertillon, police official and French criminologist of a prefecture in Paris, proposed a scientific method called "bertillonage" or anthropometries to identify criminals using detailed records of their body measurements, physical descriptions and photographs. Currently, the main methods of analysis in the field of biometrics are based on patterns of characteristics generated from the iris, fingerprint and face of a person. In 1936, Frank Burch proposed using the texture of the iris as a digital pattern to be representative of a person and to use it as a guide to recognize a person (Jain and Kumar 2010; Daugman 1993). The approach of recognizing a person based on the characteristics of the iris was established by the 1985 research conducted by ophthalmologists Drs. Leonard Flom and Aron Safir; they established that the iris pattern cannot be repeated, and in 1987, they obtained the patent on the identification of people through the iris. This moment was the origin of modern automated iris recognition; they proposed the idea but without any algorithm (Flom and Safir 1987). The shape of the iris structure forms a very particular texture for each person, which makes the iris a very specific pattern of characteristics to recognize people (Bowyer, Hollingsworth, and Flynn 2008). Dr. Flom approached Dr. John Daugman to establish the mathematical requirements that would allow the automatic identification of a person through his/her human iris. John Daugman generated the digital representation of the iris pattern, realizing for human authentication a comparison between digital iris patterns (Daugman 1993). With the joint work of Flom, Safir, and Daugman, a prototype was held in 1995 at the Nuclear Defense Agency (Shepard, Wing, Miles, and Blackburn 2006). As of 1994, with the obtaining of the patent by Daugman, the first iris recognition system was commercialized (Daugman 1993; ISO/IEC 19794-6:2005). It was based on the use of a two-dimensional digital template of the texture of the iris, obtaining the features of this texture by means of Gabor 2D filters, and for the comparison metric, the hamming distance was used. In 2005, the broad patent of Flom covering the basic concept of iris recognition expired. In 2011, once the Daugman patent expired, different iris recognition products were created by new companies. Methods for iris recognition have grown rapidly since the 1994 patent (Daugman 1993), that is 26 years.

The iris texture is a physical characteristic of human physiology that can be used for biometric authentication and identification of persons. The robustness of the iris biometric pattern is based on a set of factors (Jain, Flynn, and Ross 2007; Jain, Ross and Prabhakar

2004; Zamudio Fuentes 2010); a system of identification of people using the iris pattern meets these factors:

Universality (everyone possesses the characteristic): it means that every person possesses the iris texture features in an iris system.

Uniqueness (unique characteristic): the texture of the iris has specific and unique traits, which make the difference between individuals. Genetically, the texture pattern of the iris is not determined (epigenetic).

Distinctiveness (highly discriminative texture): the randomness of iris patterns has very high dimensionality, which is superior to 266 degrees of freedom (Daugman 1993).

Permanence (the characteristic remains invariant over life time, except for pigmentation change over time): it means that the iris texture remains invariant over time. Then, a "good" permanence will be changed reasonably over time with respect to the specific matching algorithm.

Measurability (the characteristic is easy to capture): it relates to the ease of acquisition of iris texture. The unique characteristics are extracted from the image of the texture of the iris.

Performance: it specifies the accuracy of the iris recognition technology, its speed, and robustness to uncontrolled conditions.

Acceptability (the characteristic is not invasive): it relates to how well individuals accept the iris technology. The iris image capture is non-invasive. In addition, due to its properties of the texture of the iris, the development of the field of iris biometrics has increased.

Circumvention (imitability of the characteristic): it relates to the ease or not; the human iris might be imitated using an artifact or substitute.

An iris recognition system can operate in two modes (Jain and Kumar 2010; Zamudio Fuentes 2010; Colores Vargas 2013). In *Authentication-verification mode*, the process consists of comparing a person's biometric template IrisCode (Term used by Daugman 1993 represents the texture of the iris) with the biometric template IrisCode of the person they claim to be; direct comparison of templates. The system for the *Identification mode* performs an operation that compares the biometric template of an unknown person against many templates that are in an iris database, seeking to recognize the identity of the unknown person. To identify an individual, a threshold is established, when carrying out the identification of the individual, and a result is obtained. If this result is within the area of the established threshold, there is a positive identification, but if this result is outside the area established at the threshold, the individual will not be identified (Jain and Kumar 2010).

Traditionally, the recognition of a person by their iris requires acquiring the iris image at a short distance, with an illumination and controlled characteristics, use of infrared charge-coupled device (CCD) camera, and respecting an iris radius with a minimum of 70 pixels (Jain and Kumar 2010; ISO/IEC 19794-6:2005); however, reality requires uncontrolled environments. The challenge in recent years has been focused on obtaining iris images under uncontrolled conditions and developing iris recognition systems in non-cooperative environments (Zamudio Fuentes 2010; Colores Vargas 2013; Daugman 2009; Zamudio-Fuentes, García-Vázquez, and Ramírez-Acosta 2010, 2011;

Colores, García-Vázquez, Ramírez-Acosta, and Pérez-Meana 2011; García-Vázquez and Ramírez-Acosta 2012; Rattani and Derakhshani 2017). To face this problem and improve the accuracy of recognition systems, important factors that affect the representation of the biometric model must be taken into account, which are the following: (i) the quality of the images, (ii) the lighting variations, (iii) the reduction of the blurring of the images, (iv) the reduction of the specular reflections, (v) the distance of the acquisition system to the user, and (vi) standardize the size of the iris and the pixel density of the iris texture. However, these factors can be controlled by correcting the camera selection, depth of field analysis (DOF), and field of vision (FOV) for iris recognition. The selection of an adequate quality measure of the iris image is another aspect to be taken into account.

In the last years and to the present date, the techniques of recognition of people by means of the biometric pattern of the Iris are considered the most precise and robust. However, under non-ideal conditions, a system of this type is affected in the quality and information of the captured images, having a low performance; some of these conditions are presented as non-collaborative movements and actions, as well as capturing images in visible-spectrum environments (Zamudio Fuentes 2010; Colores Vargas 2013; Zamudio-Fuentes, García-Vázquez, and Ramírez-Acosta 2011; Colores, García-Vázquez, Ramírez-Acosta, and Pérez-Meana 2011, Labati, Genovese, Piuri, and Scotti 2012).

The development and interconnection in the network of the social sectors have made the system of automatic; of people, based on their biometric patterns, develop with greater impetus. The recognition systems of people based on biometric patterns of the iris have high reliability, besides being non-invasive. The preprocessing stage in an iris recognition system has a high responsibility task, specifically in the location and delimitation of the iris texture region. With the techniques developed in recent years, the iris region has been obtained very precisely (Zamudio-Fuentes, García-Vázquez, and Ramírez-Acosta 2010, 2011; Cui, Wang, Tan, Ma, and Sun 2004).

This chapter addresses the aspects related to iris recognition systems in terms of design of recognition systems for images and video and the challenges that this implies. It also discusses problems, challenges, and proposals for iris segmentation in poorly controlled environments. The approach of iris images fusion at segmentation stage is a way to increase recognition rates in iris recognition systems. We will talk about the problem of the iris feature extraction and will analyze the main approaches that have been developed in this sense, including the recent results using deep neural networks (DNNs). We will also analyze the databases that are used by the scientific community to validate the performance of new recognition systems based on the biometric pattern of the iris. Finally, we will analyze some of our main experimental results in the development of iris recognition systems in non-cooperative environments.

6.2 Iris Recognition Systems from Images and Video

Nowadays, the available iris recognition systems impose some constraints on the position and movement of subjects during recognition to acquire a high-quality image (Phillips, Scruggs, O'Toole et al. 2007; Proença and Alexandre 2007; Newton and Phillips 2009); an ideal image is one in which the eye image is centered, without aberration errors as defocus blur or motion blur effects. This fact allows that conventional iris recognition systems achieve recognition rates higher than 99%, as shown in Figure 6.1; the user must be

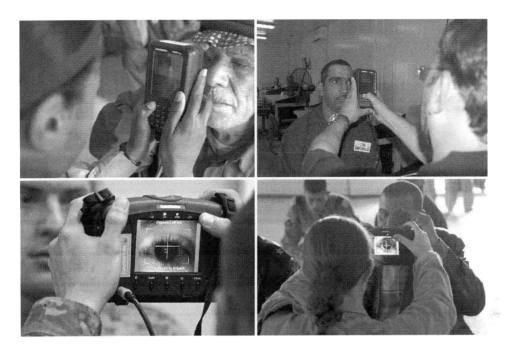

FIGURE 6.1
Conventional Iris recognition systems require users to be still as the camera locates the eye and captures the image.

remaining static in front of the camera during capture and receive visual and auditory feedback from the system.

However, the iris recognition rates may be significantly reduced if the procedures established by the system are not met, due to the considerable degradation in the quality of the iris images (Gamassi, Lazzaroni, Misino, Piuri, Sana, and Scotti 2005), as happens when the system is required to operate in unconstrained environments. Then, it is necessary to select suitable high-quality images before the next recognition processing. Otherwise, it can have a negative impact on the subsequent processing stage of the recognition system, increasing the error probability (Chen, Dass, and Jain 2006; Kalka, Zuo, Schmid, and Cukic 2006).

During recent years, the demand for secure biometric recognition systems (indoor/outdoor) at the massive level of people has increased around the world; however, it can be complicated to comply in detail with the rigorous conditions established by the capture stage in order to have obtained a trustworthy iris recognition system. Thus, with the aim of overcoming these inconveniences, new improvements are being implemented over different stages of the iris recognition system to increase its performance when operating under non-ideal conditions in the acquisition stage (unconstrained iris recognition systems), that means, the system identifies people by maintaining minimal interaction with them; nevertheless, in a less restrictive environment, the users are free to move outside the optimal distance from the camera during the capture process, which means that they may move outside the optimal "depth of field" of the system causing the blurring effects in the captured pictures, therefore, choosing an eye frame with an appropriate image quality seems to be a challenge. It is well known that defocus and motion blur effects are the most frequent causes for degradation in the quality of the capture eye frames, in addition to the

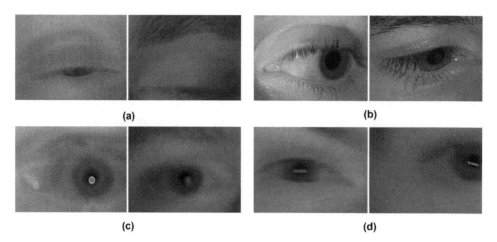

FIGURE 6.2
Non-ideal quality eye images. (a) Occlusion, (b) off-angle, (c) defocus blur and (d) motion blur.

negative effects due to improper illumination, off-axis, occlusions, specular reflections, etc. A sample set of all these problems in capturing eye images are shown in Figure 6.2.

To address the problem of image quality, related work on this subject can be placed into two categories (Zamudio Fuentes 2010; Zuo and Schmid 2009): local and global analyses. The global methods are quickly quality evaluation procedures for eliminating very poor quality images, selecting only those images that are possibly the most suitable for iris recognition. Global methods are mainly based on focus and motion blur estimation (Tenenbaum 1970; Jarvis 1976; Nayar and Nakagawa 1994).

A simple methodology for discriminating images with blur effects is based on the power spectral density of the 2D image; an image with blur effects concentrates its energy in the lower frequencies, while the energy of the Fourier spectrum in the images without blur effects will be a uniform distribution (Daugman 1993). Therefore, obtaining the spectral energy at higher frequencies can be a way to estimate and discriminate those images of the eye that have been distorted during the capture process (Wildes, Asmuth, Green et al. 1994). Colores-Vargas, García-Vázquez, and Ramírez-Acosta (2010), for image quality assessment, analyzed four methods used to obtain the high-frequency power spectrum of the eye frames. Evaluation results show that the better performance according to accuracy and computational cost is the convolution kernel proposed by Kang and Park (2005).

On the other hand, the methods for local analysis are based on measurements and extracting additional information in the pixels that compose the region of the eye and iris texture, calculating local factors such as out-of-focus blur, motion blur, and occlusion (Chen, Dass, and Jain 2006; Zhu, Liu, Ming, and Cui 2004; Chen, Hu, and Xu 2003; Ma, Tan, Wang, and Zhang 2003; Zhang and Salganicoff 1999; Belcher and Du 2007). However, the majority of previous methods require the involvement of traditional segmentation methods that are iterative and, thus, computationally expensive.

The video sequence acquisition procedure on uncontrolled environments is the current trend in iris recognition systems, where frames are acquired by the system while the user passes in front of it, at an average walking pace. Among these approaches, the long-distance eye image acquisition for iris recognition seems to be an interesting solution (Hollingsworth, Peters, Bowyer, and Flynn 2009; Colores-Vargas, García-Vázquez, Nakano-Miyatake, and Perez-Meana 2012; Matey, Naroditsky, Hanna et al. 2006) since capturing

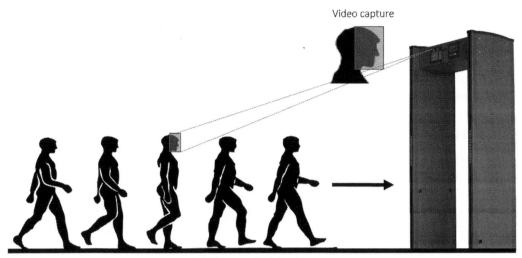

Walking speed < 1.4 m/s

FIGURE 6.3
Biometric recognition at large standoff distances on uncontrolled environments is the current trend; a person is recognized while walking through an access control point.

videos for iris sequences can provide additional information compared to a single capture of an eye image. Besides that, it is a friendly system because it is not intrusive and does not require the user's cooperation (see Figure 6.3).

Normally, in these acquisition schemes for video sequences under unconstrained environments, there are different qualities in the frames, but most of them suffer from several kinds of noisy effects such as light reflections, defocus blur, motion blur, occlusion (eyelid/eyelashes), etc.

Negative effects of noise result, sometimes, in the capture of frames with degraded iris texture information, even without any iris; so the entire recognition process must be repeated with a new frame until getting a proper result.

Therefore, it is important to implement different quality assessment methodologies in the first stage of the iris recognition system. To deal with quality problem, new stages for quality evaluation have been proposed in the iris recognition system to quickly eliminate those frames that are useless to achieve recognition, due to the degradation suffered in the iris texture information (Kalka, Zuo, Schmid, and Cukic 2006; Belcher and Du 2007). However, those quality metrics require achieving a very precise segmentation task that may be difficult to obtain in many cases.

In 2009, Hollingsworth, Peters, Bowyer, and Flynn improved the matching performance using signal-level fusion; they used a video of the iris from which they extract a sequence of frames to create a single average image, however, they considered ideal conditions in which incorrectly segmented frames are manually discarded which could limit its implementation.

Another interesting approach is proposed by Colores-Vargas, García-Vázquez, Ramírez-Acosta, Pérez-Meana, and Nakano-Miyatake (2013); they propose the inclusion of two evaluation stages on an unconstrained iris recognition system: measurement of global quality in frames and segmentation failures. The global quality assessment stage only allows frames with high quality and eliminates those with low quality. The other proposition is based on iris segmentation quality assessment. This latter stage only provides frames with good results in iris segmentation to continue with the following processes in an unconstrained iris recognition system. The evaluation stage is based on the evaluation of three segmentation accuracy metrics. These metrics produce scores that are used to determine whether the segmentation is a success or failure. The performance of the proposed scheme generates an Equal Error Rate (EER) equal to 2.48% for an optimal threshold, while the conventional scheme achieves an EER equal to 14.62%, i.e., a reduction of about 12.14%.

6.2.1 Segmentation Iris Texture Region

In order to have an iris recognition system with high performance in precision, the segmentation process of the iris annular region must establish the limits of the pupil region and the sclerotic region. The precision in obtaining the texture region of the iris is crucial to have a good performance of the recognition system; a segmentation with regions of pupil, sclera, eyelids, or eyelashes will affect the accuracy of the Iris recognition system (Llano, Vargas, García-Vázquez, Fuentes, and Ramírez-Acosta 2015; García-Vázquez, Llano, Vargas, Fuentes, and Ramírez-Acosta 2015; Proença and Alexandre 2010), as shown in Figure 6.4a–d. The segmentation stage includes three following steps:

Detection of the border between the region of the pupil and the region of the iris.

Detection of the border that delimits the established regions of the iris and the sclera.

Detection of the boundary between eyelids and iris.

In the literature, there are a lot of iris segmentation algorithms, including recent results obtained by the paradigm of deep learning which has been a very promising approach to face the task of segmentation of the iris in poorly controlled environments; but for this chapter, we have chosen four algorithms, for its public availability as open-source software for reproducibility and besides, they provide good results in the iris segmentation.

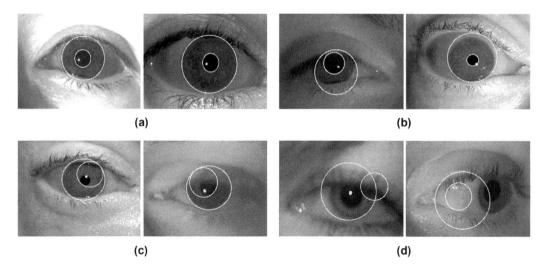

(a)   (b)

(c)   (d)

FIGURE 6.4
Segmentation results (a) successful segmentation, (b) failed iris segmentation, (c) failed pupil segmentation, (d) failed pupil and iris segmentation.

6.2.1.1 Viterbi-Based Segmentation Algorithm

The open-source reference system OSIRIS, version v4.12, in the segmentation part uses a Viterbi algorithm based iris segmentation algorithm (Sutra, Garcia-Salicetti, and Dorizzi 2012) to find the iris and pupil boundaries.

The first step of Viterbi algorithm (Sutra, Garcia-Salicetti, and Dorizzi 2012; García-Vázquez, Garea-Llano, Colores-Vargas, Zamudio-Fuentes and Ramírez-Acosta 2015) consists of a rough localization of the pupil area. As a first step, the effect of the specular reflections caused by the illumination sources on the iris by filling them is eliminated. Then, we apply an opening morphological operator with a disk-shaped structuring element to remove small dark areas. Then, considering a notable difference in the luminosity of the three regions of the eye (skin and sclera, iris and pupil), the algorithm calculates the sum of the intensity values in the pixels for different windows of the image; the minimum values correspond to the pupil that is the darkest region. When the region that comprises the pupil is detected, it proceeds to the estimation of its centroid, which is a fundamental element for the next steps of the algorithm; this detection is carried out using a morphological reconstruction. The second step uses the calculated coordinates of the centroid in the pupil to estimate its circular contour; the precision in this step is very critical for the normalization process. The contour of the pupil will be used to generate the biometric iris template for recognition purposes. The Viterbi algorithm is executed in two different resolutions corresponding to the number of points used to estimate the contour. For a higher resolution, all pixels are considered, which allows finding the pupil region contour, while at low resolution the Viterbi algorithm calculates some contours that are used to optimize the normalization circles process. Three advantages can be described in the Viterbi algorithm: does not require a threshold for the gradient map, it is a generic implementation that can be used in any database without any adaptation, finally, the facility to generalize the normalization from circular contours to elliptical contours, as well as other parametric curves in polar coordinates.

6.2.1.2 Contrast-Adjusted Hough Transform Segmentation Algorithm

The algorithm was implemented in USIT software (Rathgeb, Uhl, and Wild, 2012) from the Masek algorithm (Masek 2003), one of the classic detection methods based on the Hough transform. In this new version of the algorithm, it is necessary as a previous step to make a contrast adjustment of the iris images according to the characteristics of the database; this previous step contributes to improving the location of the edges of the iris. Additionally, it is known that this implementation implies a high computational cost; therefore, an edge detection method (Canny operator) is used to quickly identify the borders between the iris region and the pupil, and improvement techniques to remove unlikely edges.

6.2.1.3 Weighted Adaptive Hough and Ellipsopolar Transform

It is one of the methods implemented by the USIT system of the University of Salzburg (Rathgeb, Uhl, and Wild 2012; García-Vázquez, Garea-Llano, Colores-Vargas, Zamudio-Fuentes, and Ramírez-Acosta 2015). The algorithm is based on the use of weighted Gaussian functions in order to take advantage of the existing knowledge of the model. To estimate the center coordinates of the iris region, an adaptive Hough transform is applied at different resolutions of the image. Thereafter, a polar transformation is used to detect the elliptical boundary that separates the pupil from the iris region, and then an ellipso-polar transform finds the second boundary that encloses the iris region. In this way, both iris images captured in visible wavelength (with clear pupillary limits) and the iris images captured in the near infrared (clear pupillary boundaries) can be processed uniformly.

6.2.1.4 Modified Hough Transform Segmentation Algorithm

The algorithm for the segmentation of the iris texture region is based on the circular Hough transform, used in the method of Wildes, Asmuth, Green et al. (1994); this algorithm makes a combination of contour maps when using the Canny edge detector (Colores-Vargas, García-Vázquez, Ramírez-Acosta, Pérez-Meana, and Nakano-Miyatake 2013; Canny 1983; Hough 1962). Hence, it is name as Modified Hough Transform (MHT). From the edge map, votes are cast in Hough space for the parameters of circles passing through each edge point. The parameters specify the coordinates of the central point and the radius, which specify the circular region of the pupil and the iris. These parameters are the center coordinates $\left[\left(x_{cp}, y_{cp}\right),\left(x_{ci}, y_{ci}\right)\right]$ and radius $\left[r_p, r_i\right]$ for the iris and pupil outer boundaries, respectively, as shown in equations 6.1 and 6.2.

$$\left(x - x_{cp}\right)^2 + \left(y - y_{cp}\right)^2 = r_p^{\ 2} \tag{6.1}$$

$$\left(x - x_{ci}\right)^2 + \left(y - y_{ci}\right)^2 = r_i^{\ 2} \tag{6.2}$$

6.2.2 Normalization Process

The region of the texture of the iris is dynamic due to the dilatation and contraction of the pupil; to compensate for this variability of the iris texture region, a normalization process is performed. For this process, the Daugman model called linear rubber sheet model (Daugman 1993) is used. The process consists of taking each point that is in the region of the iris and mapping it to polar coordinates (r, θ), where r is in the interval [0, 1] and θ in [0, 2π]. Equation 6.3 specifies the normalized non-concentric polar representation, which is obtained by mapping the iris region from the Cartesian coordinates (x, y).

$$I\left(x(r,\theta), y(r,\theta)\right) \rightarrow I(r,\theta) \tag{6.3}$$

with

$$x(r, \theta) = (1 - r)x_p(\theta) + rx_i(\theta)$$

$$y(r, \theta) = (1 - r)y_p(\theta) + ry_i(\theta)$$

where I(x, y) represents the texture region of the Iris in the image, (x, y) represent the original points in Cartesian coordinates, (r, θ) represent the points mapped in the normalized polar coordinates, the coordinates x_p, y_p correspond to the pupil, and the coordinates x_i, y_i correspond to the iris; these points are the limits of the iris along the θ direction.

The recognition of the iris in a non-cooperative environment involves making the recognition of a person by means of his iris at a distance greater than conventional systems;

it also involves performing a series of algorithms that provide solutions to the factors that disturb the quality of the texture of the iris in an image; these disturbances occur when there are occlusions, blur, off-axis, specular reflections, and poor illumination in the iris texture region. To improve the accuracy of recognition systems of people based on the iris region, some works have focused on fusion at the level of the segmentation. The above is because a single method of segmentation does not have the robustness to provide the solution to all the factors that disturb the quality of the texture of the iris in the image.

Current methods require a considerable amount of prior and subsequent mathematical processing to carry out the detection of the iris region. These methods are based primarily on the knowledge that can be acquired from the environment in which the images were acquired. In addition, due to their characteristics experimentally obtained, they lack the flexibility to find optimal descriptors. In recent years, deep convolutional neuronal networks (CNNs) have proved to be a powerful model that allows automatically to learn and almost or optimally enough characteristics and classifiers. CNNs have been used quite successfully in image classification tasks (Szegedy, Liu, Jia et al. 2015), biometry, spoofing detection (Menotti, Chiachia, Pinto et al. 2015), face recognition (Schroff, Kalenichenko, and Philbin 2015), and iris recognition (Liu, Li, Zhang, Liu, Sun, and Tan 2016; Minaee, Abdolrashidiy, and Wang 2016; Al-Waisy, Qahwaji, Ipson, Al-Fahdawi, and Nagem 2017). In Liu, Li, Zhang, Liu, Sun, and Tan (2016), two CNN models are proposed to address the task of iris segmentation in images taken in uncontrolled and noisy environments. The results presented in this work, although limited to two databases, demonstrate the robustness of this approach that requires even a better and deeper study. The authors of this chapter have carried out some implementations, and results have shown that for the task of segmentation of the iris this is a promising way (see Table 6.3 and Figure 6.12).

6.2.3 Iris Image Fusion

Recent works have shown (Hollingsworth, Peters, Bowyer, and Flynn 2009; Colores-Vargas, García-Vázquez, Ramírez-Acosta, Pérez-Meana, and Nakano-Miyatake 2013; Llano, Vargas, García-Vázquez, Fuentes, and Ramírez-Acosta 2015; Sanchez-Gonzalez, Chacon-Cabrera, and Garea-Llano 2014; Liu and Xie 2006) that combining of information from multiple frames in the stages of segmentation and normalization could be advantageous, besides using multiple algorithms for extraction and comparison of biometric feature. The objective of an image fusion technique is to combine information from multiple images taken of the same object to get a new fused image. It has been found that using an image fusion algorithm can improve the performance of iris recognition systems, by generating a fused image that represents the biometric information from multiple images. A related example is direct linear discriminant analysis (DLDA) that in combination with wavelet transform allows the extraction of iris features.

Their results showed that recognition performance increases dramatically (Sutra, Garcia-Salicetti, and Dorizzi 2012). Other works report that one average template formed with a set of frames from an iris video is comparable with any single frame from the set (Masek 2003).

Some experimentally studies have demonstrated that the recognition accuracy for representative algorithms increases efficacy in diverse benchmark iris databases by using image fusion techniques (Colores-Vargas, García-Vázquez, Ramírez-Acosta, Pérez-Meana, and Nakano-Miyatake 2013; Llano, Vargas, García-Vázquez, Fuentes, and Ramírez-Acosta 2015; Sanchez-Gonzalez, Chacon-Cabrera, and Garea-Llano 2014; Liu and Xie 2006). These are proven methods whose results show an improvement in the quality of iris images and the biometric information associated with their texture, that is, fusion methods improve the recognition performance and reduce error rates.

In Llano, Vargas, García-Vázquez, Fuentes, and Ramírez-Acosta (2015), novelty of our proposal is the scheme to combine the measurement of the quality in the frames of video iris and a fusion method at segmentation level for simultaneous images and video iris recognition. As part of the proposed scheme, at segmentation level, we present a Modified Laplacian Pyramid based fusion method; this fusion method is based on the use of the first level of its Gaussian Pyramid (Llano, Vázquez, Vargas, Fuentes, and Acosta 2018).

Two of the best segmentation results are combined, thus obtaining the new proposed fusion scheme. We assume the experimental fusion model presented in Llano, Vargas, García-Vázquez, Fuentes, and Ramírez-Acosta (2015) and Llano, Vázquez, Vargas, Fuentes, and Acosta (2018) to evaluate the performance of experimented fusion methods. This model uses the Sum-Rule Interpolation technique with results obtained from the segmented iris images normalized by the Daugman's rubber sheet model. For the fusion stage, we experimented with four algorithms, three of them previously used to fuse iris images taken from video: LP (Laplacian Pyramid), EM (Exponential Mean); PCA (Principal Component Analysis) (Llano, Vargas, García-Vázquez, Fuentes, and Ramírez-Acosta 2015; Llano, Vázquez, Vargas, Fuentes, and Acosta 2018) (See Figure 6.5).

The proposed scheme produces an increase in decidability (measures how well separated the two distributions) for Hamming distance distributions. Intra-class (same subjects) and Inter-class (different subjects): the separation of the two distributions is very important because the recognition error is caused by their overlap.

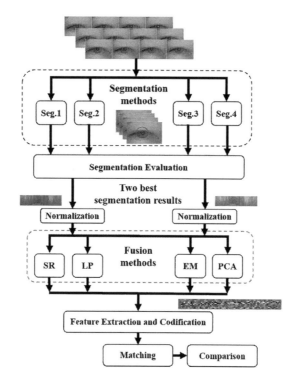

FIGURE 6.5
The proposed fusion scheme (adapted from: Llano, Vázquez, Vargas, Fuentes, and Acosta 2018); SR, Sum Rule Interpolation; LP, Laplacian Pyramid; EM, Exponential Mean; PCA, Principal Component Analysis.

The experiments have been demonstrated that fusion of multiple frames in an iris video obtains a higher recognition performance than single frame. Fusion methods are a good option that can be implemented and are computationally faster in comparison to other proposals, fusion methods at score level. It is important to establish future research in this area; the exploration of new methods for image fusion that have potential to yield further performance improvements.

6.2.4 Features Iris Texture

One of the important tasks in the process of iris recognition is the feature extraction from the texture patterns of the iris. If we analyze the different proposed approaches to extract and encode the texture of the iris, it can be said that it has been a very intensive research area in iris biometry (Bowyer, Hollingsworth, and Flynn 2008).

The general concept of the process of extraction of biometric characteristics is defined as the process applied to a biometric sample to isolate and eliminate repeatable and distinctive numbers or labels that can be compared with those extracted from other biometric samples.

In the particular case of the iris biometrics, the objective of feature extraction step is to extract discriminative features from iris texture and store them in a more compact form so that it is more effective performing pattern matching in the following stages. But Which are the most significant features and distinctive? What is the minimum amount needed? Is there any way to automatically learn what features are best for the classifier? (Duda, Hart, and Stork 1973) are the questions that are not completely solved today.

6.2.4.1 Taxonomy of Iris Feature Extraction Methods

The iris has a particularly complex and random structure that provides abundant texture information. For its description, there are analysis schemes that make use of: (i) methods that involve working in the spatial domain or the domain of the frequencies, performing the signal processing. This group of methods makes use of the digital filtering techniques to describe the iris texture, (ii) statistical processing methods, seeking to highlight the existence of spatial relationships between different regions of iris texture, (iii) the use of combined techniques, to describe the texture exhaustively, taking advantage of the power of the combination of several sources of analysis. In Figure 6.6, we represent taxonomy of the extraction methods of most used and recognized features to describe and numerically represent the texture of the iris.

6.2.4.2 Statistical Methods

The texture characteristics obtained from the second-order statistical methods (Tuceryan and Jain 1993; Sun, Tan, and Qiu 2006; Tian, Qu, Zhang, and Zong 2011; De Marsico, Nappi, and Riccio 2012; Li, Liu, and Zhao 2012) as co-occurrence and local binary pattern (LBP) have been used to exploit the spatial dependencies that characterize the texture of the image, but basically, this type of texture descriptor only provides information about events that occur on a single scale. When texture patterns are present with sizes of several scales and in different angular directions, these methods are not suitable.

However, the Independent Component Analysis (ICA) method (Huang, Luo, and Chen 2002; Dorairaj, Schmid, and Fahmy 2005; Chowhan and Shinde 2009; Bouraoui, Chitroub, and Bouridane 2012) is a statistical technique that aims to reveal relationships and hidden

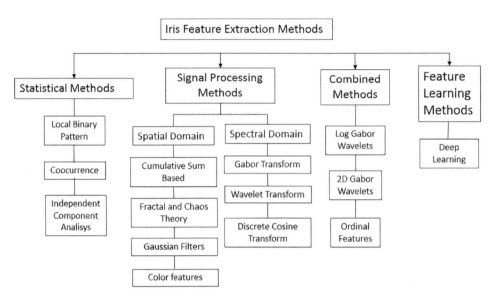

FIGURE 6.6
Taxonomy of iris feature extraction method.

factors within sets of original variables. Therefore, the ICA method can be applied to a set of texture features extracted from the iris using any texture description method. The result is a new set of revealed features with new properties that are useful for interpreting iris texture.

6.2.4.3 Signal Processing Methods

Signal processing techniques (Daugman 1993; Wildes 1997; Zhang, Sun, Tan, and Wang 2012; Zhu, Tan, and Wang 2000; Sanchez-Avila, and Sanchez-Reillo 2005; Kwanyong, Shinyoung, Kwanyong, Okhwan, and Taiyun 2001; Monro, Rakshit, and Dexin 2007) have the most powerful methods for extracting texture characteristics. They have theoretical principles that help in the modification of the spectral composition of the image using a mathematical operator in the spatial domain or using some transformation, such as the Fourier, Gabor or Wavelet transform, to extract the texture characteristics in the frequency domain. This type of technique is known as spectral techniques. It is based on the digital filtering properties and facilitates the elimination of noise in the images acquired during the capture process. This image filtering process modifies the spectral composition, low frequencies (or high frequencies) in the image. Digital filtering using Gabor filters and Wavelets filters describe the iris texture in the context of several scales. This type of texture descriptors uses filter banks to quantify the energy of the texture at different scales using various frequencies and angular directions in the image.

6.2.4.4 Combined Methods

To obtain a better representation of the texture of the iris, the combined methods are used (Sun and Tan 2009; Tan, Zhang, Sun, and Zhang 2012; Li, Wang, Sun, and Tan 2013); these methods perform the fusion of the characteristics calculated by each method used in the characteristic extraction task.

Ordinal Measures (OMs) are qualitative measures to describe the relative order of various quantities.

TABLE 6.1

Results of Recognition Accuracy on UBIRIS-V1 Dataset in EER (%) and d' at FAR ≤ 0.001%

Daugman 2D Gabor		Masek 1D Log-Gabor Wavelets		Ma 1-D Wavelet		Monro 1D-DCT	
EER	d'	EER	d'	EER	d'	EER	d'
Segmentation by Algorithm 1							
6,73	2,27	6,11	2,43	7,19	1,89	5,51	2,08
Segmentation by Algorithm 2							
9,9	2,04	10,53	2,1	11,85	1,74	12,02	2,04

Source: Llano, Vázquez, Vargas, Fuentes, and Acosta 2018.

In the case of combined methods, the OMs for iris recognition proposed by Sun and Tan (2009) showed excellent results compared to state-of-the-art algorithms. From the pioneering work of Dr. Daugman, work on iris recognition has focused on the use of local images filtering; to obtain a better performance in iris recognition, OMs' ordinal measurements of non-local differential filters are used.

Llano, Vargas, García-Vázquez, Fuentes, and Ramírez-Acosta (2015) compared the performance of four feature extraction algorithms based on Log-Gabor wavelet, Gabor Wavelet, and DCT on four benchmark iris datasets. The results (Table 6.1) clearly demonstrate the effectiveness of the combined spectral methods. The method that uses the Gabor filter under an image quality control was obtained that has better performance than the Log-Gabor wavelet and DCT methods. The authors also argue that the methods based on Log-Gabor wavelet can be tolerant to illumination variations.

6.2.4.5 Feature Learning Methods

There are not so many research works addressing iris feature extraction by machine learning techniques. Most of them are developed to obtain better recognition, some only focus on the detection of the iris texture region and extraction of representative characteristics of the texture of the iris. The methods for the representation of the texture of the iris must be effective and have a complexity that does not affect the learning and performance of the iris recognition system.

Traditional methods for extracting iris texture characteristics usually require elaborate preprocessing methods, as well as parameter updates, to obtain the representation of an iris database. The use of these traditional methods in another database of iris will not have a guarantee to obtain the same performance as with the database with which they were conceived. There have been quite a lot of research works in recent years, to obtain characteristics that represent in a general way different tasks of representation of characteristic patterns; these approaches have been made under the idea of learning features and being able to use them in different tasks. The methods of DNNs have yielded good results in this context. In recent years, a set of classic tasks of pattern recognition such as object detection and segmentation of images and textures has been approached from the focus of DNNs. Examples of networks are Alex-Net, ZF-Net, VGG-Net and Res-Net (Simonyan and Zisserman 2014; He, Zhang, Ren, and Sun 2016). Minaee, Abdolrashidiy, and Wang (2016) developed work for Iris recognition; they were based on the DNN architecture VGG-Net. To test the use of learning features, Minaee et al. make use of trained DNN architecture, for the extraction of characteristics of the texture of the iris in the images; they do not make

TABLE 6.2

Comparison of the Proposed CNN Approach by Correct Recognition Rate (CRR) with Other Existing Approaches on CASIA Iris v3 Iris Database

Approach	CRR (%)
Gabor transform and Euler numbers	97.21
Haar wavelet transform	98.45
Log-Gabor wavelet	86.00
Convolutional Neural Network	**100**

Source: Al-Waisy, Qahwaji, Ipson, Al-Fahdawi, and Nagem 2017.

any modification to the architecture. This algorithm was tested on two well-known CASIA datasets and achieved promising results, outperforming the previous best results on these databases.

In Al-Waisy, Qahwaji, Ipson, Al-Fahdawi, and Nagem (2017), authors combine the CNN with the Softmax classifier to obtain a set of inherent characteristics of the iris in an unsupervised way without having previous knowledge of the characteristics from image. These characteristics extracted in an unsupervised way will allow them to classify iris images in N classes. Then from this, they establish a system of scores for the identification of the classes and classification of the images in left or right for each individual. Finally, they establish a form of integration of the obtained classifications and create a classification list using fusion techniques to reach the final score.

The DNN approaches have shown a promising path that should still be explored because there are still important challenges in the task of extracting iris features, especially for those images captured in the visible light spectrum and under uncontrolled conditions (Table 6.2).

6.3 Databases in a Non-Cooperative Environments

The databases are important for the development of digital mathematical methods. To have a good biometric recognition system, it is necessary to have a robust database. A database with information of iris images acquired in different uncontrolled spaces and with different characteristics of digital cameras provides information to develop a high-performance biometric iris recognition system. In this section, we present four databases.

6.3.1 CASIA-V3-Interval

CASIA-V3-interval (CASIA 2002): the images of this database were acquired with a digital camera of near-infrared illumination (NIR) developed by CASIA, which gives high-quality images of resolution 320×280 pixels. The format of these images is JPEG (Joint Photographic Experts Group) in 8-bit grayscale; this database has 395 classes and a total of 2,639 images. Almost all subjects are Chinese. The database was conceived by the National Laboratory of Pattern Recognition, which belongs to the Automation Institute of the Chinese Academy of Sciences; they are also responsible for managing it (CASIA 2016).

6.3.2 CASIA-V4-Thousand

CASIA-V4-thousand (CASIA 2002): is a collection that contains 20,000 irises images from 1,000 subjects. This database contains iris images' specular reflections and individuals with eyeglasses. The system used in the acquisition creates this database; the IriskingIKEMB-100 sensor is used in the acquisition. The resolution of the images that make up this database is 640×480 pixels.

6.3.3 UBIRIS-V1

UBIRIS-V1: this database is conceived by SOCIA Lab (Soft Computing and Image Analysis Group) of the Department of Computer Science; the iris images of this database are under the RGB additive color model. This group is in the University of Beira Interior, Portugal (SOCIA Lab 2018). For the acquisition of iris images, the group used a Nikon E5700 digital camera with a 2/3-in. CCD sensor. UBIRIS-V1 (Proença and Alexandre 2005) is a data-set comprised of 1,877 images collected from 241 persons in two distinct sessions. The peculiarity of this database is that for the acquisition of iris images have environments with fewer restrictions, so, there are noise characteristics that make the iris images have contrast, luminosity, occlusion by eyelids and eyelashes, degrees of focus, and different luminosity.

6.3.4 Multiple Biometrics Grand Challenge MBGC

MBGC V2: the database of the Multiple Biometrics Grand Challenge (NIST 2018) was collected in the University of Notre Dame by the group of The Computer Vision Research Lab. MBGC V2 provided 986 near-infrared eye videos, MPEG-4 format in 8 bits-grayscale space. The acquisitions of this video-database were with LG2200 EOU iris capture system, and the resolution is 640×480 pixels.

The camera uses NIR illumination. Noise factors such as reflections and contrast, as well as the factor of brightness, and the obstruction of the iris using the eyelids and eyelashes, and the focusing characteristics; they are noise factors present in this database. These factors make the MBGC V2 database have the information that is required for the texture of the iris for the real conditions of acquisition in uncontrolled environments; in this way, it provides the necessary conditions to develop iris recognition systems of very high performance. Some images of the databases are shown in Figure 6.7.

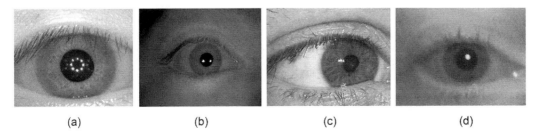

    (a)   (b)   (c)   (d)

FIGURE 6.7
Examples of databases images (a) CASIA-V3-Interval, (b) CASIA-V4-Thousand, (c) UBIRIS-V1, (d) MBGC-V2.

6.4 Experimental Results in Iris Recognition Systems in a Non-Cooperative Environments

The performance of a system of recognition of people using the iris pattern is affected when working in conditions of non-cooperative environments, such as changes in lighting or poor cooperation of the individual. The segmentation stage of the iris texture region has a critical importance for the good performance of an iris recognition system. Performing this segmentation under conditions of non-cooperative environments has a high degree of difficulty, because the iris region is small and has constant movements both voluntary and involuntary; also, the images can present noise in the region of the texture of the iris, that is to say, have occlusion of the region of the iris either by the eyelids, the eyelashes, and the reflections of illumination. It has been observed that having an incorrect segmentation affects the other stages of the recognition system, generating results with low precision and with recognitions of erroneous people.

The demand for large-scale biometric applications requires high precision to avoid possible erroneous classifications. It is widely known that the performance of cooperative biometric systems based on the control of the conditions under which the images were taken is different from the applications implemented in less controlled environments. This fact has meant that over the past 10 years there has been a lot of activity on research topics related to the optimization of preprocessing methods and segmentation methods. The quality of the captured iris images directly affects the recognition process and on which these techniques depend.

6.4.1 Fusion Segmentation and Quality Evaluation as Part of Iris Recognition System in Non-Cooperative Environments

The combination of biometric information can increase accuracy at the expense of additional resources and is traditionally used in the classification or decision level. However, these fusion strategies have presented limitations, as this requires a series of algorithms to performing similar processing that leads to a higher computational cost. The existence of conflicting information from various processing algorithms can reduce system performance.

In this sense in Uhl and Wild (2013), the concept of fusion of multiple segmentations was introduced to combine results of iris segmentations separately. The authors demonstrated experimentally that the recognition accuracy increases the effectiveness in the CASIA.V4 database when the results of two manual segmentations are combined.

Based on the proven idea that segmentation-level fusion improves iris recognition rates, in Sanchez-Gonzalez, Chacon-Cabrera, and Garea-Llano (2014), a comparative study was performed at the level of the results of the segmented normalized iris images. It was experienced the fusion of three automatic segmentation algorithms that were compared by their performance in the verification task when in the segmentation stage each segmentation algorithm is used separately and when their results are merged into four different combinations using the Sum Interpolation Rule. The evaluations on three datasets showed an improvement in the recognition rates for each of the feature extraction methods employed.

In our recent works (Llano, Vargas, García-Vázquez, Fuentes, and Ramírez-Acosta 2015, García-Vázquez, Garea-Llano, Colores-Vargas, Zamudio-Fuentes, and Ramírez-Acosta 2015; Llano, Vázquez, Vargas, Fuentes, and Acosta 2018; Llano, García-Vázquez, Zamudio-Fuentes, Vargas, and Ramírez-Acosta 2017), we have developed the idea of segmentation fusion and their use combined with other techniques to improve recognition rates in systems where iris images are captured in non-cooperative environments and with multiple sensors.

The proposed system in Llano, Vargas, García-Vázquez, Fuentes, and Ramírez-Acosta (2015) (Figure 6.8) consists of the following steps: from the capture of one or more images of the iris (one class) by the same type of sensor or several sensors simultaneously, pre-processing is performed to extract the internal and external boundaries of the iris using at least two segmentation algorithms. Once the two images of the iris are obtained, the coordinate transformation is performed to obtain the normalized iris image. Then, the combination of the two images is done using a fusion algorithm. The fusion strategy should try to obtain as results the most non-redundant information present in the merged images.

For the experimentation of the proposal, four databases were used; these datasets were taken by different sensors in order to simulate different capture conditions. The segmentation stage was based on two segmentation algorithms; they were selected from a set of four methods of the state of the art taking as selection criteria the stability of their performance under conditions of the four evaluated databases (Llano, Vargas, García-Vázquez, Fuentes, and Ramírez-Acosta 2015).

For the fusion stage, we experimented with four algorithms of the state of the art, three of them were used previously to merge iris images taken from the video (Colores-Vargas, García-Vázquez, Ramírez-Acosta, Pérez-Meana, and Nakano-Miyatake 2013).

To estimate the accuracy of the performance of the system presented, the Detection Error Tradeoff (DET) curves were used. The results show the positive impact of the proposed stage of merged segmentation. It was observed that, in all cases, the False Rejection Rate (FRR) and the Equal Error Rate (ERR) decreased compared to the results obtained by the system using a unique segmentation algorithm. As a result of the fusion process of the obtained segmentations, the FRR was reduced in a range of 0.12%–0.02% and the EER in a range of 3.1%–0.54%. These results are obtained by using the PCA as a fusion method and the 2DGabor feature-based method. (Figure 6.9).

In Llano, Vázquez, Vargas, Fuentes, and Acosta (2018), we extended our previous work and presented a new robust multisensor approach to non-ideal conditions (Figure 6.10), which is based on a strategy that combines the evaluation of the quality of iris images captured by sensors with a new fusion method at the segmentation level for simultaneous video recognition and iris image. The proposed fusion method is based on the Modified Laplacian Pyramid (MLP). The novelty of this proposal is in the combination

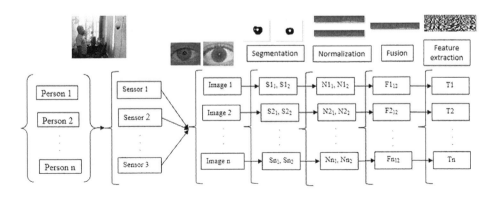

FIGURE 6.8
General scheme of the proposed system in (adapted from: Llano, Vargas, García-Vázquez, Fuentes, and Ramírez-Acosta 2015).

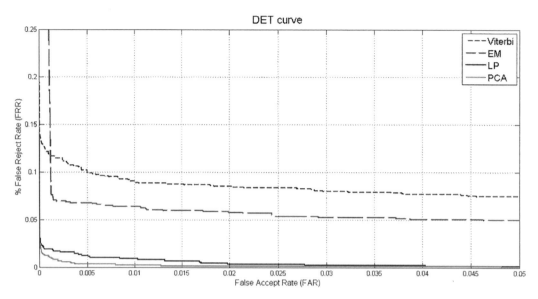

FIGURE 6.9
DET curve obtained for UBIRIS-V1 database (SOCIA Lab 2018) using 2D Gabor feature extractor.

of the processes of simultaneous capture of ocular images and video, the evaluation of its quality, the segmentation, and its fusion using a new method. This scheme guarantees the independence of the system from the type of captures (either images or video sequences of the iris), (Figure 6.10).

The experiments followed the same protocol of Llano, Vargas, García-Vázquez, Fuentes, and Ramírez-Acosta (2015) on the Casia-V3-Interval, Casia-V4-Thousand, Ubiris-V1, and MBGC-V2 databases. They showed that the proposal increases recognition accuracy, is robust for different types of iris sensors, and can work simultaneously with video and images. In Figure 6.11, an example of obtained results is shown. The inclusion of the quality evaluation stage and the new fusion method produces a decrease in FRR compared with the system proposed in Llano, Vargas, García-Vázquez, Fuentes, and Ramírez-Acosta (2015) from 0.02% to 0.01% and in an EER from 0.54% to 0.22% using the feature extraction method based on 2D Gabor Filters.

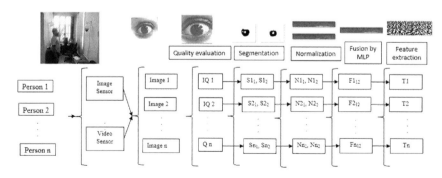

FIGURE 6.10
General scheme of the proposed system in (adapted from: Llano, Vázquez, Vargas, Fuentes, and Acosta 2018).

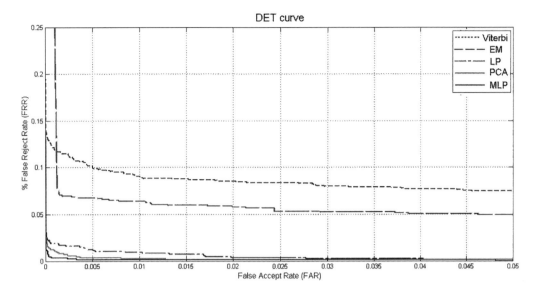

FIGURE 6.11
DET curve obtained for UBIRIS-V1 database (SOCIA Lab 2018) using 2D Gabor feature extractor. (Llano, Vázquez, Vargas, Fuentes and Acosta 2018).

The inclusion of a stage of quality assessment and a robust fusion method at the level of the segmentation stage in our proposal of iris recognition system showed that regardless of the type of captures (videos or images) the recognition rates increase, reducing the error with respect to the classical scheme that is capable of using only one segmentation method. The proposed MLP method demonstrated its superiority in the fusion task over other methods.

6.4.2 Perspectives of the Use of Deep Neural Networks in the Improvement of the Accuracy of Iris Biometric Systems

In recent years, the use of DNNs has made a qualitative leap to the solution of pattern recognition problems with rapid development and evolution of precision rates in given solutions. In the case of iris biometry and systems that work in uncontrolled environments, this benefit has been shown in the conception of methods fundamentally oriented to the segmentation stages (Liu, Li, Zhang, Liu, Sun, and Tan 2016) and more recently to the feature extraction and classification (Minaee, Abdolrashidiy, and Wang 2016; Al-Waisy, Qahwaji, Ipson, Al-Fahdawi, and Nagem 2017; Krizhevsky, Sutskever, and Hinton 2012). These experiences and some experiments carried out by us on challenging databases show that the DNN is a promising paradigm for iris recognition systems and it can be a future line of development of more robust systems in the face of unfavorable conditions of image capture with little control over the variables of illumination, pose, distance without having prior knowledge of its characteristics.

In our experiments, we trained and compare the performance of two fully convolutional neural networks (FCNN), Alexnet (Krizhevsky, Sutskever, and Hinton 2012) and fcn8s (Long, Shelhamer, and Darrell 2015), on NICE I (Proença and Alexandre 2007) iris dataset for iris segmentation task.

TABLE 6.3

Results of Training and Evaluation of FCNN on NICE I Iris Dataset
for Iris Segmentation Task

AlexNet FCNN		fcn8s FCNN	
Loss	Accuracy (%)	Loss	Accuracy (%)
0.04	98,1	0.03	98.8

FIGURE 6.12
Example of a tested image and masks obtained by trained models and its comparison with the ground truth.

For the evaluation, we used two models, one established at 50% for the training and test samples, and the other at 90% for training and 10% test samples. The training and test were carried out on the Caffe framework (Jia 2018).

Table 6.3 shows the results obtained for both models, and Figure 6.12 presents two examples that show the result of two evaluated images after training.

These preliminary results show that the fcn8s network has better qualities to face the problem of iris segmentation since it produces more refined edges compared to those obtained by Alexnet network. At the same time, it achieves better accuracy in the segmentation (Figure 6.12).

6.5 Conclusion

Research studies in feature extraction methods for iris recognition have been widely addressed to obtain techniques and algorithms for robust iris recognition over the last few years. However, research continues to advance to obtain a better representation of the texture of the iris, intending to characterize in a robust way this texture for non-cooperative environments and to achieve a compact representation for better portability.

An emerging and challenging area is non-cooperative iris recognition; it can make iris recognition more friendly and flexible to use, and this also will allow its further use in video surveillance and multi-biometric systems.

It has been demonstrated that for the systems of recognition of people using the iris pattern, which work in non-cooperative environments, the inclusion of the combination of the stages of evaluation of the quality of the iris images and the fusion of segmentation methods generates recognition rates with higher percentages.

DNNs have demonstrated their effectiveness to face the problems derived from the evolution of iris recognition systems towards work in uncontrolled conditions. These conditions have been derived from the need for its application in less demanding scenarios and where installation and maintenance costs are minimal. Then DNNs can provide the possibility that the system integrally works in a non-cooperative environment. The DNNs simultaneously learn features and classifiers from the data and do not depend on complex characteristics or statistical descriptors. In addition, the use of preprocessing and postprocessing stages of data and results would not be necessary.

Acknowledgments

This research was supported by Iris research project grant from CENATAV, Cuba and IPN-SIP2017, IPN-SIP2018 projects from IPN, México.

References

Al-Waisy, A. S., Qahwaji, R., Ipson, S., Al-Fahdawi, S., and Nagem, T. A. 2017. A multi-biometric iris recognition system based on a deep learning approach. *Pattern Analysis and Applications*, 21(3): 783802.

Belcher, C., and Du, Y. 2007. Information distance-based selective feature clarity measure for iris recognition. In *Image Quality and System Performance IV* (Vol. 6494, p. 64940E). International Society for Optics and Photonics.

Bouraoui, I., Chitroub, S., and Bouridane, A. 2012. Does independent component analysis perform well for iris recognition? *Intelligent Data Analysis*, 16(3), 409–426.

Bowyer, K. W., Hollingsworth, K., and Flynn, P. J. 2008. Image understanding for iris biometrics: A survey. *Computer Vision and Image Understanding*, 110(2), 281–307.

Canny, J. F. 1983. Finding edges and lines in images. Master's Thesis. Massachusetts Institute of Technology.

CASIA Iris Databases. 2002. CASIA iris image database [Database]. http://biometrics.idealtest.org/.

Chen, J., Hu, G., and Xu, J. 2003. Iris image quality evaluation method based on wavelet packet decomposition. *Qinghua Daxue Xuebao/Journal of Tsinghua University (China)*, 43(3), 377–380.

Chen, Y., Dass, S. C., and Jain, A. K. 2006. Localized iris image quality using 2-D wavelets. In *International Conference on Biometrics* (pp. 373–381). Springer, Berlin, Heidelberg.

Chinese Academy of Sciences Institute of Automation. 2016. Iris database [Database]. http://biometrics.idealtest.org/dbDetailForUser.do?id=4.

Chowhan, S. S., and Shinde, G. N. 2009. Evaluation of statistical feature encoding techniques on iris images. In *WRI World Congress on Computer Science and Information Engineering* (Vol. 7, pp. 71–75). IEEE.

Colores, J. M., García-Vázquez, M., Ramírez-Acosta, A., and Pérez-Meana, H. 2011. Iris image evaluation for non-cooperative biometric iris recognition system. In *Mexican International Conference on Artificial Intelligence* (pp. 499–509). Springer, Berlin, Heidelberg.

Colores Vargas, J. M. 2013. Sistema de Reconocimiento basado en Iris con técnicas de fusión de video. PhD Dissertation. CITEDI-IPN, México.

Colores-Vargas, J., García-Vázquez, M., Nakano-Miyatake, M., and Perez-Meana, H. 2012. Iris recognition system based on video for unconstrained environments. *Scientific Research and Essays*, 7(35), 3114–3127.

Colores-Vargas, J. M., García-Vázquez, M. S., and Ramírez-Acosta, A. A. 2010. Measurement of defocus level in iris images using different convolution kernel methods. In *Mexican Conference on Pattern Recognition* (pp. 125–133). Springer, Berlin, Heidelberg.

Colores-Vargas, J. M., García-Vázquez, M., Ramírez-Acosta, A., Pérez-Meana, H., and Nakano-Miyatake, M. 2013. Video images fusion to improve iris recognition accuracy in unconstrained environments. In *Mexican Conference on Pattern Recognition* (pp. 114–125). Springer, Berlin, Heidelberg.

Cui, J., Wang, Y., Tan, T., Ma, L., and Sun, Z. 2004. A fast and robust iris localization method based on texture segmentation. In *Biometric Technology for Human Identification* (Vol. 5404, pp. 401–409). International Society for Optics and Photonics.

Daugman, J. 2009. How iris recognition works. In Bovik, A. C. (ed.) *The Essential Guide to Image Processing* (pp. 715–739). Academic Press, London.

Daugman, J. G. 1993. High confidence visual recognition of persons by a test of statistical independence. *IEEE Transactions on Pattern Analysis and Machine Intelligence*, 15(11), 1148–1161.

De Marsico, M., Nappi, M., and Riccio, D. 2012. Noisy iris recognition integrated scheme. *Pattern Recognition Letters*, 33(8), 1006–1011.

Dorairaj, V., Schmid, N. A., and Fahmy, G. 2005. Performance evaluation of iris-based recognition system implementing PCA and ICA encoding techniques. In *Biometric Technology for Human Identification II* (Vol. 5779, pp. 51–59). International Society for Optics and Photonics.

Duda, R. O., Hart, P. E., and Stork, D. G. 1973. *Pattern Classification* (Vol. 2). Wiley, New York.

Flom, L., and Safir, A. 1987. U.S. Patent No. 4,641,349. U.S. Patent and Trademark Office, Washington, DC.

Gamassi, M., Lazzaroni, M., Misino, M., Piuri, V., Sana, D., and Scotti, F. 2005. Quality assessment of biometric systems: A comprehensive perspective based on accuracy and performance measurement. *IEEE Transactions on Instrumentation and Measurement*, 54(4), 1489–1496.

García-Vázquez, M. S., Garea-Llano, E., Colores-Vargas, J. M., Zamudio-Fuentes, L. M., and Ramírez-Acosta, A. A. 2015. A comparative study of robust segmentation algorithms for iris verification system of high reliability. In *Mexican Conference on Pattern Recognition* (pp. 156–165). Springer, Cham.

García-Vázquez, M. S., Llano, E. G., Vargas, J. M. C., Fuentes, L. M. Z., and Ramírez-Acosta, A. A. 2015. Cross-sensor iris verification applying robust segmentation algorithms. In *Congreso Internacional de Investigación Tijuana (CIIT)* (ISSN 2007-9478, Vol. 4, N.7, pp. 65–73).

García-Vázquez, M. S., and Ramírez-Acosta, A. Á. 2012. Avances en el reconocimiento del iris: Perspectivas y oportunidades en la investigación de algoritmos biométricos.

He, K., Zhang, X., Ren, S., and Sun, J. 2016. Deep residual learning for image recognition. In *Proceedings of the IEEE Conference on Computer Vision and Pattern Recognition* (pp. 770–778), Las Vegas, NV.

Hollingsworth, K., Peters, T., Bowyer, K. W., and Flynn, P. J. 2009. Iris recognition using signal-level fusion of frames from video. *IEEE Transactions on Information Forensics and Security*, 4(4), 837–848.

Hough, P. V. 1962. U.S. Patent No. 3,069,654. U.S. Patent and Trademark Office, Washington, DC.

Huang, Y. P., Luo, S. W., and Chen, E. Y. 2002. An efficient iris recognition system. In *Proceedings of the International Conference on Machine Learning and Cybernetics, 2002* (Vol. 1, pp. 450–454). IEEE, Sofia, Bulgaria.

ISO/IEC 19794-6:2005- Information technology—Biometric data interchange formats. Part 6: Iris image data", ISO. This standard has been revised by ISO/IEC 19794-6:2011.

Jain, A. K., Flynn, P., and Ross, A. A. 2007. *Handbook of Biometrics*. Springer, Secaucus, NJ.

Jain, A. K., and Kumar, A. 2010. Biometrics of next generation: An overview. *Second Generation Biometrics*, 12(1), 2–3.

Jain, A. K., Ross, A., and Prabhakar, S. 2004. An introduction to biometric recognition. *IEEE Transactions on Circuits and Systems for Video Technology*, 14(1), 4–20.

Jarvis, R. A. 1976. Focus optimization criteria for computer image processing. *Microscope*, 24(2), 163–180.

Jia, Y. 2018. Caffe|Deep learning framework. http://caffe.berkeleyvision.org/.

Kalka, N. D., Zuo, J., Schmid, N. A., and Cukic, B. 2006. Image quality assessment for iris biometric. In *Biometric Technology for Human Identification III* (Vol. 6202, p. 62020D). International Society for Optics and Photonics.

Kang, B. J., and Park, K. R. 2005. A study on iris image restoration. In *International Conference on Audio-and Video-Based Biometric Person Authentication* (pp. 31–40). Springer, Berlin, Heidelberg.

Krizhevsky, A., Sutskever, I., and Hinton, G. E. 2012. Imagenet classification with deep convolutional neural networks. In *Advances in Neural Information Processing Systems* (pp. 1097–1105).

Kwanyong, L., Shinyoung, L., Kwanyong, L., Okhwan, B., and Taiyun, K. 2001. Efficient iris recognition through improvement of feature vector and classifier. *ETRI Journal*, 23, 61–70.

Labati, R. D., Genovese, A., Piuri, V., and Scotti, F. 2012. Iris segmentation: State of the art and innovative methods. In Liu, C., and Mago, V. K. (eds.) *Cross Disciplinary Biometric Systems* (pp. 151–182). Springer, Berlin, Heidelberg.

Li, P., Liu, X., and Zhao, N. 2012. Weighted co-occurrence phase histogram for iris recognition. *Pattern Recognition Letters*, 33(8), 1000–1005.

Li, X., Wang, L., Sun, Z., and Tan, T. 2013. A feature-level solution to off-angle iris recognition. In *Paper presented at the Proceedings of the 6th International Conference on Biometrics*. doi:10.1109/ICB.2013.6612991.

Liu, C., and Xie, M. 2006. Iris recognition based on DLDA. In *18th International Conference on Pattern Recognition, ICPR* (Vol. 4, pp. 489–492). IEEE, Hong Kong.

Liu, N., Li, H., Zhang, M., Liu, J., Sun, Z., and Tan, T. 2016. Accurate iris segmentation in non-cooperative environments using fully convolutional networks. In *International Conference on Biometrics (ICB)* (pp. 1–8). IEEE, Halmstad, Sweden.

Llano, E. G., García-Vázquez, M. S., Zamudio-Fuentes, L. M., Vargas, J. M. C., and Ramírez-Acosta, A. A. 2017. Analysis of the improvement on textural information in human iris recognition. In *VII Latin American Congress on Biomedical Engineering CLAIB 2016*, Bucaramanga, Santander, October 26–28, (pp. 373–376). Springer, Singapore.

Llano, E. G., Vargas, J. M. C., García-Vázquez, M. S., Fuentes, L. M. Z., and Ramírez-Acosta, A. A. 2015. Cross-sensor iris verification applying robust fused segmentation algorithms. In *International Conference on Biometrics (ICB), 2015* (pp. 17–22). IEEE, Phuket, Thailand.

Llano, E. G., Vázquez, M. S. G., Vargas, J. M. C., Fuentes, L. M. Z., and Acosta, A. A. R. 2018. Optimized robust multi-sensor scheme for simultaneous video and image iris recognition. *Pattern Recognition Letters*, 101, 44–51.

Long, J., Shelhamer, E., and Darrell, T. 2015. Fully convolutional networks for semantic segmentation. In *Proceedings of the IEEE Conference on Computer Vision and Pattern Recognition* (pp. 3431–3440), Boston, MA, USA.

Ma, L., Tan, T., Wang, Y., and Zhang, D. 2003. Personal identification based on iris texture analysis. *IEEE Transactions on Pattern Analysis and Machine Intelligence*, 25(12), 1519–1533. ISO 690.

Masek, L. 2003. Recognition of human iris patterns for biometric identification. Master's Thesis.

Matey, J. R., Naroditsky, O., Hanna, K., et al. 2006. Iris on the move: Acquisition of images for iris recognition in less constrained environments. *Proceedings of the IEEE*, 94(11), 1936–1947.

Menotti, D., Chiachia, G., Pinto, A., et al. 2015. Deep representations for iris, face, and fingerprint spoofing detection. *IEEE Transactions on Information Forensics and Security*, 10(4), 864–879.

Minaee, S., Abdolrashidiy, A., and Wang, Y. 2016. An experimental study of deep convolutional features for iris recognition. In *Signal Processing in Medicine and Biology Symposium (SPMB)* (pp. 1–6). IEEE.

Monro, D. M., Rakshit, S., and Dexin, Z. 2007. DCT-based iris recognition. *IEEE Transactions on Pattern Analysis and Machine Intelligence*, 29, 586–595.

Nayar, S. K., and Nakagawa, Y. 1994. Shape from focus. *IEEE Transactions on Pattern Analysis and Machine Intelligence*, 16(8), 824–831.

Newton, E. M., and Phillips, P. J. 2009. Meta-analysis of third-party evaluations of iris recognition. *IEEE Transactions on Systems, Man, and Cybernetics-Part A: Systems and Humans,* 39(1), 4–11.

NIST. 2018. Multiple biometric grand challenge (MBGC).

Phillips, P. J., Scruggs, W. T., O'Toole, A. J., et al. 2007. FRVT 2006 and ICE 2006 large-scale results. National Institute of Standards and Technology, NISTIR, 7408(1).

Proença, H., and Alexandre, L. A. 2005. UBIRIS: A noisy iris image database. In *International Conference on Image Analysis and Processing* (pp. 970–977). Springer, Berlin, Heidelberg.

Proença, H., and Alexandre, L. A. 2007. The NICE. I: Noisy iris challenge evaluation-Part I. In *First IEEE International Conference on Biometrics: Theory, Applications, and Systems, 2007. BTAS 2007* (pp. 1–4). IEEE, Crystal City, VA.

Proença, H., and Alexandre, L. A. 2010. Iris recognition: Analysis of the error rates regarding the accuracy of the segmentation stage. *Image and Vision Computing,* 28(1), 202–206.

Rathgeb, C., Uhl, A., and Wild, P. 2012. *Iris Biometrics: From Segmentation to Template Security* (Vol. 59). Springer Science & Business Media, New York.

Rattani, A., and Derakhshani, R. 2017. Ocular biometrics in the visible spectrum: A survey. *Image and Vision Computing,* 59, 1–16.

Sanchez-Avila, C., and Sanchez-Reillo, R. 2005. Two different approaches for iris recognition using Gabor filters and multiscale zero-crossing representation. *Pattern Recognition,* 38, 231–240.

Sanchez-Gonzalez, Y., Chacon-Cabrera, Y., and Garea-Llano, E. 2014. A comparison of fused segmentation algorithms for iris verification. In *Iberoamerican Congress on Pattern Recognition* (pp. 112–119). Springer, Cham.

Schroff, F., Kalenichenko, D., and Philbin, J. 2015. Facenet: A unified embedding for face recognition and clustering. In *CVPR.*

Shepard, K., Wing, B., Miles, C., and Blackburn, D. 2006. Iris recognition-National Science and Technology Council (NSTC)–Committee on Technology–Committee on Homeland and National Security–Subcommittee on Biometrics (EUA), last updated 31 March 2006.

Simonyan, K., and Zisserman, A. 2014. Very deep convolutional networks for large-scale image recognition. arXiv preprint arXiv:1409.1556.

SOCIA Lab. 2018. Soft computing and image analysis group. University of Beira Interior: UBIRIS.v1 Database. http://iris.di.ubi.pt/ubiris1.html.

Sun, Z., and Tan, T. 2009. Ordinal measures for iris recognition. *IEEE Transactions on Pattern Analysis and Machine Intelligence,* 31, 2211–2226.

Sun, Z., Tan, T., and Qiu, X. 2006. Graph matching iris image blocks with local binary pattern. In *International Conference on Biometrics* (pp. 366–372). Springer, Berlin, Heidelberg.

Sutra, G., Garcia-Salicetti, S., and Dorizzi, B. 2012. The Viterbi algorithm at different resolutions for enhanced iris segmentation. In *5th IAPR International Conference on Biometrics (ICB), 2012* (pp. 310–316). IEEE, New Delhi, India.

Szegedy, C., Liu, W., Jia, Y., et al. 2015.Going deeper with convolutions. In *CVPR.*

Tan, T., Zhang, X., Sun, Z., and Zhang, H. 2012. Noisy iris image matching by using multiple cues. *Pattern Recognition Letters,* 33, 970–977.

Tenenbaum, J. M. 1970. Accommodation in computer vision (No. CS-182). Stanford University, Department of Computer Science, Stanford, CA.

Tian, Q., Qu, H., Zhang, L., and Zong, R. 2011. Personal identity recognition approach based on iris pattern. In Yang, J., and Nanni, L. (eds.) *State of the Art in Biometrics* (pp. 1–326). InTechOpen, London, United Kingdom, pp. 1–326. DOI: 10.5772/17110.

Tuceryan, M., and Jain, A. K. 1993. Texture analysis. In Chen, C. H., Pau, L. F., and Wang, P. S. (eds.) *Handbook of Pattern Recognition and Computer Vision* (pp. 207–248). World Scientific, Singapore.

Uhl, A., and Wild, P. 2012. Weighted adaptive Hough and ellipsopolar transforms for real-time iris segmentation. In *5th IAPR International Conference on Biometrics (ICB), 2012* (pp. 283–290). IEEE, New Delhi, India.

Uhl, A., and Wild, P. 2013. Fusion of iris segmentation results. In *Iberoamerican Congress on Pattern Recognition* (pp. 310–317). Springer, Berlin, Heidelberg.

Wildes, R. P. 1997. Iris recognition: An emerging biometric technology. *Proceedings of the IEEE*, 85, 1348–1363.

Wildes, R. P., Asmuth, J. C., Green, G. L., et al. 1994. A system for automated iris recognition. In *Proceedings of the Second IEEE Workshop on Applications of Computer Vision, 1994* (pp. 121–128). IEEE.

Zamudio Fuentes, L. M. 2010. Reconocimiento del iris como identificación biométrica utilizando elvideo. Master of Sciences Dissertation, CITEDI-IPN, México.

Zamudio-Fuentes, L. M., García-Vázquez, M. S., and Ramírez-Acosta, A. A. 2010. Iris segmentation using a statistical approach. In *Mexican Conference on Pattern Recognition* (pp. 164–170). Springer, Berlin, Heidelberg.

Zamudio-Fuentes, L. M., García-Vázquez, M. S., and Ramírez-Acosta, A. A. 2011. Local quality method for the iris image pattern. In *Iberoamerican Congress on Pattern Recognition* (pp. 79–88). Springer, Berlin, Heidelberg.

Zhang, G. H., and Salganicoff, M. 1999. U.S. Patent No. 5,953,440. U.S. Patent and Trademark Office, Washington, DC.

Zhang, H., Sun, Z., Tan, T., and Wang, J. 2012. Iris image classification based on color information. In *Paper presented at the 21st International Conference on Pattern Recognition* (pp. 3427–3430), Tsukuba, Japan.

Zhu, X. D., Liu, Y. N., Ming, X., and Cui, Q. L. 2004. A quality evaluation method of iris images sequence based on wavelet coefficients in "region of interest". In *The Fourth International Conference on Computer and Information Technology, 2004 CIT'04* (pp. 24–27). IEEE, Wuhan, China.

Zhu, Y., Tan, T., and Wang, Y. 2000. Biometric personal identification based on iris patterns. In *Paper presented at the Proceedings of the 15th International Conference on Pattern Recognition* (pp. 801–804), Barcelona, Spain.

Zuo, J., and Schmid, N. A. 2009. Global and local quality measures for NIR iris video. In *IEEE Computer Society Conference on Computer Vision and Pattern Recognition Workshops, 2009. CVPR Workshops 2009* (pp. 120–125). IEEE, Miami, FL.

7

Slap Fingerprint Authentication and Its Limitations

Puneet Gupta
Indian Institute of Technology Indore

Phalguni Gupta
National Institute of Technical Teachers' Training and Research

CONTENTS

7.1 Introduction

Rapid progress in the technology and easy accessibility of cheap hardware have proliferated the requirements of personal authentication in this digital age. Authentication refers to grant access to genuine users while restricting impostors. Few years back, keys and passwords were used for authentication purpose, but they could be easily forgotten, transferred, guessed and/or stolen. Hence, they are replaced by biometric traits that utilize the distinctive attributes attached to the user (Gupta & Gupta 2015c). Fingerprint constituting the ridge-valley pattern on the fingertip is an extensively explored biometric trait for authentication (Gupta & Gupta 2015e). Fingerprints develop when the fetus develops at around 7 months, and they remain unchanged throughout an individual's life, unless the he or she meets with a hand injury or so. As fingerprints can withstand temporal deformations and are highly distinctive, these are considered legal supporting evidence in courts and forensics. Fingerprints can be categorized in the following manner:

1. Dab fingerprint: The single fingerprint obtained by optical or capacitive fingerprint acquisition sensors is known as a dab fingerprint (Jain et al. 1999). Most frequently, if the dab fingerprints are appropriately acquired, they are used for authentication. This is because they usually contain sufficient number of features required for authentication. The area of the acquisition sensor is fixed and usually small in size, which may result in the partial fingerprint acquisition. During the matching process, it is possible that the partial fingerprints are acquired from different areas of the same fingerprint. In such a case, fewer features may be available for matching, which may eventually lead to poor authentication.

2. Rolled fingerprint: Due to partial fingerprints acquired from the limited area sensors, dab fingerprints perform poorly. This issue is mitigated by acquiring the full fingerprint in the case of rolled fingerprint that requires the nail-to-nail impression of a fingerprint (Ratha et al. 1998). It not only handles the problem of partial fingerprint acquisition but also provides more fingerprint area; hence, one gets more features for better fingerprint matching. First, it acquires an image sequence of multiple dab fingerprints by rolling the finger on the sensor. Subsequently, the sequence is consolidated to obtain a rolled fingerprint. Image mosaicking is used for the consolidation, which automatically aligns multiple images to create an aggregated image. The aggregated image is free from any visible deformation in the overlapping regions. Even though the rolled fingerprints provide better authentication than dab fingerprints, they are neglected for authentication because their acquisition is highly time-consuming and requires large operator intervention.

3. Latent fingerprint: When an individual touches an object, the impressions of his or her fingers are left on the object's surface. It is possible to lift such a finger impression and analyze it to authenticate the individual. Such inadvertent impressions are known as latent fingerprints (Jain & Feng 2011). These fingerprints are highly useful in forensics, but they are avoided for authentication required for day-to-day activities because lifting them is a highly time-consuming and expensive process. Some problems involved in matching latent fingerprints are poor quality; large, nonlinear deformations; partial fingerprints; and overlapping fingerprints.

4. Slap fingerprint: All the above-mentioned fingerprints are types of single fingerprints. The applicability of a single fingerprint for authentication is restricted when the acquired fingerprints are of a poor quality (Gupta & Gupta 2015d) because of the following reasons:

    i. Environmental factors that can result in the acquisition of dry or wet fingerprints

    ii. User factors like cuts and bruises on the fingertip

    iii. Sensor condition like the presence of dirt or latent-print on the sensor

    iv. Smoothening of ridge-valley structure because of occupation or age

    v. Introduction of shear deformation by skin elasticity because of nonuniform finger pressure on the sensor

FIGURE 7.1
Example of a slap image containing all the single fingerprints of a hand.

If multiple fingers can be used for authentication, the fingerprint-based authentication system's performance can be improved (Maltoni et al. 2009). The most obvious way to achieve this goal is by capturing the different fingerprints one after another. But such a mechanism is highly time-consuming and requires a large operator intervention. Fortunately, a slap fingerprint acquisition sensor has been designed for the acquisition of a slap image, which contains all four single fingerprints. It acquires all the four fingerprints of a hand simultaneously in a time-efficient manner. An example of slap image is depicted in Figure 7.1. Utilizing multiple fingerprints of the slap image for authentication not only improves the authentication by providing more features but also reduces the spoofing (Gupta & Gupta 2014c) because spoofing all four fingerprints is more difficult than spoofing a single fingerprint (Ulery et al. 2005). Due to these advantages, the slap fingerprint is preferred over other types of fingerprints for personal authentication (AADHAAR AUTOMATED).

The chapter is organized in the following manner. A brief description about a biometric system's working and performance evaluation is provided in the next section. Section 7.3 discusses the existing work in the realm of slap image authentication. It also discusses the problems in the existing work. Some potential research directions are provided in the last section.

7.2 State of the Art

This section provides a brief overview of the operations required in a biometric system and their performance evaluation metrics.

7.2.1 Working of a Biometric System

Usually, a biometric authentication requires the following three operations (Bolle 2004):

1. Enrollment: In order to use a biometric authentication system, a user must provide its distinctive features to the system using which matching can be performed. This first mandatory operation of storing the user's features in the database corresponding to its unique identity is known as enrollment or user registration. Initially, a biometric trait required by the system is acquired using the acquisition sensors. The noise present in the acquired image is mitigated by applying the preprocessing techniques to improve the quality. In some cases, if the acquired image is of poor quality with respect to its features, the user data are again acquired. The desired area from the acquired image is cropped to perform further processing for generating features. This is known as extraction of the region of interest (ROI). Several distinctive features are extracted from the ROI. For example, vein locations are extracted from the palm dorsal images (Gupta & Gupta 2015a) and minutiae are extracted from the fingerprint images (Gupta & Gupta 2014a). Eventually, a new entry is appended to the database, which stores the extracted features and a unique identifier, corresponding to the user.

2. Verification: This operation is required when the user claims an identity against his or her credentials and the requirement is to determine whether the user has correctly revealed his or her identity. In essence, it verifies the user on the basis of the claimed identity and accordingly decides whether the user is matched or unmatched. The verification operation also extracts the features following the same strategy as discussed in the enrollment stage. The extracted features are matched against the template of the claimed identity, which is stored in the database. A similarity or dissimilarity score is obtained after the matching. In the case of a similarity score, a genuine or an impostor user is reported when the similarity score is above or below the predefined threshold, respectively. While for the dissimilarity score, this criterion is reversed. The key point here is that verification is a one-to-one matching process where the extracted features are matched against the template of the claimed identity only. Moreover, the output of this operation is either genuine (match) or impostor (nonmatch). One of the most commonly used example of the verification operation is an ATM machine, where the identity is claimed by swapping the card.

3. Identification: In contrast to the verification operation, the user does not claim any identity in the identification operation, and the aim here is to reveal the correct identity of the user. It extracts all features from the biometric image by following the steps considered in the enrollment stage. As the user identity is unknown, the extracted features are matched against all the templates stored in the database. The template corresponding to which maximum matching score is obtained is marked as a user template, and its identity is marked as the user identity. Thus, the identification estimates the best template match to reveal the user identity. In other words, verification answers the question, "Am I the one I claim?", and identification answers the question, "Who am I?"

7.2.2 Performance Evaluation

The performance of a biometric system can be evaluated using various performance metrics. Some of the commonly used verification and identification metrics are described in this subsection.

7.2.2.1 *Performance Evaluation in Verification*

In the verification process, features of the query image are matched against the template of the claimed identity to generate a matching score. A decision threshold needs to be set prior to the matching for classifying the genuine and impostor matching scores. That is, a genuine match is reported when the matching score is greater than the decision threshold in case of similarity score; otherwise, it is treated as impostor. It is possible that the impostor matching score is greater than the decision threshold because of which the impostor is marked as a genuine user. This is known as the cases of false positives at the decision threshold. Similarly, genuine matching scores can spuriously consider impostors when the decision threshold exceeds the matching score. This is known as the cases of false negatives at the decision threshold. The following metrics are usually utilized to assess the performance of a biometric verification system:

1. False acceptance rate (FAR): It is given by the percentage of those impostors who are incorrectly marked as genuine, i.e., false positives.

$$FAR = \frac{\text{Total number of false positives}}{\text{Total number of accepted cases}} \tag{7.1}$$

2. False rejection rate (FRR): It is given by the percentage of those genuine users who are incorrectly marked as spurious, i.e., false negatives.

$$FRR = \frac{\text{Total number of false negatives}}{\text{Total number of rejected cases}} \tag{7.2}$$

3. Equal error rate (EER): It can be observed that FAR reduces while FRR increases when the decision threshold increases. Similarly, FRR reduces while FAR increases when the decision threshold decreases. That is, FAR and FRR possess competitive behavior. EER is given by

$$\left(\frac{FAR(t_1) + FRR(t_1)}{2} \right) \tag{7.3}$$

where the decision threshold t_1 is defined and where FAR and FRR are equal, i.e., $FAR(t_1) = FRR(t_1)$.

4. Receiver operating characteristics (ROC) curve: The most useful way to assess the performance of a biometric system is to compare their FAR and FRR at all possible decision thresholds. For this purpose, the graph of FAR against FRR is plotted at the possible decision thresholds. A hypothetical ROC curve is shown in Figure 7.2. The system performs accurately when there is a decision threshold at which both FAR and FRR are zero. In such a case, the EER will also be equal to zero.

7.2.3 Performance Evaluation in Identification

In the identification process, the extracted features are matched against all templates stored in the database and eventually they are ranked according to the matching score. Correct recognition refers to the case when the template corresponding to Rank 1 actually belongs to the user. The rate of correct recognition is known as CRR, i.e.,

$$CRR = \frac{\text{Total number of genuis matches having Rank 1}}{\text{Total number of user attempts for identification}} \times 100 \tag{7.4}$$

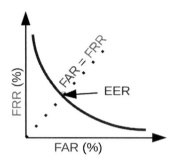

FIGURE 7.2
A hypothetical example of an ROC.

7.3 Slap Image Authentication

A multi-instance system refers to the biometric system that consolidates multiple instances of the same biometric trait, but these are acquired from different body parts using the same sensor (Ross et al. 2006). A slap image authentication system is one such multi-instance system where multiple instances of the fingerprint are used. These fingerprint instances are acquired from different fingers but using the same slap acquisition sensor. Usually, a slap image authentication system consists of the following three stages: slap segmentation, single fingerprint matching, and fusion of the single fingerprint matching scores. The flow-graph of the slap authentication system is shown in Figure 7.3. In the first stage of slap segmentation, all four single fingerprints existing in the acquired slap image

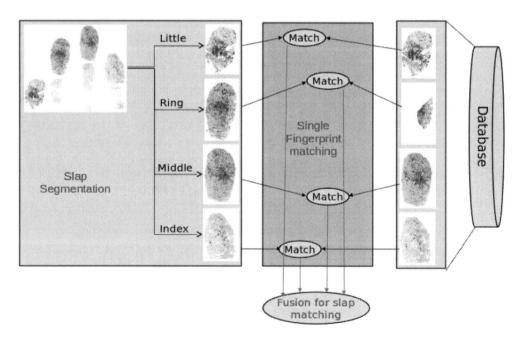

FIGURE 7.3
Flow-graph of slap image authentication.

are determined, cropped, and labeled. Labeling a fingerprint refers to determine that the fingerprint belongs to index, middle, ring, or little finger of either left or right hand. In the next stage, features are extracted from each single fingerprint, and they are matched against the corresponding templates stored in the database. It is important to understand that the single fingerprint matching only matches the fingerprints belonging to the same finger. Hence, if fingerprints are labeled incorrectly, wrong fingerprints are matched leading to high FRR. Moreover, one matching score is obtained corresponding to each fingerprint; thus, total four matching scores are obtained in this stage. In the last stage, all four matching scores are consolidated to generate a single slap image matching score that is used for performing the authentication. Correct authentication can be guaranteed only when all these three stages perform their task correctly. In this section, working and limitations of the existing slap image authentication systems are discussed.

7.3.1 Slap Segmentation

The first stage in a slap authentication system is slap segmentation, in which all single fingerprints existing in the acquired slap image are first determined and subsequently labeled as index, middle, ring, or little finger of either left or right hand. Correct authentication requires that it should satisfy the following requirements:

1. Correct fingerprint localization: The fingerprints existing in the slap should be located correctly. If the erroneous area is marked as fingerprint, spurious features can be generated and matched during the next stage of single fingerprint matching, which eventually results in poor authentication.

2. Correct fingerprint labeling: Fingerprint labeling refers to correctly classify the hand present in the slap image (i.e., slap image contains either left or right hand) and determine that the single fingerprint belongs to which finger (i.e., index, middle, ring, or little finger). If the labeling is incorrect, wrong fingerprints can be matched that will eventually lead to poor authentication.

3. Near-real time computations: Most of the day-to-day life applications require near-real time authentication, which can impose the condition that the slap segmentation should be performed in near-real time.

7.3.1.1 Framework of Slap Segmentation

The basic steps required in slap segmentation are illustrated in Figure 7.4. The slap image is first preprocessed using: (i) down-sampling to reduce the time computations; (ii) noise reduction (Ramaiah & Mohan 2011); and (iii) detection of finger pixels. In the realm of slap image, pixels belonging to fingers are considered as foreground pixels and the remaining pixels are considered as background pixels. Since the foreground pixels belong to finger areas, they contain ridge-valley structure that possesses large intensity fluctuations in their local neighborhoods. On the other hand, background pixels possess nearly uniform intensity fluctuations in their local neighborhoods. This provides the motivation behind the exploration of anisotropic measures for the detecting the finger pixels. The detected foreground pixels are congregated according to the spatial locations to create components. These components belong to different finger areas and some of them are derived from the fingertips. The components derived from fingertips contain fingerprints, and hence they are referred as fingerprint components. Such components are represented by their geometrical properties like mean and orientation. These geometrical properties are

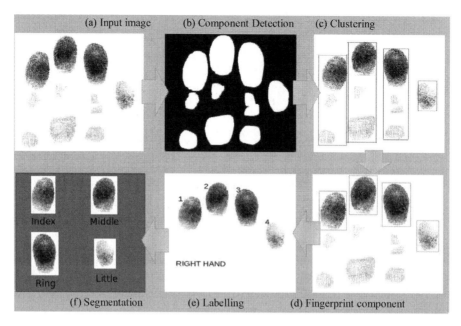

FIGURE 7.4
Framework of slap segmentation. (a) Input image. (b) Component detection. (c) Clustering. (d) Fingerprint component. (e) Labelling. (f) Segmentation.

used to cluster the components such that all components derived from a common finger lie in the same class, but the components derived from different fingers lie in different classes. In essence, each class denotes a finger in the component clustering. Along with the hand geometry information, sensor information can also be utilized to improve the clustering. It is obvious that a fingerprint lies at the top of a finger; hence, the components present at the top in each class (which represents finger) are marked as fingerprint components. The topmost component or fingerprint component can be determined by analyzing the geometrical properties like mean of components. Subsequently, hand geometry constraints are utilized for labeling the extracted fingerprint components.

7.3.1.2 Existing Approaches

A well-known publicly available slap segmentation algorithm, NFSEG, was proposed in Watson et al. (2004). It down-samples the slap image to reduce the time computation. Components are extracted from the resultant image using global thresholding (Otsu 1975). It is quite possible that it can split up the fingerprint into multiple components, which degrades the performance due to poor fingerprint localization. One such example is shown in Figure 7.5, where a fingerprint component is wrongly localized. The components are clustered using the intuition that components belonging to different fingers can be separated using several parallel lines containing equal spacing between them. Thus, parallel lines are drawn at various heuristically chosen angles and spacings. The configuration of parallel lines and spacings that best separate the components is chosen. After clustering the components using the chosen configuration, fingertip components are determined. This is highly computationally expensive as it is a multi-pass algorithm. It assumes that all fingers are in one common direction and thus it fails when the slap

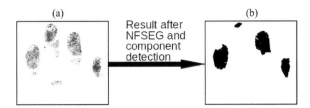

FIGURE 7.5
Example of erroneous component detection due to global thresholding. (a) Input slap image. (b) Component detection using NFSEG.

FIGURE 7.6
Example of a slap image containing open fingers.

image contains open fingers. One such example of slap image containing open fingers is shown in Figure 7.6.

Anisotropic measures like mean and local variance are explored for detecting the finger pixels (Zhang et al. 2010). These measures do not provide the correct detection when the slap image contains finger pixels of varying intensities. Thus, local entropy can also be employed to handle this issue (Gupta & Gupta 2012). In addition, neighboring components from different fingers can be merged due to down-sampling. One such example is shown in Figure 7.7. This issue can be handled by proposing a multi-resolution approach (Gupta & Gupta 2014b). It prunes most of the image area using low-resolution image and, eventually, analyzes the remaining area in the original acquired slap image. Several hand geometry constraints can be utilized for extracting the fingerprint components lying in the upper half of the slap image (Lo & Sankar 2006). Such constraint is often violated during the slap image acquisition, and one such example is shown in Figure 7.8. Usually, the components are extracted by utilizing neighborhood connectivity (Zhang et al. 2010). Another way employed for this purpose is the mean-shift algorithm that is based on elliptical fitting (Hödl et al. 2009). This algorithm is computationally expensive, and it fails when the slap image contains non-elliptical shape components. One such example of slap image containing non-elliptical shape components is shown in Figure 7.8.

(a) Input slap image   (b) Merged component

FIGURE 7.7
Example of a slap image where neighboring components are merged due to down-sampling. (a) Input slap image (b) Merged component.

FIGURE 7.8
Example of a slap image containing non-elliptical fingerprints in the lower half of the slap image.

Extracted components are clustered into four classes that denote four fingers. Such clustering utilizes the hand geometry information (Yong-liang et al. 2010; Li et al. 2010; Singh et al. 2012a). Mostly topmost component from each class is marked as fingerprint component. Component clustering can be avoided and only four topmost components can be marked as fingerprint components (Singh et al. 2012b). It provides erroneous fingerprint components when the slap image contains large rotation. One such example is shown in Figure 7.9. This algorithm is modified in Singh et al. (2012b), where the slap image is rotated at various heuristically chosen angles, and the angle that best describes the hand geometry is used for the fingerprint component extraction. This multi-pass algorithm is computationally expensive. Relative placements of fingers can be analyzed along with the hand geometry constraints for fingerprint labeling (Gupta & Gupta 2014b). This fingerprint labeling outperforms the fingerprint labeling proposed in Hödl et al. (2009), which utilizes the mean of fingerprint components and global orientation of hand. The labeling in Hödl et al. (2009) performs erroneously when the slap image contains partial and non-elliptical-shaped fingerprint components.

FIGURE 7.9
Example of a slap image containing large rotation.

7.3.1.3 Challenges in Component Detection

Spurious generation of a component or missing a genuine component is referred as the cases of wrong component detection in a slap image. One reason behind the partially or completely missing a genuine component is the large intensity variations of finger pixels in the slap image. Sometimes the global threshold is set in such a way that some finger areas are erroneously removed. Various well-known algorithms given in Singh et al. (2012a,b) and Watson et al. (2004) utilize this threshold for finger pixel detection and, hence, reported several cases of wrong component detection. One such example is shown in Figure 7.5. Another reason for the wrong component detection is the down-sampling of slap image, which can merge the neighboring components. Its example is shown in Figure 7.7. Different fingerprint components can be merged or spurious components can be generated when the slap image contains halo and sweat (Gupta & Gupta 2018). This can be visualized from Figure 7.10, where different components are erroneously merged.

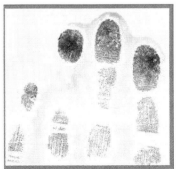 

(a) Slap image having halo   (b) Erroneous component

FIGURE 7.10
Example of wrong component detection due to halo. (a) Slap image having halo. (b) Erroneous component.

7.3.1.4 Challenges in Fingerprint Component Detection

It is possible that fingerprint components are erroneously detected due to:

1. Noise in background: If slap sensor is not properly cleaned during acquisition, noise or latent fingerprints can be simultaneously acquired with slap image (Yong-liang et al. 2010).

2. Halo or sweat: Skin and environment conditions can introduce problems like halo or sweat in the slap image. Halo and sweat can merge different components or introduce spurious components.

3. Improper orientation: Despite the correct component extraction, fingerprint component can be incorrectly detected if hand is placed in such a way that can violate hand geometry constraints. An unavoidable reason of the inappropriate hand placement is the hand disease like osteoarthritis and swan neck deformity.

7.3.1.5 Challenges in Hand Detection

Hand detection can be erroneous when:

1. Hand geometry is inappropriately altered due to the hand disease.

2. User places the hand in such a way that the correct hand cannot be detected even by the visual inspection.

3. Geometrical properties are incorrectly estimated because the fingerprint components are of poor quality.

7.3.2 Single Fingerprint Matching

In the literature, more than 100 types of local ridge deformations or minutiae have been discovered and utilized for single fingerprint based authentication, but the most prominent among these are ridge endings and bifurcations (Maltoni et al. 2009). Minutiae-based fingerprint matching consists of the following two tasks: minutiae extraction and matching. Usually, Mindtct and Bozorth are utilized for minutiae detection and fingerprint matching, respectively (Watson et al. 2004). It is because these algorithms are publicly available and they accomplish their tasks correctly. But this approach performs erroneously if the fingerprint is of poor quality.

7.3.2.1 Existing Approaches

Single fingerprint matchers existing in the literature are mainly categorized as:

1. Minutiae-based matchers: Major factors that limit the fingerprint matching are the shear (or non-linear) deformations introduced due to varying pressure applied by the user and partial fingerprints that are acquired due to inappropriate finger placement and limiting sensor size (Tico & Kuosmanen 2003). To handle these unavoidable scenarios, the distinctive features known as minutiae are mostly used for fingerprint matching (Lee et al. 2002). In order to match two fingerprints, minutiae are extracted from them and the extracted minutiae are compared to estimate the similarity between the fingerprints. Various kinds of minutiae can be employed for the matching, but usually minutiae ending and bifurcation are

employed for the fingerprint matching because these are usually available in abundance in fingerprint images (Peralta et al. 2015). A minutia is defined by its spatial location and rotation (Feng 2008). Various kinds of features are extracted using these minutiae like relative length and orientation. It is important to define these features in such a way that the geometric transformations are minimized. These geometric transformations are introduced due to the translation, rotation, and non-linear deformation due to skin and hand placement. In addition, these features should be represented such that the problems of the spurious generation of false minutiae and missing of genuine minutiae can be minimized. One such minutiae-based fingerprint matcher is Bozorth (Watson et al. 2004) that uses Mindtct for extracting the minutiae ending and bifurcations. A minutiae-based fingerprint matcher can correctly match the fingerprints if the fingerprints involved in the matching contain sufficiently large number of minutiae (Jain et al. 2001). This requirement is often violated when the fingerprint contains poor quality.

2. Correlation-based matchers: In this category of fingerprint matchers, the fingerprints are first superimposed and the similarity between the fingerprints is determined using their overlapping and non-overlapping areas (Meraoumia et al. 2012). Thus, these matchers compare the fingerprints globally in contrast to the minutiae-based fingerprint matchers that analyze the local features for the matching. Minutiae-based fingerprint matchers outperform the correlation-based matchers when partial fingerprints are acquired or global fingerprint structure is deformed due to skin elasticity (Gupta & Gupta 2016b). But correlation-based matchers outperform minutiae-based fingerprint matchers when few genuine minutiae can be extracted due to poor quality fingerprints (Maltoni et al. 2009). Both correlation-based and minutiae-based fingerprint matchers can also be simultaneously used for improving the fingerprint matching (Jiang & Yau 2000).

3. Ridge feature-based matchers: A fingerprint structure is given by the coherent flow of ridges and valleys. Directions of ridges and valleys deviate slightly due to shear deformations introduced by skin elasticity, but these directions provide valuable information for poor quality fingerprints. This provides the motivation to employ the ridge features for determining the similarity between two fingerprints. Some of the frequently used ridge features in these matchers are defined using the local ridge orientations (Gupta & Gupta 2016a), texture (Jain et al. 2000), and frequency of ridges (Alonso-Fernandez et al. 2009). Just like correlation-based matchers, these matchers fail when the fingerprints are partially acquired.

In the realm of slap image authentication, minutiae-based fingerprint matcher known as Bozorth (Watson et al. 2004) is most frequently utilized. It extracts the minutiae ending and bifurcations using Mindtct that consists of the following stages:

1. Preprocessing: Noise presents in the fingerprint image is mitigated, and the image is enhanced by applying filtering. It is followed by binarizing and skeletonizing the enhanced image so as to reduce the time computations.

2. Minutiae detection: Several filters that can mimic the possible minutiae endings and bifurcations are applied on the skeletonized image to localize the minutiae. Some such filters are depicted in Figure 7.11.

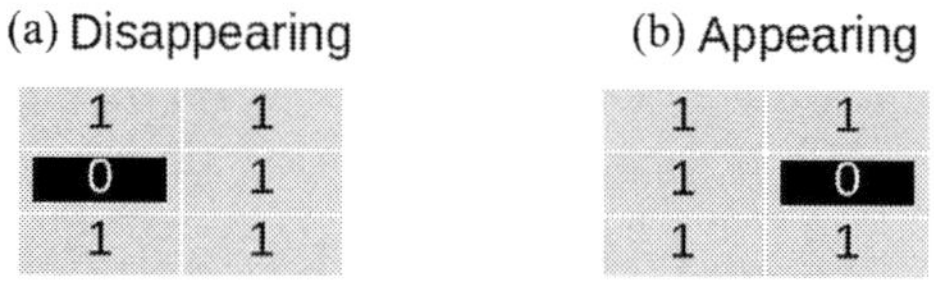

(a) Disappearing

1	1
0	1
1	1

(b) Appearing

1	1
1	0
1	1

FIGURE 7.11
Filters mimicking ridge ending minutiae points. (a) Appearing. (b) Disappearing.

3. Remove false minutiae: Other local deformations like island, hole, and overlaps can be spuriously detected as minutiae endings and bifurcations. Thus, several constraints are utilized to remove these false minutiae endings and bifurcations. The constraints are defined using the domain knowledge of fingerprints, for example, the distance between two minutiae should be greater than the predefined threshold.

After the minutiae extraction, the similarity score between two fingerprints is determined using Bozorth (Watson et al. 2004) that matches the extracted minutiae. For correct fingerprint matching, this matcher should not only be rotation and translation invariant, but it should also mitigate the issues related to partial fingerprints, missing genuine minutiae, and spurious minutiae generation. Hence, Bozorth performs the fingerprint matching in three steps. First, it defines two separate tables known as Intra-Fingerprint Minutia Comparison Tables. One table is defined for the probe fingerprint, while another one is defined for the gallery fingerprint. In the subsequent step, these tables are used to determine the matching minutiae pairs, and they are stored in a new table, known as Inter-Fingerprint Compatibility Table. In the last step, matching pairs stored in Inter-Fingerprint Compatibility Table are consolidated.

7.3.2.2 Challenges

Single fingerprint matching can be spurious due to the non-linear deformation introduced by the elastic skin on the fingertip; partial fingerprints acquired due to improper hand placement and limited sensor area; and poor quality fingerprints that result in the generation of spurious minutiae and missing of genuine minutiae. Poor quality fingerprints can be generated due to environmental factors such as humidity that results in wet fingerprints; sensor condition like dirt on the sensor; and user factors like fingerprint may get worn out or smoothen due to occupation and age.

7.3.3 Fusion

This section provides a brief overview of those existing approaches that are used for consolidating all the four single fingerprint matching scores. In Lo and Bavarian (2009), authentication is performed by applying serial fusion on the single fingerprint matching scores. It aims to iteratively match the extracted single fingerprints such that some potential non-matched fingerprints are eliminated in each iteration. The ordering in which single fingerprints are utilized for matching is fixed according to the domain knowledge of slap image matching. It requires that the single fingerprints involved in the slap matching should be of good quality. This requirement is often violated in the case of slap images. A better approach is to utilize score level fusion for consolidating the single fingerprint matching scores, as in Li et al. (2010); Singh et al. (2012b); Gupta and Gupta (2016b). In Li et al. (2010), fixed weights

TABLE 7.1

Description of the Existing Systems

System	Fusion Strategies	Description of Fusion
Lo and Bavarian (2009)	Serial fusion	Fixed ordering
Bendale and Boult (2010)	Feature level fusion	Inaccurate fusion strategy
Li et al. (2010)	Score level fusion	Fixed weights
Singh et al. (2012b)	Quality level fusion	Inaccurate fingerprint quality
Gupta and Gupta (2014a)	Decision level fusion	Fail for poor quality fingerprints
Gupta and Gupta (2016b)	Quality level fusion	Insufficient fingerprint quality

obtained by minimizing the error rate are utilized. In contrast, weights are adaptively assigned in Singh et al. (2012b) according to the quality of single fingerprints. Quality in this approach is provided by the local orientation of fingerprint, and such a quality measure is inaccurate in the fingerprint areas containing large curvature (like the areas near singular points) (Gupta & Gupta 2015b). A better adaptive weight assignment is discussed in Gupta and Gupta (2016b), where symmetric filters based fingerprint quality estimation is used for weight assignment. Such a quality measure performs correctly for singular point areas. Decision level fusion is introduced in Gupta and Gupta (2014a) for slap image matching. It assumes that in the case of genuine slap matching, minimum of two single fingerprints should be matched. Similarly, feature level fusion is proposed in Bendale and Boult (2010) for the slap image matching. The epitome of these slap matching systems is presented in Table 7.1. It is important to note that no single fusion strategy performs correctly in all the possible scenarios.

7.4 Future Research Directions

The efficacy of the slap image authentication can be improved by mitigating these shortcomings in the following manner:

1. Indexing of slap images: One of the major challenges involved in the slap image based authentication is a large amount of time incurred during identification. The user identity is revealed in a biometric authentication system after sequentially matching each template of the database. This is computationally expensive. For the time-efficient matching, it is preferable to arrange the templates in such a manner that only few potential templates need to be matched corresponding to the query image. This phenomenon is referred to as database indexing. Even though a large number of indexing techniques are available, slap image indexing is not properly studied and requires rigorous exploration.

2. Fusing with other traits: Slap fingerprint can be used along with other biometric traits to provide better authentication. The main problem with the multi-modal fusion is that it requires multiple biometric traits whose acquisition will be time consuming. Thus, it is advisable to consider those biometric traits that can be simultaneously acquired (Gupta et al. 2016). Keeping this in mind, the possibility of fusing the slap image with other traits can be studied. As an instance, slap image can be fused with palm-print.

3. Utilizing fingerprints from non-distal phalanges: In the literature, fingerprints from non-distal phalanges have been utilized for authentication (Kayaoglu et al. 2015). Fortunately, slap image provides fingerprints derived from both distal and non-distal phalanges, but only distal phalanges are used for authentication. Fingerprints from non-distal phalanges can be incorporated in the existing slap image authentication to improve the authentication.

4. Multiple fusions framework: Slap image authentication can be improved if single fingerprint matching and the fusion of single fingerprint matching scores can be improved. A particular class of fingerprint matcher is insufficient to provide correct single fingerprint matching in all possible scenarios. Hence, the possibility of employing multiple fingerprint matchers can be studied to provide better slap image authentication (Gupta & Gupta 2017). In a similar manner, the fusion of the matching scores can be improved for enhancing the authentication.

7.5 Conclusions

Slap fingerprint (i.e., all fingerprints of a hand) is highly useful for authentication, which is one of the major requirements in the current digital age. It is because: (i) just like any other biometric trait, they cannot be forgotten, transferred or guessed; (ii) they can withstand temporal deformations, provides highly distinctive features, and acceptable as legal supporting evidence in courts; and (iii) difficult to spoof as spoofing all four fingerprints are difficult. Despite these advantages, the efficacy of slap fingerprint has been restricted for authentication. The overview of existing work in the realm of slap image authentication along with the factors that limit the efficacy of slap fingerprint for authentication has been discussed in this chapter. In addition, this chapter has provided some potential research directions in the realm of slap fingerprint based authentication.

References

AADHAAR AUTOMATED. UID enrolment proof-of-concept report. uidai.gov.in/images/FrontPageUpdates/uid_enrolment_poc_report.pdf.

Alonso-Fernandez, F., Bigun, J., Fierrez, J., Fronthaler, H., Kollreider, K., & Ortega-Garcia, J. (2009). Fingerprint recognition. In D Petrovska-Delacrétaz, G Chollet, B Dorizzi (eds.) *Guide to Biometric Reference Systems and Performance Evaluation* (pp. 51–88). Springer, London.

Bendale, A. Z., & Boult, T. E. (2010). id-privacy in large scale biometric systems. In *International Workshop on Information Forensics and Security* (pp. 1–6). IEEE.

Bolle, R. (2004). *Guide to Biometrics*. Springer, New York.

Feng, J. (2008). Combining minutiae descriptors for fingerprint matching. *Pattern Recognition*, 41, 342–352.

Gupta, P., & Gupta, P. (2012). Slap fingerprint segmentation. In *IEEE International Conference on Biometrics: Theory, Applications and Systems* (pp. 189–194).

Gupta, P., & Gupta, P. (2014a). A dynamic slap fingerprint based verification system. In DS Huang, V Bevilacqua, P Premaratne (eds.) *Intelligent Computing Theory* (pp. 812–818). Springer.

Gupta, P., & Gupta, P. (2014b). An efficient slap fingerprint segmentation and hand classification algorithm. *Neurocomputing*, 142, 464–477.

Gupta, P., & Gupta, P. (2014c). Extraction of true palm-dorsa veins for human authentication. In *Proceedings of the 2014 Indian Conference on Computer Vision Graphics and Image Processing* (p. 35). ACM.

Gupta, P., & Gupta, P. (2015a). An accurate finger vein based verification system. *Digital Signal Processing*, 38, 43–52.

Gupta, P., & Gupta, P. (2015b). Fingerprint orientation modeling using symmetric filters. In *Winter Conference on Applications of Computer Vision* (pp. 663–669). IEEE.

Gupta, P., & Gupta, P. (2015c). Multi-modal fusion of palm-dorsa vein pattern for accurate personal authentication. *Knowledge-Based Systems*, 81, 117–130.

Gupta, P., & Gupta, P. (2015d). A robust singular point detection algorithm. *Applied Soft Computing*, 29, 411–423.

Gupta, P., & Gupta, P. (2015e). Slap fingerprint segmentation using symmetric filters based quality. In *2015 Eighth International Conference on Advances in Pattern Recognition (ICAPR)* (pp. 1–6). IEEE.

Gupta, P., & Gupta, P. (2016a). An accurate fingerprint orientation modeling algorithm. *Applied Mathematical Modelling*, 40, 7182–7194.

Gupta, P., & Gupta, P. (2016b). An accurate slap fingerprint based verification system. *Neurocomputing*, 188, 178–189.

Gupta, P., & Gupta, P. (2017). An efficient slap image matching system based on dynamic classifier selection and aggregation. *Information Sciences*, 417, 113–127.

Gupta, P., & Gupta, P. (2018). A slap fingerprint based verification system invariant to halo and sweat artifacts. *Applied Mathematical Modelling*, 54, 413–428.

Gupta, P., Srivastava, S., & Gupta, P. (2016). An accurate infrared hand geometry and vein pattern based authentication system. *Knowledge-Based Systems*, 103, 143–155.

Hödl, R., Ram, S., Bischof, H., & Birchbauer, J. (2009). Slap fingerprint segmentation. In *Computer Vision Winter Workshop*.

Jain, A., Ross, A., & Prabhakar, S. (2001). Fingerprint matching using minutiae and texture features. In *International Conference on Image Processing* (volume 3, pp. 282–285). IEEE.

Jain, A. K., & Feng, J. (2011). Latent fingerprint matching. *IEEE Transactions on Pattern Analysis and Machine Intelligence*, 33, 88–100.

Jain, A. K., Prabhakar, S., & Hong, L. (1999). A multichannel approach to fingerprint classification. *IEEE Transactions on Pattern Analysis and Machine Intelligence*, 21, 348–359.

Jain, A. K., Prabhakar, S., Hong, L., & Pankanti, S. (2000). Filterbank-based fingerprint matching. *IEEE Transactions on Image Processing*, 9, 846–859.

Jiang, X., & Yau, W.-Y. (2000). Fingerprint minutiae matching based on the local and global structures. In *Proceedings. 15th international conference on Pattern recognition, 2000* (volume 2, pp. 1038–1041). IEEE.

Kayaoglu, M., Topcu, B., & Uludag, U. (2015). Biometric matching and fusion system for fingerprints from non-distal phalanges. arXiv preprint arXiv:1505.04028.

Lee, D., Choi, K., & Kim, J. (2002). A robust fingerprint matching algorithm using local alignment. In *Proceedings. 16th International Conference on Pattern Recognition, 2002* (volume 3, pp. 803–806). IEEE.

Li, Y., Zhang, Y., Lu, J., Liu, C., & Fang, S. (2010). Robust rotation estimation of slap fingerprint image for e-commerce authentication. In *IEEE International Conference on Information Theory and Information Security* (pp. 66–69).

Lo, P., & Sankar, P. (2006). Slap print segmentation system and method. US Patent 7,072,496.

Lo, P. Z., & Bavarian, B. (2009). Adaptive fingerprint matching method and apparatus. US Patent 7,515,741.

Maltoni, D., Maio, D., Jain, A. K., & Prabhakar, S. (2009). *Handbook of Fingerprint Recognition*. Springer Science & Business Media.

Meraoumia, A., Chitroub, S., & Bouridane, A. (2012). Multimodal biometric person recognition system based on fingerprint & finger-knuckle-print using correlation filter classifier. In *International Conference on Communications* (pp. 820–824). IEEE.

Otsu, N. (1975). A threshold selection method from gray-level histograms. *Automatica*, 11, 285–296.

Peralta, D., Galar, M., Triguero, I., Paternain, D., García, S., Barrenechea, E., Benítez, J. M., Bustince, H., & Herrera, F. (2015). A survey on fingerprint minutiae-based local matching for verification and identification: Taxonomy and experimental evaluation. *Information Sciences*, 315, 67–87.

Ramaiah, N. P., & Mohan, C. K. (2011). De-noising slap fingerprint images for accurate slap fingerprint segmentation. In *10th International Conference on Machine Learning and Applications and Workshops (ICMLA), 2011* (volume 1, pp. 208–211). IEEE.

Ratha, N. K., Connell, J. H., & Bolle, R. M. (1998). Image mosaicing for rolled fingerprint construction. In *Proceedings. Fourteenth International Conference on Pattern Recognition, 1998* (volume 2, pp. 1651–1653). IEEE.

Ross, A. A., Nandakumar, K., & Jain, A. K. (2006). *Handbook of Multibiometrics*. volume 6. Springer, Boston, MA.

Singh, N., Nigam, A., Gupta, P., & Gupta, P. (2012a). Four slap fingerprint segmentation. In DS Huang, J Ma, KH Jo, MM Gromiha (eds.) *Intelligent Computing Theories and Applications* (pp. 664–671). Springer.

Singh, N., Tiwari, K., Nigam, A., & Gupta, P. (2012b). Fusion of 4-slap fingerprint images with their qualities for human recognition. In *IEEE World Congress on Information and Communication Technologies* (pp. 925–930).

Tico, M., & Kuosmanen, P. (2003). Fingerprint matching using an orientation-based minutia descriptor. *IEEE Transactions on Pattern Analysis and Machine Intelligence*, 25, 1009–1014.

Ulery, B., Hicklin, A., Watson, C., Indovina, M., & Kwong, K. (2005). Slap fingerprint segmentation evaluation, SlapSeg04 analysis report, 2004.

Watson, C. I., Garris, M. D., Tabassi, E., Wilson, C. L., McCabe, R. M., & Janet, S. (2004). Users guide to NIST fingerprint image software 2 (NFIS2). National Institute of Standards and Technology.

Yong-liang, Z., Yan-miao, L., Hong-tao, W., Ya-ping, H., Gang, X., & Fei, G. (2010). Principal axis and crease detection for slap fingerprint segmentation. In *IEEE International Conference on Image Processing* (pp. 3081–3084).

Zhang, Y., Xiao, G., Li, Y., Wu, H., & Huang, Y. (2010). Slap fingerprint segmentation for live-scan devices and ten-print cards. In *International Conference on Pattern Recognition* (pp. 23–26). IEEE.

8

The Reality of People Re-Identification Task, Where We Are and Where We Are Going: A Review

Daniela Moctezuma

CONACyT—Consejo Nacional de Ciencia y Tecnología

Centro de Investigación en Ciencias de Información Geoespacial

Mario Graff, Eric S. Tellez, and Sabino Miranda-Jiménez

CONACyT—Consejo Nacional de Ciencia y Tecnología

INFOTEC Centro de Investigación e Innovación en Tecnologías de la Información y Comunicación

CONTENTS

8.1 Introduction

Recently, many countries increased their security needs to prevent terrorist attacks, thefts, and other disturbances. In a supporting reaction, the scientific community focused their attention to several research areas related to security like Intelligent Video Surveillance (IVS).

In the past few years, several impressive works have been published in this area, proving that IVS systems are a useful tool to detect and assess potentially insecure situations in public spaces such as airports, train stations, malls, universities, trains, or airplanes (Saghafi, Hussain, Zaman, & Saad, 2015). IVS comprises a broad area, including several tasks such as people re-identification, action recognition, scene understanding, and trajectory analysis (Wang, 2013).

In literature, there is a consensus over the major challenge in person re-identification problem; the challenges lie in the preservation of the identity of the person against large variations caused by complex backgrounds, occlusions, and illuminations changes. The big question is how to deal with this challenge while discriminating different individuals.

People re-identification can be defined as identifying a particular person over a set of cameras located in environments with different conditions of lighting, zoom, angle, space, and time. Therefore, it is the process of assigning a reliable identifier to a person in different situations (i.e., images) (Bedagkar-Gala & Shah, 2014). To understand the re-identification process the European Commission in EUROSUR-2011 (Frontex, 2011) defined several relevant tasks such as detection, classification, identification, recognition, and verification. Under this standardization, the more related definitions to re-identification are identification and recognition. According to EUROSUR-2011, identification is the task of establishing the unique identity of an object without any prior knowledge. Recognition is the task of confirming that a detected object is a specific predefined unique object. Therefore, re-identification process can be defined as the task of recognizing an individual who has been observed, for example, in a surveillance cameras circuit.

Figure 8.1 illustrates the general workflow of a classical person re-identification system. First, people detection is done; this is a critical starting point, only after background segmentation. Broadly speaking, once the person is segmented in an image, the next step is to extract its features. This extraction is done using the appearance or the motion such that a feature vector is computed. Finally, a similarity metric or a trained classifier is used for assigning the ID. Numerous studies have been carried out over the years (e.g., Luna et al., 2017; Li, Yang, Zhang, & Xu, 2014); however, people detection still remains an open problem. Although the people detection task is a hard problem to solve, it is crucial to improve people re-identification.

8.1.1 Person Re-Identification Formalization

Formally, because person re-identification can be modelled as a matching problem, we can assume that we have N samples of C classes (where each class corresponds to a person), N_i^c denotes sample i from class c; that is, image i from person c in our dataset. Furthermore, we have a multi-camera problem, so $N_{i_{cam}}^c$ where i_{cam} denotes image i in camera cam. From these specifications, we obtain a vector X which is the result of the extraction of features in each sample $N_{i_{cam}}^c$, so we can have $X_{i_{cam}}^c$ which denotes the feature vector from image i in camera cam for person c. A matching problem is about comparing two samples, where the objective is that the samples from the same person are close to each other, while those

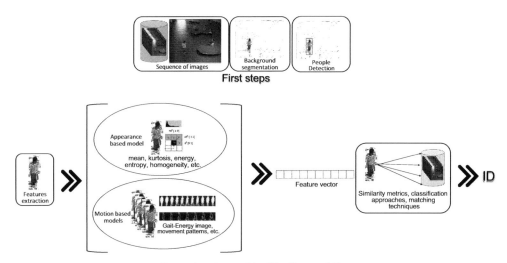

FIGURE 8.1
General scheme of a person re-identification system.

from different persons differ. This problem can be addressed from two viewpoints, which are as follows:

1. Proposing new extraction and descriptor methods to improve the quality of generated features for each image $N_{i_{cam}}^c$, with the idea that this quality improvement is transitive to vector $X_{i_{cam}}^c$.

2. Learning metrics that, usually through a transformation matrix, reduce the distance between samples in the same class and in samples of different classes increase the distance.

For instance, let us suppose we have two persons in our dataset, person a and person b, giving us two vectors $X_{i_{cam}}^a$ and $X_{i_{cam}}^b$, which means we have the feature vector from image i of camera cam for each person. The Euclidean distance is a straightforward way to compare these two vectors:

$$d\left(X_{i_{cam}}^a, X_{i_{cam}}^b\right) = \left(X_{i_{cam}}^a - X_{i_{cam}}^b\right)^T \left(X_{i_{cam}}^a - X_{i_{cam}}^b\right).$$

Nevertheless, in most cases, Euclidean distance does not satisfy the discrimination necessities; this issue is solved through metric learning. The core idea is to use a learned matrix or kernel M to transform the distance function as follows:

$$\left(X_{i_{cam}}^a, X_{i_{cam}}^b\right) = \left(X_{i_{cam}}^a - X_{i_{cam}}^b\right)^T M \left(X_{i_{cam}}^a - X_{i_{cam}}^b\right).$$

Both new descriptors and metric learning approaches have been proposed recently. This chapter is dedicated to review and analyze the current literature of the person re-identification task. The analysis utilizes several recognized benchmark datasets, and all the works are placed in similar evaluation conditions. The rest of this chapter is organized as follows. Section 8.2 describes the major available benchmarks of people re-identification task. Section 8.3 analyzes several approaches used to cope with the re-identification problem. The performance of classification approaches to the problem is studied in Section 8.4. Finally, Section 8.5 summarizes and concludes this chapter.

8.2 Evaluation Datasets

The evaluation step is critical in a recognition system because it is how we determine the performance of our approaches. In people re-identification task, there are several efforts dedicated to evaluating proposed approaches in the community. These efforts try to provide realistic scenarios with several conditions, such as one-shot (one image) per person, multi-shot (several images) per person, mono or multi-camera environments, overlapping or non-overlapping circuits, crowded or uncrowded scenarios, etc. These works also provide variations of other severe problems such as lighting, perspective, and zoom. This section presents an overall point of view of the most popular and current cited datasets. Table 8.1 gives a benchmark list published between the years 2007 and 2016. It also shows some of the central aspects to be analyzed such as the number of images and cameras,

TABLE 8.1

Most Popular Benchmark Datasets a Long Time for People Re-Identification Task

Name	# Persons	# Cameras	Non-Overlapping	# Images	Image Resolution	Publication Year
ViPER	632	2	NA	1,264	128×48	2007
ETHZ	146	1	No	8,580	32×64	2009
i-LIDS	119	2	Yes	476	NA	2009
GRID	250	8	Yes	1,025	Not standardized	2009
Indoor Multi-Camera Pedestrian Datasets (just 1)	1	3	No	15,966	384×288	2010
Indoor Multi-Camera Pedestrian Datasets (3–5 persons)	max 5	3	No	14,163	385×288	2010
Indoor Multi-Camera Pedestrian Datasets (5+ persons)	5+	3	No	13,053	386×288	2010
CAVIAR4REID	72	2	Yes	2,440	17×39 and 72×144	2011
Person Re-ID (PRID2011) Dataset 2011	200	2	Yes	94,987	64×128	2011
CUHK01	971	2	Yes	3,384	160×60	2012
SAIVT Soft Biometric Database	152	8	Yes	64,778	704×576	2014
Market-1501	1,501	6	No	32,668	128×64	2015
PKU-ReID	114	2	yes	1,824	128×48	2016
Person Re-ID Dataset 450S (PRID450S)	450	2	Yes	2,700	64×128	2016
DukeMTMC-reID	702	8	Yes	19,000	NA	2016
ALERT Airport Re-Identification Dataset	9,651	6	Yes	39,902	128×64	2016

the resolution of images, the overlapping or non-overlapping views, the number of persons to identify (or identities), and the year of publication.

The oldest dataset considered here is the VIPeR (Gray, Brennan, & Tao, 2007), one of the most-cited benchmark, despite the fact it only contains a single shot per camera (two cameras in total); it has 632 identities which is a huge number even today. Another dataset is CAVIAR4REID (Cheng, Cristani, Stoppa, Bazzani, & Murino 2011); it contemplates a real video surveillance scenario with issues like resolution changes, light conditions, occlusions, and pose variations.

The Person Re-ID (PRID2011) dataset (Hirzer, Beleznai, Roth, & Bischof, 2011) consists of 245 identities taken from two cameras. The PRID450S dataset was generated 5 years later based on the PRID2011 dataset. PRID450S increases the number of identities considered in PRID from 245 to 450. This dataset provides an automatically generated motion-based background segmentation, as well as a manual segmentation of several parts of the human body.

The ETHZ dataset (Schwartz & Davis, 2009) was generated through a moving camera. It contains three sequences of images acquired from a single camera. Not only ETHZ was designed originally for human detection it is also broadly used in person re-identification tasks.

The i-LIDS dataset (Zheng, Gong, & Xiang, 2009) was collected in an airport with two non-overlapping cameras. It contains images with scenarios with heavy occlusion and pose variation. The GRID dataset (Loy, Liu, & Gong, 2013) was acquired in a busy underground station through eight non-overlapping cameras, the quality of its images is poor, implying low image resolution and some kind of noise in the images (see Figure 8.2). The CUHK01 dataset (Li, Zhao, & Wang, 2013) contains images from two non-overlapping cameras with good clarity in images; in general, this dataset is considered a really good one.

There are other datasets although less used nowadays, for example, the ALERT Airport Re-Identification dataset (Karanam et al., 2016) which is one of the newest and complex benchmarks. This dataset includes images from six cameras in an indoor surveillance environment (airport), and it contains images of 9,651 persons, but not necessarily each person is captured by all six cameras. Other datasets are PKU-ReID (Ma, Liu, Hu, Wang, & Sun, 2016), DukeMTMC-reID (Zheng, Zheng, & Yang, 2017), SAIVT soft-biometric (Bialkowski, Denman, Sridharan, Fookes, & Lucey, 2012), and indoor multi-camera pedestrian dataset (Roth, Leistner, Berger, & Bischof, 2010).

Figure 8.2 shows several images from VIPER, PRID2011, ETHZ, GRID, and CAVIAR4REID. Here, it can be seen how the images look, for example, some images from GRID have noise and low clarity and look blurred.

A review of several benchmark datasets is presented in Vezzani and Cucchiara (2014); this study analyzes datasets older than 2014.

All the above datasets are used to test most current people re-identification approaches; nevertheless, there are some issues to be discussed. Aspects such as space and time are barely analyzed in the evaluation scenarios included in these datasets; the datasets are composed only of short video data, which is a limitation for re-identification in longer periods of time.

Although the mono-camera scenario is less important than multi-camera, there are still important issues to solve in this scenario, like crowded scenes, with high overlapping between people, etc.

There are several datasets with more than four cameras, but in most of them, people do not appear in more than four cameras that means although the dataset considers eight cameras if there are no images from these eight views of each person, it is hard to evaluate a real multi-camera environment.

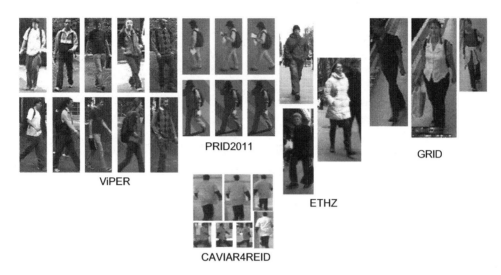

FIGURE 8.2
Sample images of several datasets.

When datasets containing long-time videos (weeks or months), the current approaches of person re-identification task must be changed. This consideration is important since many techniques use temporal-dependent features like appearance (mainly based on clothes), which is not a stable feature in long periods of surveillance.

Other benchmark datasets are focusing on depth information such as 3D images, etc. This work focuses on 2D image datasets; techniques dealing with depth images datasets are given in Baltieri, Vezzani, and Cucchiara (2011); Barbosa, Cristani, Bue, Bazzani, and Murino (2012).

8.3 Current Approaches

The contributions around person re-identification research can be divided into two main categories: those proposing new feature representations and those working around learning distance metrics that reduce the distance between samples of the same classes while increasing it among the samples of different classes. In this section, we briefly discuss both approaches.

8.3.1 Feature Representation

Feature representation or extraction is a fundamental problem in person re-identification task. This procedure can be designed to capture features from several different sources like the appearance based on clothes, the appearance based on movement, the body structure, or even the appearance based on biometrics traits. Despite these extensive efforts, finding an ideal feature representation or descriptor for person re-identification is still an unsolved problem (Fang, Hu, Hu, Liao, & Li, 2018). Here, we study appearance-based models, motion-based models, spatiotemporal features, and learning metrics techniques.

8.3.1.1 Appearance-Based Models

As the classical biometric traits (e.g., face) have been considered as impractical and not useful in real-world video surveillance scenarios, the use of clothing appearance has been broadly studied over the years to tackle people re-identification problem. The appearance-based methods have a huge problem because the appearance of people changes considerably due to many factors, such as pose and zoom variations, lighting conditions, etc.

Fang et al. (2018) presented a descriptor that combines three different types of features. The authors start with several low-level features of the image, that is:

$$f = nR, nG, nB, H, S, Y, \nabla_Y, \nabla_Y^2, \theta_Y^2$$

where (nR, nG, nB) correspond the normalized RGB; H, S, and Y are the channels used for the HSV and YCbCr color spaces; ∇_Y represents the first-order gradient's magnitude; and ∇_Y^2, θ_Y^2 represent the second-order gradient's magnitude and orientation. From these features, a Hash Feature Map is calculated through the average vector value of each feature and its difference between each feature. Then, three more values are generated: a histogram, a mean vector, and a co-occurrence matrix. The concatenation of all these features produces the final descriptor vector (called MSHF-multi-statistics on hash feature map). Moreover, they use the Gabor Filters to generate an additional vector to obtain the enriched MSHF.

A hierarchical region descriptor based on the Gaussian distribution of the features is proposed in Matsukawa, Okabe, Suzuki, and Sato (2016). The above is through a dense extraction of local patches, thus each region is described based on the information of this set of local patches. Several features are defined to represent pixels, magnitudes of the gradient along four orientations (0°, 90°, 180°, and 270°) using the RGB, Lab, HSV, and nRGB (normalized RGB) space color values. These features are summarized through a Gaussian distribution with mean and covariance parametrization to describe each patch in the image.

In Liao, Hu, Zhu, and Li (2015), it introduced a new feature representation called Local Maximal Occurrence (LOMO). Despite its relative novelty, LOMO has become a point of reference for recent work, see for instance Li, Shen, Wang, Guan, and Tang (2018); Dai, Zhang, Lu, & Wang (2018); Wang, Wang, Liang, Gao, and Sang (2018); Ren, Li, and Lu (2018); this suggests its potential. LOMO analyzes the horizontal occurrence of local features, and it takes care of considering the camera views changes. The Retinex algorithm is applied to handle the color changes from one camera view to another; then, the HSV color histogram is computed to extract features, and furthermore, the Scale Invariant Local Ternary Pattern as texture's descriptor. All these features are computed through sliding windows over the whole image.

Lisanti, Masi, Bagdanov, and Del Bimbo (2015) introduced a method to rank candidate targets. The contribution combines a new feature extraction with an iterative extension of Sparse Discriminative Classifiers. The descriptor, called WHOS (Weighted Histograms of Overlapping Stripes), is based on coarse and striped pooling of local features. An HOG descriptor is calculated from each stripe, with and without overlapping. Also, the HS (Hue and Saturation) and RGB histograms are calculated and concatenated to the HOG features to get the final vector. Finally, in the classification stage, an iterative sparse algorithm is proposed; it realizes an interactive process to rank features with various weighting schemes.

An extensive analysis of features about people-appearance features is provided in Doretto, Sebastian, Tu, and Rittscher (2011).

8.3.1.2 Motion-Based Models

Alternatively, it is possible to characterize a person using its movements; it is commonly used to avoid the limitations of appearance traits. Such is the case of the method proposed in Zhu et al. (2017); here, the authors introduce an image-to-video person re-identification system. The core idea is to combine a projection matrix along with a heterogeneous dictionary pair-learning (PHDL) approach. Despite the similarity between tasks, the task of finding a person in a video sequence is distinct from the image-to-image matching. In the video case, the walking cycle is constructed with both the visual appearance and movement features. Notice that a walking sequence only can be extracted from videos while appearance features are extracted from images. For appearance features, the authors employ the WHOS feature (Lisanti et al., 2015), and for video, the SFTV3D (Zhang, Ma, Liu, & Huang, 2017) feature descriptor is used.

8.3.1.3 Biometrics-Based Models

The use of biometric characteristics bypasses several problems like color-constancy, temporal traits like clothing, and other appearance features. Biometric traits are the most used features to identify a person, and there has been an analysis in the community to identify the usefulness of face, iris, palm print, fingerprints, etc., as distinctive and high potential features. However, it is hard to use this kind of traits in IVS because of the number of potential environmental problems like long-distance acquisition, large lighting changes, perspective and zoom changes, poor quality images, among others. Under these conditions, the person identification is still hard to identify even for a human being. The use of biometric traits like the face is a new and open hard problem in the community.

There are indeed few works related to person re-identification using biometric traits, such as Koide, Menegatti, Carraro, Munaro, and Miura (2017), where a face recognition system is used to perform people-tracking and re-identification. This system uses RGB-D images from a Kinect sensor. It is based on the combination of deep neural network and Bayesian inference-based face classification method. The authors use two image sequences of six different persons reaching an accuracy rate of 60%. Nevertheless, there are two issues with performance; first, the images used are acquired by themselves; and second, the distance and conditions of the scenes are more straightforward than a real video surveillance environment.

Despite the few works, we believe that the usage of biometrics for person re-identification is promising; in fact, we found important achievements on those contributions using biometric traits to solve the people re-identification task. For instance, Huang, Wang, and Kuo (2016) and Yang, Wei, Yeh, and Wang (2015) focus on face recognition at distant targets; this can be seen as a crucial starting point. Most contributions in this area are restricted to proprietary datasets limiting the broad comparison of these techniques.

8.3.1.4 Spatiotemporal Features

The usage of this kind of features is increasing; such is the case of the work presented in Ukita, Moriguchi, and Hagita (2016) where a new spatiotemporal trajectory feature (transit times between cameras) of each person regarding her/his group is used to perform people re-identification. The basic spatiotemporal features are related to several metrics between each people, for example, the distance between *person i* and *person j*, their absolute difference in speed and direction, the time-overlap radio which is the time steps in which *person i* is observed by the camera, etc. These features are calculated only in frames where the velocity is overpassing a given threshold; the purpose is to avoid issues produced by noise. Finally, these proposed group features are classified (using Support Vector Machines) as "group" or "non-group." Furthermore, more traditional features are employed like color histograms, transition time between cameras, that means when a person enters and exits in the camera's field of view. On the other hand, groups can be characterized using several traits like the appearance-color, the number of members, and the relationship between the trajectories. Several spatiotemporal features are proposed later, wherein a feature vector represents each frame; in these circumstances, the time overlap is redefined as the difference between overlap in time t and a predefined parameter. Although this idea is interesting, there are a lot of issues that must be carefully attended, for example, the group must be well detected as well as the trajectories of each people, and also, the definition of *a person belongs to the group* must be clear.

In Li et al. (2018), the spatiotemporal information is considered through the use of activity cues as features to reduce the search space to a correct re-identification over a large candidate options list. An activity is defined as different movements' paths realized by the person over the camera's network (in this case, three cameras). Also, the LOMO and XQDA descriptors are tested in combination with their scheme.

In Zhang et al. (2017), the whole walking person is considered to generate a spatiotemporal body-action model. For this purpose, blobs based on spatial and temporal segmentation were obtained from each video. Then, each blob is split into several small chunks, and from each fragment, a Fisher vector is calculated to generate a bag-of-words feature vector (BoW). That is, each image of the walking sequence of a person is transformed into the BoW scheme to get the final feature vector. The metric learning stage uses the KISSME method.

8.3.2 Distance Metrics or Learning Metrics

An alternative to feature representation research is to deal with learning metrics, or improving a classifying technique. This section is devoted to studying such approaches.

An effective solution based on feature transformation is presented in Dai et al. (2018); here, the authors introduce the cross-view semantic projection learning algorithm. The core idea is to project all features acquired by non-overlapping into a new feature space where they must share a semantic structure. At the same time, they learn the association function between non-overlapping views. This semantic projection uses hyper-parameter optimization to improve the performance of the transformation function. In the people matching stage, the similarity between two images is measured with the Euclidean distance. LOMO and GOG (Hierarchical Gaussian Descriptor) descriptors are used as feature vectors. In this work, the contribution focuses on finding the equivalence between two feature vectors acquired by different non-overlapping views.

To improve the similarity metric between the extracted features, Wu, Wang, Gao, and Li (2018) propose to learn local-adaptive similarity metrics to facilitate deep embedding learning. Authors use a deep neural network to transform the input data into a deep feature space, where the main objective is seeing the data distribution and structure within classes. The final is used to improve the measure of similarity between images.

Another approach dealing with minimizing the inter-class distance of the samples to solve person re-identification problem is presented in Wang, Zhou, Wang, and Hou (2018). This approach works with a deep convolutional neural network to generate the feature representation and later proposes an adaptive loss function which shows that intra-class distance is smaller than the inter-class distance. The deep neural network comprises three parts:

a global feature learning, which consists of some convolutional and max-pooling layers; a local feature learning, which consists of four convolutional and max-pooling layers with a set of specific filters; and a feature learning fusion with four teams of fully connected layers.

A modified Mahalanobis distance, called EquiDistance constrained Metric Learning (EquiDML) for person re-identification is proposed in Wang et al. (2018a). This EquiDML applies least-square regression to map features from the same person to the same vertex of a regular simplex while maps feature vectors from different persons to different vertices. The above with the objective of learning an optimal distance metric assigns small values to images from the same person and large values, otherwise. The LOMO and GoG are used for feature representation. Both of them have high-dimensional representations, e.g., 26,960 features for LOMO and 27,622 for GoG. For two of the five used datasets, the ID-discriminative Embedding features are also tested that produce a lower dimension descriptor.

In addition to proposing the LOMO feature (see Section 8.3.1.1), Liao et al. (2015) introduce a metric learning method called Cross-view Quadratic Discriminant Analysis (XQDA). XQDA is an extension of the Bayesian face technique of Kostinger, Hirzer, Wohlhart, Roth, and Bischof (2012) and KISSME (Moghaddam, Jebara, & Pentland, 2000) approaches, but XQDA learns sub-spaces through generalized eigenvalue decomposition.

With the purpose to avoid the overfitting problem in metric learning, Ren et al. (2018) propose to learn multi-metric subspaces. The subspace learning aims to learn a projection matrix W by minimizing the intra-class distance and maximizing the inter-class distance. A multi-metric projection with a feedback of top ranks matching pairs is used with the purpose to separate the negative pairs. The authors report that this method helps them to avoid the overfitting caused by small sample size (SSS) and high feature dimension. The authors use several independent subspaces to learn their metrics. Finally, the LOMO descriptor is used to construct the feature vector.

In Zhao, Zhao, and Su (2018), the Kernelized Random Keep It Simple and Straightforward metric is proposed. This metric learning is based on the well-known KISS metric learning, but the authors consider that not all the pair differences follow a Gaussian distribution; thus, instead of using Gaussian distributions, first the features are transformed into a set of kernelized features to fit the Gaussian distribution better. Furthermore, to solve the SSS problem which can cause the lack of inverse covariance matrices, they apply a random subspace ensemble method to obtain an estimation of the covariance matrix. As features, the color, texture, and shape traits LOMO (Liao et al., 2015) and kCCA (Lisanti, Masi, & Bimbo, 2014) are extracted from the image.

The approach proposed in Zhao, Ouyang, and Wang (2014) was designed to learn mid-level filters to discover clusters of patches that can capture a visual pattern related to a particular body trait. For this, a dense grid is used over the image (in LAB color space) to capture a 32-bin color histogram and 128-dimensional SIFT features. These features are compared with the nearest neighbor (NN) search to establish an NN set. Then, partial AUC quantization is proposed to see the discriminative and generalization power of patches, this through the accumulation of distances between each patch and its NN set. Finally, a hierarchical patch clustering is applied to identify both subtle and coarse aspects of the image.

A two-stage transfer learning approach for person re-identification is proposed in Zhang, Kato, Wang, and Mase (2015). In the first stage, an adaptive metric learning method is used to transfer the knowledge from the dataset to assist the learning of target (i.e., the person who must be re-identified). In the second stage, they transfer the learned distance to each probe-specific person using side-information; here, the side-information consists in the image data from another person who is, or was, in the same camera. The SIFT, HOG, and LBP descriptors were used to create the feature vector.

In Mirmahboub, Mekhalfi, and Murino (2017), a method is proposed that penalizes the distance scores obtained with several descriptors in function to their capability to do a correct matching. The authors use the LOMO, WHOS, and GOG descriptors. First, the learning model is fed with the descriptors extracted, which learns a weighting scheme; then, each vector is replaced by the resulting vector of computing the distances between the test and the gallery samples. The final feature vector consists of applying the learned weighting scheme to the distance vector.

8.4 Systems Evaluation and Discussion

A crucial aspect of the evaluation is to determine which performance metric should be used. In fact, a real system may perform according to which error or success is used. For example, maximizing the true positive rate or avoid any false alarm even if the true positive detection is low. These aspects must be taken seriously into account in any IVS system. In this research area, the most used metrics are precision, recall, False Acceptance rate, False Reject rate, Receiver Operating Characteristic curve, and the Cumulative Match Characteristic (CMC) curve (Vezzani & Cucchiara, 2014).

The recent literature in the area uses the CMC curve as the evaluation metric. The CMC curve measures the success performance, and it also judges the ranking capabilities of the identification system, where the result is obtained with the first best-scores (Bolle, Connell, Pankanti, Ratha, & Senior, 2005). This condition is determined by the value of k parameter. That is, when a system returns the correct identity inside the k-first best scores, it is considered a success.

This section analyzes in depth the most relevant work presented above. This chapter is organized by the analysis of each benchmark dataset; we compare benchmarks used by three or more references in our state-of-the-art review.

8.4.1 VIPeR Dataset Results

The VIPeR benchmark is one of the most used in person re-identification task, in spite of it was proposed several years ago, in 2007; it is still very used in the community.

Despite the fact that the initially reported results in Gray et al. (2007) have been greatly improved, it is recognized as a challenging dataset among the community. In Table 8.2, the results of the reviewed works over the VIPeR dataset are presented. Here, it can be observed that the best result obtained with k = 1 is achieved recently (Mirmahboub et al., 2017), and surprisingly when k = 5, a great result over 97% of recognition rate was reached; this surpasses by far the recognition rate of the alternative works.

To see the general performance of the methods over VIPeR dataset, Figure 8.3 shows the average position (with several values of k) and the standard deviation of each compared method. Here, it is easy to see that the best method is provided in Mirmahboub et al. (2017). Unexpected behavior is that the best method with k = 1 is not necessarily the best with k = 5, or even with a higher value of k; for instance, the work presented in Ren et al. (2018) achieves a competitive result for k = 1, but as k increases its performance is not even close to the best ones. Figure 8.4 illustrates this behavior as a large variance.

In general, VIPeR dataset is a complex dataset, and there are still many challenges to solve with it because a 68.89% of recognition rate is still a low result.

TABLE 8.2

Comparison of Methods in VIPeR Dataset with the CMC Curve

	CMC			
	k = 1	**k = 5**	**k = 10**	**k = 20**
Dai et al. (2018)	47.31	75.92	86.39	94.21
Fang et al. (2018)	53.5	82.6	91.5	96.6
Liao et al. (2015)	49.04	77.13	86.26	96.2
Lisanti et al. (2016)	51.46	79.56	89.08	95.85
Matsukawa et al. (2016)	49.7	79.7	88.7	94.5
Mirmahboub et al. (2017)	40	-	80.51	91.08
Ren et al. (2018)	59.21	76.11	84.21	93.2
Wang et al. (2018a)	54.2	83.5	91.3	96.4
Wu et al. (2018)	43.39	72	85	93
Zhao et al. (2014)	27	-	-	-
Zhao et al. (2018)	68.89	97.63	99.91	100

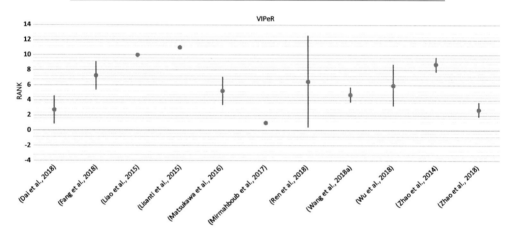

FIGURE 8.3
Average position and standard deviation of each method over VIPeR.

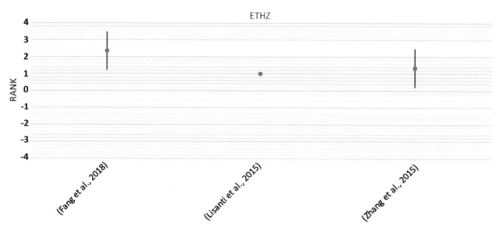

FIGURE 8.4
Average position and standard deviation of each method over ETHZ.

8.4.2 ETHZ Dataset Results

ETHZ is a dataset not too used as VIPeR. Nevertheless, there are interesting works that employ it to test their approaches. Table 8.3 shows the results reported by at least three of our reviewed works. Here, several excellent results with k = 1 can be obtained, all of them higher than 80% of recognition rate. The above is a very different result in comparison with the ViPeR dataset.

Figure 8.4 compares the methods using this dataset. Here, Lisanti et al. (2015) achieve the best result with k = 1; however, they only report experiments for that k. Under these terms, Figure 8.4 should be analyzed differently for k = 1 and k > 1. In k = 1, the best method is provided in Lisanti et al. (2015), yet for k > 1, Zhang et al. (2015) perform the better.

8.4.3 iLIDS Dataset Results

The iLIDS dataset is another widely used benchmark; nowadays, five of all the reviewed works use this dataset. It contains a lot of images, more than 42,000 images with 300 different identities or persons. Table 8.4 lists the results obtained with this benchmark. Here, the best result with k = 1 is still low (55.11% of recognition rate). Nevertheless, a significant improvement has been reached in Mirmahboub et al. (2017) with a value of k = 5.

To observe the whole performance in this dataset, Figure 8.5 shows the average position and the standard deviation of each method using iLIDS. Here, Mirmahboub et al. (2017) perform the best since it reaches the highest recognition rates over all k values. All methods exhibit a stable performance on the iLIDS dataset; this could mean that if a method performs well with k = 1, then it should perform well for k > 1.

8.4.4 CAVIAR4REID Dataset Results

CAVIAR4REID is another complex dataset, which provides images of 72 identities acquired from the previous and well-known CAVIAR dataset. In Table 8.5, the results obtained with

TABLE 8.3

Results in ETHZ Dataset with the CMC Curve

	CMC		
	k = 1	k = 5	k = 10
Fang et al. (2018)	85	96	98
Lisanti et al. (2015)	89	99	99
Zhang et al. (2015)	99	-	-

TABLE 8.4

Comparison of Results in iLIDS Dataset under the CMC Curve

	CMC			
	k = 1	k = 5	k = 10	k = 20
Fang et al. (2018)	34.46	53.73	61.92	72.57
Lisanti et al. (2015)	28.15	50.37	65.88	80.35
Mirmahboub et al. (2017)	39.5	-	-	-
Zhang et al. (2017)	49.7	78.3	84.7	91.7
Zhu et al. (2017)	55.11	81.99	90	96.45

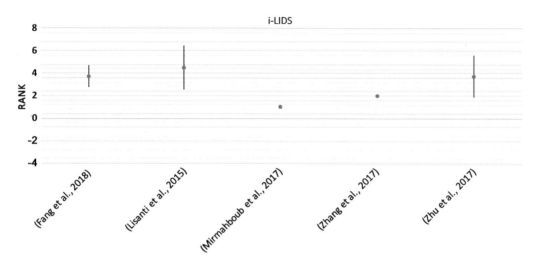

FIGURE 8.5
Average position and standard deviation of each method over iLIDS.

TABLE 8.5

Results in CAVIAR4REID Dataset with the CMC Curve

	CMC			
	k = 1	**k = 5**	**k = 10**	**k = 20**
Fang et al. (2018)	31.44	47.01	56.99	69.35
Lisanti et al. (2015)	45	65	77	90
Mirmahboub et al. (2017)	29	-	-	-
Zhang et al. (2015)	45.2	85.2	94.8	100

CAVIAR4REID dataset are showed. Here, Mirmahboub et al. (2017) achieve the best results ranging from a 45% of recognition rate with k = 1, to 100% of recognition rate with k = 20. Although the 100% of recognition rate is reached in this dataset, the most important value is the result obtained with k = 1, or even k = 5; in general, the performance on k = 1 is still low.

Figure 8.6 shows the overall performance of methods evaluating the CAVIAR4REID dataset. The rank of all methods is preserved for all k; therefore, the standard deviation is zero.

8.4.5 GRID Dataset Results

The GRID dataset contains a lot of identities, 1,025 different persons, acquired in a circuit of eight cameras; these characteristics convert GRID into a challenging benchmark dataset. For this reason, the identification rates achieved are low compared with other datasets analyzed in this chapter. Table 8.6 shows the results obtained with five methods using this dataset. In general, a low recognition rate between 60% and 70% is reached. The previous result corresponds to k = 20 while k = 1 produces a recognition rate of 25.84% with Dai et al. (2018).

In Figure 8.7, the performance of all the methods using this dataset with the different values of k can be observed. The best performance is reached by Dai et al. (2018). The least performing in this dataset is that presented in Liao et al. (2015). This dataset is one of the most challenging public benchmark datasets in the literature.

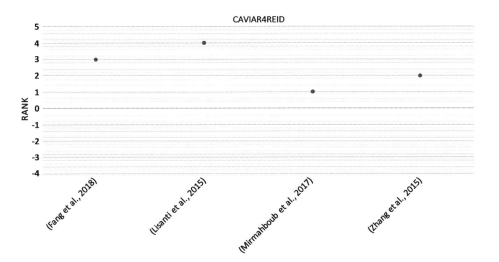

FIGURE 8.6
Average position and standard deviation of each method over CAVIAR4REID.

TABLE 8.6

Results in GRID Dataset with the CMC Curve

	CMC			
	k = 1	**k = 5**	**k = 10**	**k = 20**
Dai et al. (2018)	23.2	44.48	53.04	64.16
Fang et al. (2018)	25.84	47	59	70
Liao et al. (2015)	24.7	47	58.4	69
Matsukawa et al. (2016)	16.56	-	41.84	52.4
Ren et al. (2018)	19.68	-	49.04	59.52

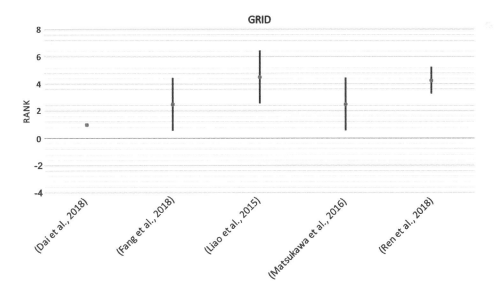

FIGURE 8.7
Average position and standard deviation of each method over GRID.

8.4.6 CUHK01 Dataset Results

CUHK01 is one of the most utilized benchmark datasets; in this analysis, it is the second more cited, below the VIPeR dataset. The high quality of its images contrasts with other datasets.

Table 8.7 shows a performance comparison of several k values. Here, Wang et al. (2018a) achieve the best performance for k = 1. For k > 1, Wu et al. (2018) and Wang et al. (2018b) achieve better performances.

Figure 8.8 shows the overall performance of the methods using this dataset. Under the average rank metric, the best methods are Wang et al. (2018a,b) and Wu et al. (2018), i.e., best positions with low standard deviation.

In general, the results achieved in CUHK01 are good even with k = 1.

TABLE 8.7

Results in CUHK01 Dataset with the CMC Curve

	CMC			
	k = 1	k = 5	k = 10	k = 20
Dai et al. (2018)	66.75	85.8	91.75	95.16
Fang et al. (2018)	72.01	88.56	92.84	96.38
Liao et al. (2015)	73.02	91.57	96.73	98.58
Matsukawa et al. (2016)	71.9	91.8	95.8	97.2
Ren et al. (2018)	74.22	90.51	94.42	96.91
Wang et al. (2018a)	57.8	79.1	86.2	92.1
Wang et al. (2018b)	63.21	83	90	94
Wu et al. (2018)	69.65	88.33	92.98	96.03
Zhao et al. (2014)	66.9	83	88.7	92.7
Zhao et al. (2018)	34.3	55	72	75

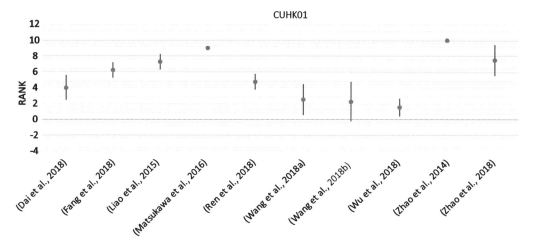

FIGURE 8.8
Average position and standard deviation of each method over CUHK01.

8.4.7 Market-1501 Dataset Results

The Market-1501 dataset contains more than 1,500 entities acquired over a circuit of six cameras, and it includes several images per person, in different views and scales. The majority of the contributions in the literature use this benchmark for k = 1, only Wu et al. (2018) report for k > 1. Table 8.8 shows the results of this dataset. In general, the results reported are high, most of them achieve recognition rates bigger than 80% for k = 1.

The lack of values for k > 1 limits the comparison between methods; then, the overall performance figure is omitted.

8.4.8 PRID2011 Dataset Results

The PRID2011 dataset contains images from two cameras, with a total of 245 identifies in both. Table 8.9 shows the results of the methods using the PRID2011 dataset where it can be seen that best results are reached by the work presented in Wang et al. (2018b), which shows a superior recognition rate in comparison with other works.

In Figure 8.9, it can be better appreciating the superiority of the performance of Wang et al. (2018b) which always reaches the best. The second-best method is reported in Zhang et al. (2017), which maintained good results but still far away from the results of the best method using this dataset. Based on the recognition rates achieved even with recent approaches, the PRID2011 dataset can be considered as a difficult dataset.

8.4.9 PRID450S Dataset Results

The PRID450S dataset is based on the PRID2011 dataset; it contains 450 image pairs (identities) acquired from two cameras and seems similar to VIPeR dataset but more

TABLE 8.8

Results in the Market-1501 Dataset Using the CMC Curve

	CMC			
	k = 1	k = 5	k = 10	k = 20
Fang et al. (2018)	51.6	-	-	-
Wu et al. (2018)	84.14	93.25	97.33	98.07
Wang et al. (2018b)	89.5	-	-	-
Wang et al. (2018a)	80.02	-	-	-

TABLE 8.9

Results in PRID2011 Dataset with the CMC Curve

	CMC			
	k = 1	k = 10	k = 20	k = 50
Ukita et al. (2016)	13.7	71.1	-	-
Wang et al. (2018b)	73.3	97.5	98.3	100
Zhu et al. (2017)	41.92	85.47	92.44	-
Zhang et al. (2015)	25	62	75	92
Zhang et al. (2017)	66.2	87.3	88.4	89.4

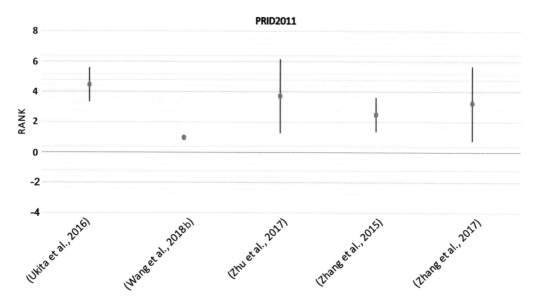

FIGURE 8.9
Performance comparison over PRID2011; the points show the average rank and its standard deviation.

current and with a better quality of images, but VIPeR has more image pairs of identities. In spite of that, it was launched in 2014; it is not popular among recent contributions, compared with VIPeR for example. In Table 8.10, the results obtained with PRID450S are showed.

Figure 8.10 shows a general view of the performance over the PRID450S dataset. Here, the method that achieves the best result is not the same that produces the best result in higher values. More detailed, Zhao et al. (2018) achieve the best result for k = 1 and k = 20. For k = 5 and k = 10, the best performing method is Dai et al. (2018).

8.5 Discussion and Conclusions

In this chapter, several works related to person re-identification problem were reviewed and analyzed. Moreover, each work is compared with others under similar conditions.

The diversity of benchmarks and the difficulty of them are stated in the above comparison. Figure 8.11 summarizes the best results as achieved in each dataset; we observe that

TABLE 8.10

Results in PRID450S Dataset with the CMC Curve

	CMC			
	k = 1	k = 5	k = 10	k = 20
Dai et al. (2018)	69.2	90.36	95.51	96.6
Matsukawa et al. (2016)	68.4	88.8	94.5	97.8
Zhao et al. (2018)	70.4	89	94.1	97.8

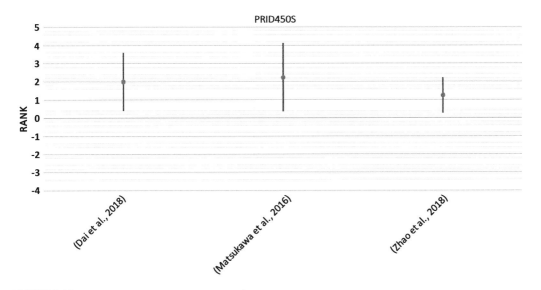

FIGURE 8.10
Average position and standard deviation of each method over PRID450S.

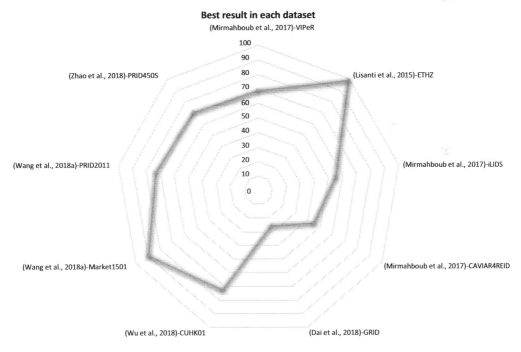

FIGURE 8.11
Best result in each dataset reviewed.

most datasets reach a recognition rate higher than 60%. In the case of ETHZ, a close to perfect recognition is available, and Market1501 with 90%; we can say that the problem is feasible to be solved, under some conditions. In contrast, the GRID dataset lacks a higher performant solution. Nowadays, the GRID, CAVIAR4REID, and i-LIDS datasets still pose a difficult challenge in the area.

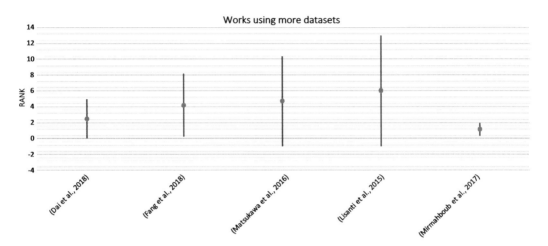

FIGURE 8.12
Works using more than four benchmark datasets.

On the other hand, Figure 8.12 ranks contributions concerning the reached performance on the evaluated benchmarks; only those works being applied to at least four datasets were considered. In this rank, Mirmahboub et al. (2017) perform well in all benchmarks, and it is the best ranked, in general. The work presented in Lisanti et al. (2015) achieves the least-performing method under our evaluation methodology. It is important to mention that the best approach is a recent work, published in late 2017; this behavior tells the activity and advances in the field are still growing and although the problem is far to be solved, the methods are getting better over the years.

From a general point of view, the best method is presented by Mirmahboub et al. (2017). We recommend practitioners to start evaluating the available techniques from here, and then adapt their systems with other techniques to get the desired performance and characteristics. Also, new researchers in the area cannot ignore to include this work in their comparisons.

From the dataset perspective, we consider that a lot of work is still necessary to tackle the challenging VIPeR, GRID, CUHK01, and PRID2011 benchmarks. Along with the inner difficulty of them, these datasets are popular enough in the area to provide a fair comparison playground for new techniques and those already recognized in the literature.

The soft-biometrics features (mainly appearance-based) are quite popular to tackle this problem. But this kind of features cannot deal with long-time videos, namely, days, weeks, or months, so this kind of feature turns out to be useless on such conditions.

Most of the current reviewed works follow the same evaluation scheme. This procedure consists of the usage of CMC curve with k = 1, 5, 10, and 20. Furthermore, in most of the cases, the way of splitting or using the images is the same.

At this moment, just a few works use biometrics like face recognition to solve the person re-identification task. This approach is quite promising, yet the face recognition at distance task is an open problem that needs to be solved to obtain reliable features. Finally, although important results are achieved with current approaches, the task of person re-identification could still be considered as an open problem. This claim is supported by the low performances produced for k = 1 (under the CMC curve scheme).

References

Baltieri, D., Vezzani, R., & Cucchiara, R. (2011). 3DPeS: 3D people dataset for surveillance and forensics. In *Proceedings of the 2011 Joint ACM Workshop on Human Gesture and Behavior Understanding* (pp. 59–64). Scottsdale, AZ. doi:10.1145/2072572.2072590.

Barbosa, I. B., Cristani, M., Bue, A. Del, Bazzani, L., & Murino, V. (2012). Re-identification with RGB-D sensors. In *Computer Vision ECCV 2012* (pp. 433–442). Springer. Berlin, Heidelberg.

Bedagkar-Gala, A., & Shah, S. K. (2014). A survey of approaches and trends in person re-identification. *Image and Vision Computing, 32*(4), 270–286. doi:10.1016/j.imavis.2014.02.001.

Bialkowski, A., Denman, S., Sridharan, S., Fookes, C., & Lucey, P. (2012). A database for person re-identification in multi-camera surveillance networks. In *2012 International Conference on Digital Image Computing Techniques and Applications, DICTA 2012*. Fremantle, WA. doi:10.1109/DICTA.2012.6411689.

Bolle, R. M., Connell, J. H., Pankanti, S., Ratha, N. K., & Senior, A. W. (2005). The relation between the ROC curve and the CMC. In *Fourth IEEE Workshop on Automatic Identification Advanced Technologies (AutoID'05)* (pp. 15–20). IEEE. Buffalo, NY. doi:10.1109/AUTOID.2005.48.

Dai, J., Zhang, Y., Lu, H., & Wang, H. (2018). Cross-view semantic projection learning for person re-identification. *Pattern Recognition, 75*, 1339–1351. doi:10.1016/j.patcog.2017.04.022.

Cheng, D. S., Cristani, M., Stoppa, M., Bazzani, L., & Murino, V. (2011). Custom pictorial structures for re-identification. In *British Machine Vision Conference (BMVC)* (pp. 68.1–68.11). London. doi:10.5244/C.25.68.

Doretto, G., Sebastian, T., Tu, P., & Rittscher, J. (2011). Appearance-based person reidentification in camera networks: Problem overview and current approaches. *Journal of Ambient Intelligence and Humanized Computing, 2*(2), 127–151. doi:10.1007/s12652-010-0034-y.

Fang, W., Hu, H. M., Hu, Z., Liao, S., & Li, B. (2018). Perceptual hash-based feature description for person re-identification. *Neurocomputing, 272*, 520–531. doi:10.1016/j.neucom.2017.07.019.

Frontex. (2011). Application of surveillance tools to border surveillance—Concept of operations. Retrieved August 20, 2001, from https://ec.europa.eu/research/participants/portal/doc/call/fp7/fp7-space-2012-1/31341-2011_concept_of_operations_for_the_common_application_of_surveillance_tools_in_the_context_of_eurosur_en.pdf.

Gray, D., Brennan, S., & Tao, H. (2007). Evaluating appearance models for recognition, reacquisition, and tracking. In *10th International Workshop on Performance Evaluation for Tracking and Surveillance (PETS), 3*, 41–47. Boston, US..

Hirzer, M., Beleznai, C., Roth, P. M., & Bischof, H. (2011). Person re-identification by descriptive and discriminative classification. In Heyden, A., & Kahl, F., (Eds.), *Image Analysis*, Lecture Notes in Computer Science (Including Subseries Lecture Notes in Artificial Intelligence and Lecture Notes in Bioinformatics), *6688 LNCS*, 91–102. doi:10.1007/978-3-642-21227-7_9.

Huang, C. T., Wang, Z., & Kuo, C. C. J. (2016). Visible-light and near-infrared face recognition at a distance. *Journal of Visual Communication and Image Representation, 41*, 140–153. doi:10.1016/j.jvcir.2016.09.012.

Karanam, S., Gou, M., Wu, Z., Rates-Borras, A., Camps, O., & Radke, R. J. (2016). A systematic evaluation and benchmark for person re-identification: Features, metrics, and datasets, *2115*, 1–14. Retrieved from http://arxiv.org/abs/1605.09653.

Koide, K., Menegatti, E., Carraro, M., Munaro, M., & Miura, J. (2017). People tracking and re-identification by face recognition for RGB-D camera networks. In *2017 European Conference on Mobile Robots (ECMR)* (pp. 1–7). IEEE. Paris. doi:10.1109/ECMR.2017.8098689.

Kostinger, M., Hirzer, M., Wohlhart, P., Roth, P. M., & Bischof, H. (2012). Large scale metric learning from equivalence constraints. In *2012 IEEE Conference on Computer Vision and Pattern Recognition* (pp. 2288–2295). IEEE. Providence, RI. doi:10.1109/CVPR.2012.6247939.

Li, B., Yang, C., Zhang, Q., & Xu, G. (2014). Condensation-based multi-person detection and tracking with HOG and LBP. In *2014 IEEE International Conference on Information and Automation, ICIA 2014, July* (pp. 267–272). Hailar. doi:10.1109/ICInfA.2014.6932665.

Li, M., Shen, F., Wang, J., Guan, C., & Tang, J. (2018). Person re-identification with activity prediction based on hierarchical spatial-temporal model. *Neurocomputing, 275*, 1200–1207. doi:10.1016/j.neucom.2017.09.064.

Li, W., Zhao, R., & Wang, X. (2013). Human reidentification with transferred metric learning. In Lee, K. M., Matsushita, Y., Rehg, J. M., & Hu, Z. (Eds.), *Computer Vision — ACCV 2012*, Lecture Notes in Computer Science (Including Subseries Lecture Notes in Artificial Intelligence and Lecture Notes in Bioinformatics), *7724 LNCS* (Part 1), 31–44. doi:10.1007/978-3-642-37331-2_3.

Liao, S., Hu, Y., Zhu, X., & Li, S. Z. (2015). Person re-identification by local maximal occurrence representation and metric learning. In *Proceedings of the IEEE Computer Society Conference on Computer Vision and Pattern Recognition, 7–12 June* (pp. 2197–2206). Boston, MA. doi:10.1109/CVPR.2015.7298832.

Lisanti, G., Masi, I., Bagdanov, A. D., & Del Bimbo, A. (2015). Person re-identification by iterative re-weighted sparse ranking. *IEEE Transactions on Pattern Analysis and Machine Intelligence, 37*(8), 1629–1642. doi:10.1109/TPAMI.2014.2369055.

Lisanti, G., Masi, I., & Del Bimbo, A. (2014). Matching people across camera views using kernel canonical correlation analysis. In *ICDSC*, (MICC), 1–6. Venezia Mestre, Italy. doi:10.1145/2659021.2659036.

Loy, C. C., Liu, C., & Gong, S. (2013). Person re-identification by manifold ranking. In *2013 IEEE International Conference on Image Processing, ICIP 2013—Proceedings* (pp. 3567–3571). Melbourne, VIC. doi:10.1109/ICIP.2013.6738736.

Luna, C. A., Losada-Gutierrez, C., Fuentes-Jimenez, D., Fernandez-Rincon, A., Mazo, M., & Macias-Guarasa, J. (2017). Robust people detection using depth information from an overhead Time-of-Flight camera. *Expert Systems with Applications, 71*, 240–256. doi:10.1016/j.eswa.2016.11.019.

Ma, L., Liu, H., Hu, L., Wang, C., & Sun, Q. (2016). Orientation driven bag of appearances for person re-identification, 13. Retrieved from http://arxiv.org/abs/1605.02464.

Matsukawa, T., Okabe, T., Suzuki, E., & Sato, Y. (2016). Hierarchical Gaussian descriptor for person re-identification. In *2016 IEEE Conference on Computer Vision and Pattern Recognition (CVPR)* (pp. 1363–1372). Las Vegas, NV. doi:10.1109/CVPR.2016.152.

Mirmahboub, B., Mekhalfi, M. L., & Murino, V. (2017). Person re-identification by order-induced metric fusion. *Neurocomputing, 275*, 667–676. doi:10.1016/j.neucom.2017.09.019.

Moghaddam, B., Jebara, T., & Pentland, A. (2000). Bayesian face recognition. *Pattern Recognition, 33*(11), 1771–1782. doi:10.1016/S0031-3203(99)00179-X.

Ren, Y., Li, X., & Lu, X. (2018). Feedback mechanism based iterative metric learning for person re-identification. *Pattern Recognition, 75*, 1339–1351. doi:10.1016/j.patcog.2017.04.012.

Roth, P. M., Leistner, C., Berger, A., & Bischof, H. (2010). Multiple instance learning from multiple cameras. In *2010 IEEE Computer Society Conference on Computer Vision and Pattern Recognition—Workshops* (pp. 17–24). San Francisco, CA. doi:10.1109/CVPRW.2010.5543802.

Saghafi, M. A., Hussain, A., Zaman, H. B., & Saad, M. H. M. (2015). Review of person re-identification techniques (December), 1–20. doi:10.1049/iet-cvi.2013.0180.

Schwartz, W. R., & Davis, L. S. (2009). Learning discriminative appearance-based models using partial least squares. In *Proceedings of SIBGRAPI 2009 - 22nd Brazilian Symposium on Computer Graphics and Image Processing* (pp. 322–329). Rio de Janeiro. doi:10.1109/SIBGRAPI.2009.42.

Ukita, N., Moriguchi, Y., & Hagita, N. (2016). People re-identification across non-overlapping cameras using group features. *Computer Vision and Image Understanding, 144*, 228–236. doi:10.1016/j.cviu.2015.06.011.

Vezzani, R., & Cucchiara, R. (2014). *Person Re-Identification*. Springer, London. doi:10.1007/978-1-4471-6296-4.

Wang, J., Wang, Z., Liang, C., Gao, C., & Sang, N. (2018a). Equidistance constrained metric learning for person re-identification. *Pattern Recognition, 74*, 38–51. doi:10.1016/j.patcog.2017.09.014.

Wang, J., Zhou, S., Wang, J., & Hou, Q. (2018b). Deep ranking model by large adaptive margin learning for person re-identification. *Pattern Recognition, 74*, 241–252. doi:10.1016/j.patcog.2017.09.024.

Wang, X. (2013). Intelligent multi-camera video surveillance: A review. *Pattern Recognition Letters, 34*(1), 3–19. doi:10.1016/j.patrec.2012.07.005.

Wu, L., Wang, Y., Gao, J., & Li, X. (2018). Deep adaptive feature embedding with local sample distributions for person re-identification. *Pattern Recognition*, *73*, 275–288. doi:10.1016/j.patcog.2017.08.029.

Yang, M. C., Wei, C. P., Yeh, Y. R., & Wang, Y. C. F. (2015). Recognition at a long distance: Very low resolution face recognition and hallucination. In *Proceedings of 2015 International Conference on Biometrics, ICB 2015* (pp. 237–242). Phuket. doi:10.1109/ICB.2015.7139090.

Zhang, G., Kato, J., Wang, Y., & Mase, K. (2015). People re-identification using two-stage transfer metric learning. In *Proceedings of the 14th IAPR International Conference on Machine Vision Applications, MVA 2015* (pp. 588–591). Tokyo. doi:10.1109/MVA.2015.7153260.

Zhang, W., Ma, B., Liu, K., & Huang, R. (2017). Video-based pedestrian re-identification by adaptive spatio-temporal appearance model. *IEEE Transactions on Image Processing*, *26*(4), 2042–2054. doi:10.1109/TIP.2017.2672440.

Zhao, R., Ouyang, W., & Wang, X. (2014). Learning mid-level filters for person re-identification. In *Proceedings of the IEEE Computer Society Conference on Computer Vision and Pattern Recognition* (pp. 144–151). Columbus, OH. doi:10.1109/CVPR.2014.26.

Zhao, Z., Zhao, B., & Su, F. (2018). Person re-identification via integrating patch-based metric learning and local salience learning. *Pattern Recognition*, *75*, 1339–1351. doi:10.1016/j.patcog.2017.03.023.

Zheng, W.-S., Gong, S., & Xiang, T. (2009). Associating groups of people. In *British Machine Vision Conference*, *5*(1), 6. London. doi:10.5244/C.23.23.

Zheng, Z., Zheng, L., & Yang, Y. (2017). Unlabeled samples generated by GAN improve the person re-identification baseline in vitro. doi:10.1109/ICCV.2017.405.

Zhu, X., Jing, X. Y., You, X., Zuo, W., Shan, S., & Zheng, W. S. (2017). Image to video person re-identification by learning heterogeneous dictionary pair with feature projection matrix. *IEEE Transactions on Information Forensics and Security*, *13*(3), 717–732. doi:10.1109/TIFS.2017.2765524.

9

Optimization of SVM-Based Hand Gesture Recognition System Using Particle Swarm Optimization and Plant Growth Simulation Algorithm

K. Martin Sagayam, Sankirthana Suresh, and D. Jude Hemanth
Karunya Institute of Technology and Science

Lawrence Henessey
Blekinge Institute of Technology

Chiung Ching Ho
Multimedia University

CONTENTS

9.1 Introduction

Gestures are the most important actions in communication. Among the various body actions in humans, hand gesture is one of the most predominant naturally. It is used in recent fields such as computer vision, image processing, and pattern recognition for real-time applications. The context explores easy ways to interact through human–computer interaction [1,2].

9.1.1 Background

Gesture recognition is widely used in the development of virtual reality applications. These are physical actions that are used for nonverbal communication, e.g., sign language. Moreover, it involves much of human–computer interaction which primarily brings in the replacement of mouse in computer interface by these gestures. This technique also allows more natural interaction with gadgets, such as tablets, smartphones, etc., and also with newer devices like Google glasses.

In today's world, gesture recognition has vast usage in various fields. Considering the world of animation and gaming, devices like Microsoft Kinect has already brought in the existence of hand gesture interfaces to the market and also the navigation of the three-dimensional (3-D) virtual environment has become more natural and easy way of interaction through gestures in the 3-D space. These gestures are also considered as great usage in robotic applications. For specially abled people with zero percentage hearing and/or speech ability, gesture recognition was a great source of help. They could interact with the computer and other devices which in fact could also be applied in the field of health-care, thereby allowing for much more natural control of the diagnostic data and surgical devices. Nowadays, gesture recognition used as vehicle interfacing strategies has been noticed [3].

9.1.2 Significance

The technique of hand gesture recognition is of great significance in the world of intelligent human–computer technology, which is a little complicated to deal with in an environment which is quite complex. It is an important part of communication in real-time appliances [4]. This is a provocative problem which has arisen due to which many different approaches have been introduced. Wearable devices have been used earlier for various applications (like gloves), instead, the vision-based approaches are now being able to capture the hand gestures which are purely based on natural interaction with the computers and without any requirement of any sort of physical device that has to be worn. It has been invoked with virtual reality systems to control machinery activities, and also it seems that it may range from simple manipulative gestures which could be used to point at and also move objects around; there were numerous such approaches introduced to be applied to interpret hand gestures.

9.2 Literature Survey

The improvisation of the human–computer interaction had become popular by the research domain which mainly introduced the hand gesture recognition system. It basically involved a set of finger movements that were extracted from the fingers and hands. This required adequate space and also a proper sustained range of resolution that was further supported by related algorithms and hardware.

The sound wave technology brought in by the Microsoft presented the utilization of the loudspeakers and microphones that were distinctively embedded in the computer system to recognize the hand gesture. Later this method was enforced to be limited by the spatial resolution of a sound wave and it could thereby only be used for meek hand gesture detection [5,6]. Another such work was authorized by Google. They designed an Frequency-Modulated Continuous-Wave (FMCW) radar system which could further recognize more complicated hand gestures. The radar system introduced operated on 60 GHz and consumed range-doppler map algorithm in order to acquire the velocity and collection of different movements and accomplish the recognition of the hand in the desired manner. On the other hand, this system was found expensive particularly for human–computer interaction [5,7].

Computer vision mainly deals with object recognition in the 3-D field. It is thereby very much essential, for example, robots have to interact in an unknown environment with computational intelligence. Thus it is found to be useful in various fields as shown in Table 9.1 [8].

9.2.1 Survey on Database

The database can be termed as a set of data such as images, texts, time series, etc. which are marginally essential for the initial step of all sorts of recognition systems. There are various sets of data existing for the research purpose and also any set of information that the scholar wants can be modified into a database [9].

9.2.2 Survey on Feature Extraction

In pattern recognition application, the feature extraction was always referred to as an initial stage of measuring a set of data which would help derive values as they were intended to be instructive and nonredundant. This also enabled the consequent learning and generalization of steps and apart from that, in rare cases, it leads to better human interpretations. Dimensionality reduction is the key feature to which feature extraction is usually related to. It had also been defined as the reduced set of transformed features which were assumed to be redundant data input of an algorithm which was again too large to get processed (example: having exact measurements in both feet and meter or the repetitiveness of images presented as pixels). These transformed features were coined as feature vector. Thus the determined subset of the primary features is known as feature selection. Thereby the selected and essential features are extracted from original data, which reduces the storage capacity of the system.

TABLE 9.1

Machine Learning versus Methods Used

Machine Learning	Methods Used
Artificial neural networks	Biological cybernetics
Adaptive signal processing	Cognitive sciences
Fuzzy and genetic systems	Mathematical statistics
Robotics and vision	Computational neuroscience
Detection and estimation theory	Nonlinear optimization
Exploratory data analyzes	Structural modeling

9.2.3 Survey on Classification

The analysis of the hand gesture which acts as input is further used for gesture classification for recognition purposes. The classification algorithm has varied for different feature points from the original data [10,11]. For recognizing dynamic hand gestures, statistical tools like finite state machines [12], principal component analysis [13] and learning vector quantization [14] are very often used. Moreover, the tool hidden Markov modeling has been found as a very effective tool for the classification system [15,16]. Classifier like deep belief network (DBN) has got its normalized parameters in terms of scale, translation, and rotation in order to train the postures of the image and is thus been used. Also, the study tells that convolutional neural network (CNN) widely has an effective parallel computation to those parameters than deep belief network [17,18]. Moreover, after processing through these various techniques of the neural network, the final output of an image is obtained for which the deep learning (DL) techniques were used. This shows the improved performance measures in terms of accuracy and recognition rate [17].

9.2.4 Survey on Optimization-Based Hand Gesture Recognition

Knowing that the methodology of hand gesture recognition involves the stage of classification, there are classifiers used in order to find the performance measure of the desired data set which is been inculcated. Thus to improvise this performance measure, there are n-number of optimization techniques which are introduced understanding the fact that on utilization of the optimizers there is a definite rise in the percentage level of the derived performance measure. Recently, Rillo F. et al. (2014) and Sotirios P. Chtzis et al. (2011) presented an idea on evolutionary-based heuristic algorithm for pattern recognition applications. Artificial bee colony algorithm is explored in the context of search problem in spatial domain in an input data which has been reported by Jayanth J. et al. [19]. Tereshko [20] has proposed a new technique based on an intelligent behavior for classifying the data effectively in remote sensing applications pointed out by Cuevas et al. [21] and Zhang et al. [22]. Many researchers have the same perspective on the hybrid meta-heuristic approach. It produces an exemplary result for the image classification, data clustering and tracking with better performance measures as reported by Pan et al. [23], Sabat et al. [24] and Stathakis and Vasilakos [25]. Banerjee et al. [26] have presented the accuracy of artificial bee colony algorithm with an existing technique with LISS III data. It was also investigated and compared with other traditional classifiers; artificial bee colony optimization has less complexity in normal distribution of training data. Therefore, it is ascertained that these types of procedures have a greater potential in improving classification accuracy.

9.3 Methodology

Hand gesture recognition has been explored widely by various philosophers using various algorithms. The framework of the proposed work contains preprocessing, feature extraction and classification shown in Figure 9.1. On that term, there are various different techniques which are brought into implementation on the basis of the above three basic elements. The classifier has basic two processes, i.e., training and testing.

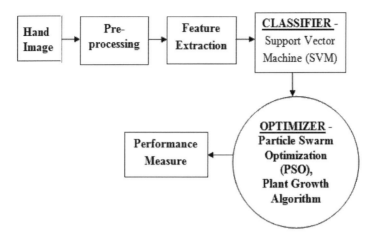

FIGURE 9.1
Basic model on the proposed methodology.

The training process can be done by forward–backward algorithm or Baum–Welch algorithm. The testing process can be done by Viterbi algorithm. During the training process, recursiveness and self-looping problems reduce system performance. Instead of existing approach, meta-heuristic approach has been used to train the process with the same data. The algorithms for the recognition system are thus influenced on the basis of the type of label output.

9.3.1 Image Database

The datasets comprising various images of diverse hand gestures were obtained from the Website of Imperial College of London [27]. The Cambridge Hand Gesture Data set is a legal and publically authorized set of data which is been improvised by the scholars for the usage of pre-existing data acquisition purpose for research analysis and study purpose. In this chapter, five datasets are considered, each equivalent to different gestures measured. The five hand gestures positioned are (i) center-to-left, (ii) compress hand, (iii) V-shape center-to-left, (iv) V-shape center-to-right and (v) left-to-right [17].

9.3.2 Image Preprocessing

Preprocessing is a part of the methodology used to relinquish disorder or the undesirable reappearance of the segment from the given input data. This procedure includes the identification of edges by distinguishing the sharpness in the input data which would clearly portray the boundaries of the patterns in the hand gesture image [17].

The comprised detection method is the Sobel edge detection method introduced in the year 1970 [28]. The major goal of the proposed work was to partition the image. Thus the Sobel edge detection technique has distinguished the edges utilizing Sobel operator which approximates to the subordinate. It was shown by the edges at the specific focus where the gradient is supreme noteworthy. The detector also represents the space of large spatial recurrence which helps relate the edges by making a 2D spatial slope sort on the particular image. It is also used to determine the total gradient size in 'n' input greyscale image at every single point [17].

9.3.3 Feature Extraction

Feature extraction is a type of reduction in the dimensionality of input data which professionally signifies and compresses input data into small data (feature vector). The input data have different features based on shape, color, texture, etc. It has been categorized based on the characteristic of input data. The robust feature set is extracted based on the cluttered backgrounds under demanding illumination [29]. Thereby, the issue of feature sets for hand has been detected using normalized histogram of oriented gradient (HOG) descriptors. It is responsible for exceptional performance than other existing feature extraction techniques.

The usage of the HOG method is beneficial when the image sizes are large and the execution is quickly completed within the duration of the process [30]. This method also involves dropping the number of resources essential to designate a large set of data [29]. The features of histogram of gradient were familiarized by Navneet Dalal and Bill Triggs who have advanced and tested several variants of HOG. These methodologies basically deal with moving body parts and are well paralleled to other feature extractors.

From the study, the evaluation of having different normalization schemes had brought improved normalization percent with less redundancy. The grouping of cells leads to larger spatial blocks and normalizing each block separately. The descriptor vector from the normalized cells had responded in the detection window. Typical overlapping of the block was done for each scalar cell component so that the final descriptor vector showed a minimum percent of improvement in its performance [31]. For the evaluation purpose, it was found that the basic two geometries were taken into consideration rectangular/square (R-HOG) and circular (C-HOG) and these were partitioned into cells in log-polar fashion.

On the basis of Ref. [31], the block normalization schemes can be derived as below:

This scheme can be initially evaluated using four various block normalization structures for both the above HOG geometries. Having considered v the un-normalized descriptor vector, 'V_k' represents k-norm for k = 1, 2 and 'e' is a small constant.

Schemes:

  a. L2-norm,

$$V \rightarrow \frac{V}{V_2^2 + e^2} \tag{9.1}$$

  b. L2-Hys, L2-norm followed by clipping (limiting the maximum values of V to 0.2) and renormalizing, as in Ref. [32]

  c. L1-norm,

$$V \rightarrow \frac{V}{V_1 + e} \tag{9.2}$$

  d. L1-sqrt, L1-norm followed by the square root

$$V \rightarrow \sqrt{\frac{V}{V_1 + e}} \tag{9.3}$$

The probability distribution and distance between the blocks have been represented for the descriptor vector as shown in the above equation. It was earlier proven that usage of L2-Hys, L2-norm, and L1-sqrt all accomplish equally well while the simple L1-norm minimizes the performance by nearly 5% and helps omit the normalization entirely by reducing it by 27% (approx.), at 10^{-4} FPPW [31].

The steps to be followed for implementing this process using the HOG descriptor algorithm (*see* Figure 9.2) are as follows:

- The image pixel is divided into small regions called cells. Each cell determines histogram directions or edge orientations of the pixels within the cell.
- The angular bin stores the value of orientation of the gradient in each cell.
- The resultant angular bin based on a weighted gradient in each cell.
- A group of adjacent cells in spatial sections called blocks. Thus by grouping these cells into a block is based on alignment and normalization of HOG.
- A group of normalized histograms symbolizes the block histogram. The set of these block histogram features represent descriptor of the object (Table 9.2).

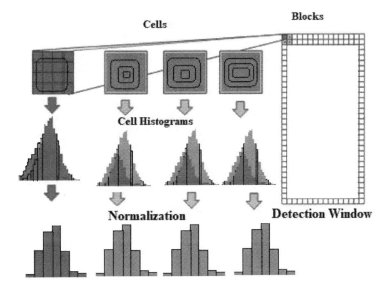

FIGURE 9.2
Working of HOG descriptor.

TABLE 9.2

Steps for HOG Algorithm

S. No.	Steps	Description
1	Gradient computation	• This method applies 1D centered point in both horizontal and vertical direction. • Also involves filtering of grey scale image
2	Orientation binning	• Involves creating cell histograms • Based on the values found in the gradient computation, each pixel of a cell comprises a weighted vote
3	Descriptor blocks	• Here, gradient strength has to be localized • Includes grouping of cells (as large, spatial, connected blocks) • Two main blocks: (i) Rectangular HOG (R-HOG) (ii) Circular HOG (C-HOG)

9.3.4 Classification

Classification is the process of identifying the correct class from the group. It can be used to predict a higher precision rate. A few approaches are used to do the classification process such as linear and nonlinear predictive analysis. Support vector machine (SVM) is one of the approaches used to check the maximum matched feature vector of the given data. Basically, this is a linear classifier that capitalizes the distance between the decision lines. The focal advantage of this classifier is that it can use kernels for nonlinear data transformation. The main objective of SVM is to classify the data provided into two diverse categories and then to get the hyper-plane to isolate the given classes [33]. Then, it is utilized for performing classification and regression.

The standard SVM is a two-class SVM which proceeds with a set of input data and foretells the possible class for each input. It also emphasizes that among the two possible classes the input is a member of the class which makes it a nonprobabilistic binary linear classifier [34]. The training process of the SVM is better than the other conventional techniques used for classification. This classifier has in turn proven to be an efficient method of finding an ideal hyper-plane for separating nonlinear data as well [34]. It can be used for classifying the classes from the multi-class level of input data. This significantly improves the performance measures by two-class SVM.

SVM is one among the most known classification methods which are stimulated from the statistical learning theory [35,36]. It helps accomplish linear classification based on supervised learning as depicted in Figure 9.3. This methodology can be utilized to avoid parameter sensitivity. Given a labeled training data $(a_i, b_i, i) = 1, \dots, p, b_i \in (-1, 1); (a_i \in \mathcal{R}^d)$, it is also found that the algorithm builds a hyperplane with a principal margin that separates the positive from the negative examples. The difficulty of finding the optimal hyperplane was found to be dealt with solving the succeeding optimization problem:

$$\min_w \frac{1}{2} w^t w + c \sum_{i=1}^{p} \xi(w, a_i, b_i) \tag{9.4}$$

where $c > 0$ is a penalty parameter and w is normal to the hyperplane.

$c \sum_{i=1}^{p} \xi(w, a_i, b_i)$ was considered to reflect a total of misclassification errors. The testing process has been carried out by determining the distance from the data obtained from testing to hyperplane. It has been found that performance measure increased by 3% at 10^{-4} false positive per window (FPPW) using SVM kernel function [31].

$$e^{\left(-\gamma |x_1 - x_2|^2\right)}. \tag{9.5}$$

The above equation had helped develop increased performance (Figure 9.4).

FIGURE 9.3
The basic model of stepping into the classification stage.

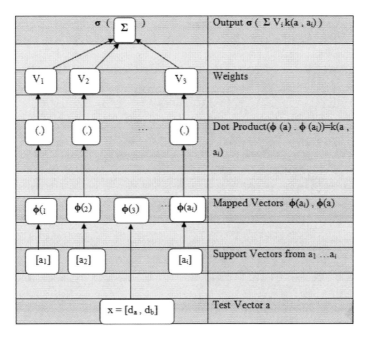

FIGURE 9.4
Sample SVM module.

9.3.5 Optimization

The optimization is a stage inculcated in order to enhance the performance measure of hand gesture recognition using two optimization techniques: particle swarm optimization (PSO) and plant growth optimization as a case study report has shown in the results section.

9.3.5.1 Particle Swarm Optimization (PSO)

Eberhart and Kennedy [37] have described the construction and working of PSO, which is yet again based on the norm of the recommended modifications in Coello Coello and Mezura-Montes [38] and in Pedersen [39]. In most of the recent approaches, PSO and genetic algorithm (GA) have been applied in various fields such as data clustering, network optimization and linear-constrained function optimization [39,40]. PSO has been initiated by crafting incident particle and allot with initial velocity in the given search space problem [36].

It thereby evaluates the objective function at each particle location and also determines the best (bottommost) function value and also its best location. Here it chooses new velocities on the bases of the particles' individual best locations, the best whereabouts of their neighbors and also on the bases of the current velocity.

It then iteratively updates the locations of the particle (the new location is the velocity plus the old one, which is modified to keep particles within the bounds), its velocities and the neighbors. The iterations proceed until the algorithm approaches a stopping criterion.

Initialization of Algorithm:

By default, this PSO optimization technique generates particles at random which are uniformly within the bounds, and if there are any unbounded component the method creates particles with a disorganized unvarying circulation from −1,000 to 1,000. If in case there is just one bound then the particle swarm shifts the construction to have the bound as an endpoint and a creation interval to be 2,000 wide. Particle j has position y(j) which is a row vector with (nvar) as its elements. The span of the initial swarm is controlled using the Initial_Swarm Span_option.

Similarly, the particle swarm creates initial particle velocities V at random and also uniformly within the range [–R, R] where R is the vector of the initial *ranges*. The range of component j is min(ub(j) – lb(j),Initial Swarm Span(j)).

This particle swarm technique evaluates the objective function at all particles. It accounts the current position y(j) of each particle j. In successive iterations, y(j) will be the location of the finest objective function that the particle j has originated with. And 'a' is the best overall particles: [a = min(fun(p(i)))]. The location is denoted by 'l' such that [a = fun(l)].

1. The technique also initializes the neighborhood size H to
   Min neighbourhood size = max (1,floor (swarm size × min neighbour fraction)).
2. The technique initializes the inertia I = max (inertia range), or if the inertia range is negative, it sets I = min (inertia range).
3. The technique initializes the stall counter c = 0.

For the convenience of the notational aspect, let set the variable

z1 = Self-adjustment weight and

z2 = Social adjustment weight, where self-adjustment weight and social adjustment weight are options.

Iteration Steps

The algorithm apprises the swarm as follows:
For a particle j which is at position y(j):

i. A random subset S of H particles is chosen other than j.
ii. Find the fbest(S) where the best objective function among the neighbours is found and g(S) which is the position of the neighbour with the best objective function.
iii. For u1 and u2 uniformly being (0, 1) distributed as random vectors of length (nvar), the velocity is updated.
   V = I*V + z1*u1*(p – y) + z2*u2*(g – y).
   This update uses a weighted sum of
   • The previous velocity V.
   • The distance between best position and current position in the particle has represented as (p – y).
   • The distance between best position and current position in the current neighbourhood is represented as (g – y).
iv. The position is updated from y to y = y + V.
v. Impose the bounds and in case if any component of y is outside any bound then set it equal to that particular bound.
vi. Now evaluate the objective function f = fun(y).
vii. If f < fun(p), then set p = y. Here, this step ensures that p has the best position the particle has seen.

viii. If f < a, then set a = f and l = x. Here, this step ensures that a has the best objective function in the swarm and l has the best location.

ix. If in the step above, the best function value had been lowered then set flag = true. Otherwise, let it be flag = false. The value of the flag is utilized in the next step.

x. Update the neighborhood.
   If flag = true : Set c = max(0,c − 1)
   (A) et N to min neighborhood size
   (B) if c < 2, then set I = 2*I
   (C) if c > 5, then set I = I/2
   (D) Ensure that I is in the bounds of the inertia range options.
   If flag = false:
   (E) Set c = c + 1.
   (F) Set H = min (H + min neighborhood size, swarm size)

Stopping Criteria

The particle swarm stops the iteration when maximum time or maximum iteration or objective functions. Then the entire process gets stopped with exit flag 1, optionally calls for hybrid function for restarting the operation.

9.3.5.2 Plant Growth Simulation Algorithm (PGSA)

The plant growth simulation algorithm (PGSA) is basically bionic random algorithm hired for the solutions of the combinational optimization problems reported by Li et al. (2005), Li and Wang (2008). The modified PGSA algorithm was later proposed by Wang and Cheng (2008). The modified PGSA algorithm was finely streamlined from an existing algorithm without making much difference in existing approach [41]. Since it has simplified and reliable parameters for optimization problem, it can be found in most of the applications like facility location, power network planning, antenna arrays configuration, etc.

The PGSA is an algorithm inspired by the growth mechanism of plant phototropism. The theory behind the algorithm is that, largely, during the plant growth, a node with the higher morphactin concentration in a plant has the higher probability to sprout a new branch. The distribution of morphactin concentration on a plant is never constant. It is something which is allotted over and over again according to the novel environment information it acquires after it sprouts a new branch. The study relates that, in the probability model of PGSA, g(Y) is a function which describes the growth environment of the node Y on a plant. This algorithm shows that node Y has better sprouting for new branch if g(Y) has a smaller value [41]. Using that concept the concentration (say C_{Mi}) or the depth of the database can be calculated as follows:

$$C_{Mi} = \frac{g(B_0) - g(B_{Mi})}{\Delta_1}$$

$$\Delta_1 = \sum_{i=1}^{P} \left(g(B_0) - g(B_{Mi}) \right).$$

(9.6)

Equation 9.6 shows the relation between the morphactin concentration distribution on a plant and growth environment at the specified point. This also means that the concentration is dependent on the growth environment. If the depth C_{Mi} consists of p nodes then

B_{Mi} (i = 1,2, ... , p) is considered to be a state space. If a random number β is obtained in the interval [0, 1], then it is thrown to the state space like a ball and will fit into one of the p nodes [42]. The corresponding node is called the preferential node that has decided to take a next step by prioritizing a new sprout set. In other words, the node B_{MT} had taken the priority if the random number β had satisfied equation (9.7) as follows:

$$\left[\begin{array}{l} 0 \leq \beta \leq \displaystyle\sum_{i=1}^{T} C_{Mi} \\[2em] \displaystyle\sum_{i=1}^{T} C_{Mi} < \beta \leq \sum_{i=1}^{T} C_{Mi} \end{array} \right. \qquad T = 1 \text{ and } T = 2,3,...,p. \qquad (9.7)$$

Basically, the PSO has explained through the activity of birds flocking in space problem [43,44]. Each agent birds are positioned that is represented in the X-Y plane and velocity represents the moving from one point to other point. Each agent knows its best value so far (pbest) and its current position. It is an analogy statement who is an agent bird. However, each agent has to know their best value gbest and pbests in that group [42]. This information is an analogy of an agent knowing how other agents have performed in the surround environment to it. Each agent has modified its position with respect to velocity.

$$v_i^k = w v_i^k + c_1 \text{rand}_1 \times \left(\text{pbest}_i - S_i^k \right) + c_2 \text{rand}_2 \times \left(\text{gbest} - S_i^k \right) \qquad (9.8)$$

Equation 9.8 shows the updating of the position with respect to velocity, where v_i^k is velocity of agent i at iteration k, w is weighting function, c_1 and c_2 are weighting factors, r_1 and and r_2 and are random numbers between 0 and 1. The following weight function is usually used in equation 9.8:

$$w = w_{max} - \left(\left(w_{max} - w_{min} \right) / \left(\text{iter}_{max} \right) \right) \times \text{iter} \qquad (9.9)$$

where w_{max} is the initial weight, w_{min} is the final weight, iter_{max} is the maximum iteration number, and iter is the current iteration number. Using the previous equations, a certain velocity, which gradually brings the agents close to pbest and gbest, can be calculated. The current position (search point in the solution space) can be modified by the following equation:

$$s_i^{k+1} = s_i^k + s_i^{k+1} \qquad (9.10)$$

The model using equation 9.9 is called gbest model. The model using equation 9.10 in equation 9.9 is called inertia weights approach (IWA).

9.4 Result and Discussion

In this chapter, Cambridge hand gesture dataset has been used. It consists of three different motions (leftward, rightward and contract) in three different shapes flat, spread and V-shape. So totally nine classes of dataset with hundred frames are stored. This experiment has been carried out with five classes of hand gesture data for validation of hidden

Markov modeling with particle swarm optimization and plant growth optimization. First, preprocess the input data using the edge detection method; Sobel edge detection technique has distinguished the edges utilizing Sobel operator which approximates to the subordinate. It was shown by the edges at the specific focus where the gradient is supreme noteworthy. Sobel operator has provided the better edge of moving objects. It avoids false detection point than other kernel operator used in edge detection. The feature point has been extracted by using histogram features and has created the label for each class of the hand gesture data. It can be validated for training and testing using SVM. The below Table 9.3 shows the training and testing of data from the dataset using the validation process.

After validation procedure, each sample of data has to be tested with SVM. It can be determined how much correct matching of feature point with the original data has been classified.

Table 9.4 shows performance measures using SVM. The average depth of accuracy for five set of data is 89.624 of execution time is 0.888 s. To improve the performance measures using natural inspired optimization techniques has been discussed with the results obtained in next two tables.

Tables 9.5 and 9.6 show the performance measures using SVM with particle swarm optimization and Plant Growth Algorithm (PGA), respectively. Although the results are almost

TABLE 9.3

Training and Testing Data Set Using Validation Procedure

Input Data from Dataset	Training Set	Testing Set
Center to left	600	300
V-shaped center to right	600	300
Compress	600	300
Center to right	600	300
V-shaped center to left	600	300

TABLE 9.4

Performance Measures Using SVM without Optimization

S. No.	Data Set	Execution Time (s)	Depth of Accuracy
1	Center to left	0.86	86.88
2	V-shaped center to right	0.91	90.15
3	Compress	0.90	89.47
4	Centre to right	0.89	86.91
5	V-shaped center to left	0.88	87.71

TABLE 9.5

Performance Measures Using SVM with PSO Optimization

S. No.	Data Set	Execution Time (s)	Depth of Accuracy
1	Center to left	0.87	87.93
2	V-shaped center to right	0.92	91.37
3	Compress	0.911	91.27
4	Centre to right	0.902	89.15
5	V-shaped center to left	0.89	88.94

TABLE 9.6

Performance Measures Using SVM with PGA

S. No.	Data Set	Execution Time (s)	Depth of Accuracy
1	Center to left	0.88	87.0237
2	V-shaped center to right	0.927	90.5018
3	Compress	0.912	90.3701
4	Centre to right	0.909	88.2924
5	V-shaped center to left	0.898	87.0487

the same for all classes of hand gesture data in both optimization, the difference of average depth of accuracy in PSO and PGA is 0.0827% with 0.0072 s. Compare the proposed system with the performance calculation using SVM, execution time for entire process is significantly less but degraded by depth of accuracy.

9.5 Conclusions

In this proposed work, hand gesture recognition is done using the SVM approach. The Cambridge hand gesture data are used in this context. Histogram features from original data are stored. Each feature produces magnitude and angular bin values of the hand gestures. It has been labeled and trained by the existing SVM approach. The same process has been re-trained by PSO and PGA. While testing process with PSO and PGA approach the performance has been improved. SVM with PSO has improved accuracy of 1.51% and PGA has improved accuracy of 0.42%. In the future scope, this work will be processed by using deep learning algorithm for real-time applications.

References

1. F. Karray, M. Alemzadeh, J. A. Saleh, and M. Nours, Human-computer interaction: Overview on state of the art, *International Journal on Smart Sensing and Intelligent System*, March 2008, Vol. 1, No. 1, 37–159.
2. A. S. Ghotkar and G. K. Kharate, Vision based hand gesture recognition techniques for human computer interaction, *International Journal of Computer Application, Computer Science Foundation, New York, USA*, May 2013, Vol. 70, No. 6, ISSN: 0975-8887, 1–6.
3. F. Dominio, M. Donadeo, and P. Zanntligh, Combining multiple depth-based descriptors for hand gesture recognition, *Pattern Recognition Letters*, 2014, Vol. 50, 101–111.
4. W. Liang, L. Guixi, and D. Hongyan, Dynamic and combined gestures recognition based on multi-feature fusion in a complex environment, doi:10.1016/S1005-8885(15)60643-4.
5. S. Lan, Z. He, and H. Tang, A hand gesture recognition system based on 24 GHz radars, *IEEE 2017 International Symposium on Antennas and Propagation (ISAP)*, Phuket, 2017, doi:10.1109/ISANP.2017.8228827.
6. S. Gupta, D. Morris, S. Patel, and D. Tan, Sound wave: Using the Doppler effect to sense gestures, *SIGCHI Conference on Human Factors in Computing Systems*, pp. 1911–1914, 2012.

7. J. Linen and N. Gillia, Soli: Ubiquitous gesture sensing with millimeter wave radars, *ACM Transactions on Graphics (TOG)*, 2016, Vol. 35, No. 4, 142.

8. V. Dutt, V. Chaudhry, and I. Khan, Pattern recognition: An overview, *American Journal of Intelligent Systems*, 2012, Vol. 2, 1, 23–27, doi:10.5923/J.AJIS.20120201.04.

9. K. Martin Sagayam and D. Jude Hemanth, Hand posture and gesture recognition techniques for virtual reality applications: A survey, *Virtual Reality*, 2017, Vol. 21, 2, 91–107.

10. R. Z. Khan and N. A. Ibraheem, Hand gesture recognition: A literature review, *International Journal of AI and Application*, July 2012, Vol. 3, 161.

11. S. Mitra and T. Acharya, Gesture recognition: A survey, *IEEE Transactions on Systems, Man and Cybernetics*, 2007, doi:10.1109/TSMCC.2007.893280.

12. R. Verma and A. Dev, Vision based hand gesture recognition using finite state machines and fuzzy logic, *IEEE International Conference on Ultramodern Telecommunications and Workshops*, St. Petersburg, 2009, doi:10.1109/ICUMT.2009.5345425.

13. Y. Minghai, Q. Xinyu, G. Qinlong, R. Taotao, and L. Zhong Wang, Outline PCA with adaptive subspace method for real time hand gesture learning and recognition, *Journal World Scientific Engineering Academy and Society*, 2010, Vol. 9, 6.

14. L. Lamberti and F. Camastra, Real time hand gesture recognition using color glove, *Springer Proceedings of the 16th International Conference on Image Analysis and Processing Part 1 (ICIAP)*, Ravenna, 2011.

15. B. Min, H. Yeon, J. Seh, Y. Yango, and T. Ejuna, Hand gesture recognition using hidden Markov model, *IEEE International Conference Computational Cybernetics and Simulation*, Orlando, FL, 1997, doi:10.1109/ICSMC.1997.637364.

16. E. Mahmoud, A. Ayoub, and M. Bernd, Hidden Markov model based isolated and meaningful hand gesture recognition, *World Academy of Science, Engineering and Technology*, 2008, vol. 2, No. 5, 985–992.

17. K. M. Sagayam, T. V. Viyas, C. C. Ho, and L. E. Henesey, Virtual robotic arm control with hand gesture recognition and deep learning strategies, IOS Press, 2017, doi:10.3233/978-1-61499-822-8-50.

18. A. Tang, K. Lu, Y. Wang, J. Huang, and H. Li, A real time hand posture recognition system using deep neural networks, *ACM Transaction Intelligence System Technology*, 2013, Vol. 9, 4, Article 29.

19. J. Jayanath, S. Koliwad, and T. Ashok Kumar, Classification of remote sensed data using artificial bee colony algorithm, *The Egyptian Journal of Remote Sensing and Space Sciences*, 2015, Vol. 18, 119–126.

20. V. Tereshko, Reaction diffusion model of a honeybee colony's foraging behaviour, In: Schoenauer, M. (Ed.), *Parallel Problem Solving from Nature, VI*, Lecture Notes in Computer Science, 17–19, Springer, Berlin; New York, 2000.

21. E. Cuevas, F. Secció n-echauri, D. Zaldivar, and M. Pérez-cisneros, Multi-circle detection on images using artificial bee colony (ABC) optimization, *Soft Computing*, 2011, Vol. 22, 15–26.

22. C. Zhang, D. Ouyang, and J. Ning, An artificial bee colony approach for clustering, *Expert Systems with Applications*, 2010, Vol. 37, 4761–4767.

23. Q. K. Pan, M. Fatih-tasgetiren, P. N. Suganthan, and T. J. Chua, A discrete artificial bee colony algorithm for the lot-streaming flow shop scheduling problem, *Information Sciences*, 2010, Vol. 26, 65–78.

24. S. L. Sabat, S. K. Udgata, and A. Abraham, Artificial bee colony algorithm for small signal model parameter extraction of MESFET, *Engineering Applications of Artificial Intelligence*, 2010, Vol. 23, 689–694.

25. D. Stathakis and A. Vasilakos, Comparison of computational intelligence based classification techniques for remotely sensed optical image classification, *IEEE Transactions on Geoscience and Remote Sensing*, 2008, Vol. 44, 8, 2305–2318.

26. S. Banerjee, A. Bharadwaj, D. Gupta, and V. K. Panchal, Remote sensing image classification using artificial bee colony algorithm, *International Journal of Computer Science and Information Technologies*, 2012, Vol. 2, 3, 67–71.

27. T. -K. Kim and R. Cipolla, Canonical correlation analysis of video volume tensors for action categorization and detection, *IEEE Transaction on Pattern Analysis and Machine Intelligence*, 2009, Vol. 31, 8, 1415–1428.

28. R. C. Gonzalez, R. E. Woods, and S. L. Eddins, *Digital Image Processing Using MATLAB*, Pearson Education Ltd., Singapore, 2004.

29. R. N. Nagashree, S. Michahial, G. N. Aishwarya, B. H. Azeez, M. R. Jayalakshmi, and R. Krupa Rani, Hand gesture recognition using support vector machine, *The International Journal of Engineering and Science (IJES)*, June 2015, Vol. 4, 6, 42–46.

30. H. -J. Lee and J. -H. Chung, Hand gesture recognition using orientation histogram, *Proceeding of the IEEE Region 10 Conference on TENCON 99*, Cheju Island, 2002, doi:10.1109/TENCON.1999.818681.

31. N. Dalal and B. Triggs, Histograms of oriented gradients for human detection, INRIA Rhone-Alps, 655 avenue de l'Europe, Montbonnot 38334, France {Navneet.Dalal,Bill.Triggs}@inri-alpes.fr, http://lear.inrialpes.fr.

32. D. G. Lowe, Distinctive image features from scale-invariant key points, *IJCV*, 2004, Vol. 60, 2, 91–110.

33. S. Gupta, J. Jaafar, and W. F. W. Ahmad, Static hand gesture recognition using local Gabor filter, *International Symposium on Robotics and Intelligent Sensors 2012 (IRIS 2012)*, pp. 827–832, 2012, doi:10.1016/j.proeng.2012.07.250.

34. G. Sharma, and Rudrakshi, Survey on prediction using back propagation neural network, *International Journal of Advanced Research in Computer Science and Software Engineering* (ISSN: 2277-128X), April 2014, Vol. 4, No. 3, 917–920.

35. V. N. Vapnik, An overview of statistical learning theory, *IEEE Transactions on Neural Networks*, 1999, Vol. 10, 5, 988–999, doi:10.1109/72.788640.

36. K. Lekdioui, R. Messoussi, Y. Ruichek, Y. Chaabi, and R. Touahn, Facial decomposition for expression recognition using texture/shape descriptors and SVM classifier, *Signal Processing: Image Communication*, October 2017, Vol. 58, 300–312.

37. J. Kennedy and R. Eberhart, Particle swarm optimization, *Proceedings of the IEEE International Conference on Neural Networks*, Perth, pp. 1942–1945, 1995.

38. E. Mezura-Montes and C. A. Coello Coello, Constraint-handling in nature-inspired numerical optimization: Past, present and future, *Swarm and Evolutionary Computation*, 2011, Vol. 1, 173–194.

39. M. E. Pedersen, *Good Parameters for Particle Swarm Optimization*, Hvass Laboratories, Luxembourg, 2010.

40. R. M. Ramadan and R. F. Abdel-Kader, Face recognition using particle swarm optimization-based selected features, *International Journal of Signal Processing, Image Processing and Pattern Recognition*, June, 2009, Vol. 2, No. 2, 51–66.

41. R. C. Eberhart and Y. Shi, Comparison between genetic algorithms and particle swarm optimization, *Proceedings of the 7th international Conference on Evolutionary Programming*, pp. 611–616, 1998.

42. S. Lu and S. Yu, A fuzzy k-coverage approach for RFID network planning using plant growth simulation algorithm, Research Article, *Journal of Network and Computer Applications*, March 2014, Vol. 39, 280–291.

43. R. Garduno, J. S. Heo, and K. Y. Lee, Multiobjective control of power plants using particle swarm optimization techniques, *IEEE Transactions on Energy Conversion*, July 2006, doi:10.1109/TEC.2005.858078.

44. C. Reynolds, Flocks, herds, and schools: A distributed behavioral model, *Computer Graphics*, 1987, Vol. 21, No. 4, 25–34.

10

Internet of Biometric Things: Standardization Activities and Frameworks

Zoran S. Bojkovic and Dragorad A. Milovanovic
University of Belgrade

CONTENTS

10.1 Introduction

Biometrics deals with the identification of a person from his measurable physical characteristics based on automatic verification. Thus, the role of biometrics in future computer applications is huge. One of the best examples is electronic commerce. Another example is the case when your fingerprint is analyzed by a computer in order for you to obtain different levels of access and significant information. The beginning of biometric system introduction coincides with the development of the computer world. The explosion of activity and everyday applications started in the early 2000s.

10.1.1 Biometrics Technology Evolution

In order to follow biometrics technology evolution, first of all, it will be useful to go back to history and track it until the start of huge development at the very beginning of this century. This time line with some important events is shown in Table 10.1.

TABLE 10.1

Biometrics Timeline

Year	Action
1896	A fingerprint classification system is developed
1903	NY State Prisons begin using fingerprints
1936	Identification based on iris pattern is proposed
1960	Face recognition becomes semi-automated
1960	The first model of acoustic speech production is developed
1969	Fingerprint recognition becomes an automated process in FBI
1970	Face recognition takes another step toward automation
1970	The first model of behavioral components of speech is developed
1974	First commercial hand geometry system becomes available
1976	The first system of speaker recognition is developed
1977	An acquisition of dynamic signature information is patented
1980	The NIST Speech Group is established
1985	Identification concept based on the fact that no two irises are alike is proposed
1986	The iris-based identification patent is awarded
1988	The first system based on semi-automatic facial recognition is deployed
1988	Face recognition technique based on eigenface is developed
1991	Real-time face recognition based on face detection is deployed
1992	Biometric consortium is established within US Government
1993	Face recognition technology program is initiated
1994	The first iris recognition algorithm is patented
1994	Palm system is benchmarked
1995	The iris-based identification prototype is commercialized
1999	Study of compatibility between biometric and machine-readable travel documents begins

In the early 2000s, an explosion of biometric computing technology appears, as shown in Table 10.2.

As a technological system, biometrics involves running data using algorithms in order to obtain specific results for the identification of a user/individual. Users can identify themselves to a computer by devices including but not limited to face scanner, scanner, figure scanner, retina or iris scanner, voice scanner.

10.1.2 Biometrics Signals Processing

The main advantage of digital signal processing (DSP) is the fact that it allows the biometric system to be small, portable and not expensive. It should be mentioned that a biometric system is a pattern recognition system which has the potential to extract a feature from the raw data and compare it to the corresponding feature in the feature set stored in the data base (Jain, 2007). The function of DSP is to perform multiplication/addition in a single cycle using multiply/accumulate hardware together with its arithmetic–logic unit. Two-dimensional FFT (fast Fourier transform) and FIR (finite impulse response) filters are applied for enhancing the resolution of the captured image.

Today there are competitive, sponsored algorithm test programs in speaker, fingerprint, and facial recognition. Fingerprint pattern recognition was one of the first applications of DSP. An important method of forensic science is the recovery of fingerprints from a crime scene (Huynh et al., 2015). Electronic fingerprint verification is used in mobile devices,

TABLE 10.2

Biometrics Timeline from the Year 2000

Year	Action
2000	NIST face recognition vendor test is ongoing
2000	A university biometrics degree program is introduced
2000	A recognition concept based on vascular patterns is published in a research paper
2002	ISO/IEC committee on biometrics starts standardization work
2003	US Government starts coordination of biometric activities
2003	European Biometric Forum is established
2003	ICAO (International Civil aviation Organization) adopts integration of biometric identification into travel documents
2004	DoD (Department of Defense) implements automatic biometric identification
2004	Face recognition ground challenge starts
2008	US Government starts coordination of biometric database use
2010	US national security implements biometrics for terrorist identification
2013	Apple Inc. includes fingerprint scanners in smartphones
2017	Apple Inc. includes 3D facial scanners in smartphones

smart homes and buildings, and transportation systems (Shah, Sangoi, and Visharia, 2014). That is the reason why automated fingerprint identification is of interest from not only academic, but also commercial point of view.

Some stages such as image acquisition, segmentation, preprocessing, feature extraction, and person verification are included in automatic fingerprint identification system. This system understands image acquisition, fingerprint segmentation, image preprocessing, and feature extraction. Image acquisition is a factor that contributes when considering fingerprints. For the recognition system, in choosing the device to capture fingerprints, parameters such as performance, cost, and size of the demanded image are considered; segmentation serves to separate the image of the finger from the background. This is an initial step performed on the observed finger by the print system (Sayeemuddin, Pithadia, and Vandra, 2014). The procedure of preprocessing comprises modifications to the image in the sense of enhancement in order to increase the accuracy of the system. In the step of feature extraction, only similar fingerprints are retrieved and compared with the target one (Raghavendra, Busch, and Yang, 2013; Zhang, 2014).

One of the main characteristics of speakers is voice print. It is well known that a voice is unique. The reason lies in the fact that our vocal activities and the way we move our mouth are different, which results in our speech. When using a voice print system, we have to say the exact words that are required. Also, we can give an extended sample of our speech. In that way, the system will identify a person.

A sound spectrogram is the data used in a voice print. It is a graph with different speech sounds characterized by determined shapes. Spectrograms are used by speaker recognition systems in order to represent human voices.

Palm print is a unique and reliable biometric characteristic of the image acquired from the palm region of the hand. It can be used in a forensic or commercial case (Fisher and Fisher, 2012). As in the case of fingerprint, it also contains some other information (texture, indents) on marking which can be useful when comparing one palm print to another. A biometric hand scanner identifies the person by the palm of their hand, while a finger scanner identifies the person by their fingerprint. A biometric retina or iris scanner can identify a person by scanning the iris or retina of their eyes.

Biometric face scanner identifies a person by measurements of a person's face. It is assumed that scanners are smart enough to differentiate between a picture and real person. Face recognition algorithm can be included in applications used for humans. The results obtained from *Face Recognition Vendor Test* in 2006 and also *Challenge Evolution* in the same year show that face recognition can have a performance which is comparable to iris recognition for verification applications, for example, access control.

10.2 Biometrics Technology and Internet of Things

As previously stated, there are several types of biometric identification schemes analyzing different features:

- facial characteristics
- individual's fingerprints
- capillary vessels located at the back of the eye (retina)
- shape of the hand and the length of the fingers
- colored rings that surround the eye's pupil
- way the person signs himself (signature)
- tone, pitch, cadence, and frequency (voice).

Today biometric systems are one of the most effective technologies for human identification with two main advantages. The first one is the fact that they offer greater security and convenience compared to the traditional methods of person recognition. Secondly, they give users greater convenience, while maintaining, on the other hand, sufficiently high accuracy. In addition, the biometric systems ensure that the user is present at the place and time of identification. On the other hand, adoption of biometric technology depends on biometric technologies' security as well as the privacy implication when using biometrics in a variety of applications such as forensics, transportation, healthcare, finance, safety and education (Jain, Nandakumar, and Nagar, 2007) (Bhattacharyya et al., 2009). Identification of a person based on biometric recognition is a significant task in any information security system.

Biometric recognition deals with establishing the identity of an individual based on the following data: fingerprint, hand geometry and images of iris, face and ear (Peer and Bule, 2013). The major obstructions in applying biometrics are accessibility and scalability of existing biometric technology and the cost of the biometric system. That is the reason why it is of importance to move the existing biometric technology to a cloud area. In this case, scalability, sufficient storage, capability for parallel processing as well as cost reduction will be confirmed. The second step which could be recommended is to use a low-cost Internet of Things (IoT) device (Bojkovic, Bakmaz, and Bakmaz, 2017).

Research in the field of IoT, which emerged as a massive technology, is enabled by sensors, networking, cloud computing, edge computing, big data, machine learning intelligence, security and privacy. This makes the domain of the field quite interdisciplinary. Applications of IoT digital and physical entities would be linked in future by information and communication technologies (ICT). The final goal is to provide a new class of applications and services. Today, the IoT can be treated as a dynamic and distributed network system with a number of smart objects to produce information. The related technologies are big data, cloud computing and so on. The IoT allows objects to be sensed and controlled remotely across existing network infrastructure. In that way, we are on the road to fully integrating the physical world and computer-based system.

In the time of IoT, biometric technology will reveal a new significant period. The advantages of incorporating IoT are low cost, low space occupancy, low power consumption as well as portability of the system. Also, the IoT will redefine management using biometrics for unlocking banking applications, e-mail accounts, and personal health databases.

10.2.1 Influence of IoT on the Development of Biometrics Technology

It was estimated that IoT would contribute an increasingly large number of 9 billion connections by the end of 2018. Information transmission is based on information distributed and ubiquitous computing facilities. In that way, industries and markets are connected and, at the same time, economic growth increases. On the other hand, users are enlarging their computing and computation capabilities, with a range of computing devices. The network becomes accessible all the time. As for the resources, they are mobile and have wired as well as wireless access. Computer services and devices are using information processing through microprocessor devices. A new generation of Internet services is using the same infrastructure that enables the widespread use of web technologies.

In the biometric community, iPhone fingerprint scanner contains a special feature called TouchID which is built right into the home button. This technology is used as secure technology for device sharing, payments, tickets as well as applications where identity is of importance. iWatch belongs to a type of mobile devices with applied biometric technology. Here, the most evident biometric authentication is a thumbprint scanner. An iris scan and facial recognition are other options. Specific hardware such as a camera is required to be built into iWatch. Another option is listening to a user's heartbeat.

10.2.2 Role of Cloud Computing

The cloud computing, which encapsulates a business model of providing services across the Internet, and mobile cloud computing concept have been used from 2007. Information processing could be completed much faster on a cloud infrastructure with two limitations: sensor deployment and battery constraints. Today, mobile devices with camera are suitable to be used in face and iris recognition applications. For the process of recognition, the computing power of mobile devices is capable of resolving the problem.

Accessibility and scalability of existing biometric technology and cost of the biometric system are the major obstructions for an effective solution. That is the reason why deployment of existing biometric technology on cloud platforms can solve issues of scalability, sufficient storage, parallel processing as well as cost reduction.

10.2.3 Biometric Sensors

A device that converts human biometric characteristics, such as fingerprint, voice, face and so on, into electrical signals is called a biometric sensor. In fact, the sensor reads pressure, temperature, speed, light, and electrical capacity. To achieve this goal, different technologies are used including display cameras or combinations of networks of sensors. It should be pointed out that the output signal of a sensor is a representation of the real-world biometrics. A biometric device is a system in which a biometric signal is embedded. In addition, communication, processing, and memory modules provide functionalities that the biometric sensor cannot.

In order to design biometric sensors, the following parameters have to be analyzed: user acceptance, portability, lifetime, calibration, operating conditions, sensor interface, power supply, failure rate, cost, sensor resolution, image depth or dynamic range, modulation transfer function, geometric image accuracy, field of view, depth of field, intensity linearity, signal-to-noise ratio, frame rate, optics. All these factors are significant depending on the application of the biometric sensor.

User acceptance. This parameter is observed when dealing with the choice of a corresponding technology as well as the design of a biometric device. There are two main families for biometric sensor and devices, i.e., intrusive and non-intrusive. The closer the device to the person, the more intrusive the device. When an operator is needed, the device can be classified as a supervised device; otherwise it can be classified as an unsupervised device.

Portability. The most sensitive step in the biometric authentication is the so-called biometric capture. One of the solutions for different applications is the existing possibility to process the captured biometrics. The process can be performed remotely. In this case, a communication interface is a part of the complex device. Unfortunately, this technology is limited by the fact that it needs supply of the energy for portable batteries.

Lifetime. Very often a biometric device has to be installed or carried in difficult environments. In this case as well as when moving parts are included, one of the features to be considered is lifetime. By difficult environments, we mean very low or high temperature, high humidity, dust and noisy location, and by moving parts, we mean line-scan and auto-focusing cameras, auto-position sensors, etc. Of course, these facts have an evident influence on the maintenance costs of the corresponding device. Special materials used in sensor manufacturing are welcome.

Calibration. There are two kinds of calibration: mechanical and electrical. The first one is performed when a device has got moving parts, while electrical calibration is used to restore the initial electrical conditions. On the other hand, mechanical calibration is carried out when mechanical friction appears in the period the sensor was designed for. When there is a need for refocusing lenses or re-establishing the initial illumination conditions, optical calibration is used.

Operating conditions. This are the set of conditions, such as voltage, temperature, pressure, humidity etc., for some specified parameters to maintain the performance rating. The operating conditions are chosen based on the applications.

Sensor interface. When choosing a sensor for a biometric application, this interface of a sensor or device with a processing unit has to be taken into consideration. In the case where the quantity of data the sensor has to transfer to a processing unit is large, the task of the interface is to transfer the corresponding data very fast. In this way, long latency will be avoided.

Power supply. Cameras and microphones as basic sensors do not need too much power. On the other hand, illuminators such as emission diode or optical fibers, line scanning cameras, and heating generators require extra power. Low power absorption is an important feature because it facilitates sensors to be embedded in devices. Communication interfaces such as USB, Ethernet, etc. can supply power to the sensor without extra wires.

Failure rate. Mean time between failures (MTBF) is the average time between failures of an observed system without including its end of life. Calculations of MTBF assume that a system is fixed after each failure.

Cost. The costs of biometric sensor or device depend on the availability of the technology used, the production material, the manufactured number of samples, and maintenance.

Sensor resolution. Increasing the resolution increases the accuracy of the sensor and cost. By sensor resolution, we mean the ability of a device to obtain, scan, or distinguish details of the biometric treat. As for the sensor resolution, it can be of the following types: spatial, frequency, time, and radial. Spatial resolution, expressed in pixel-per inch, represents the number of pixels in unit length. Frequency resolution is the measure of ability of a device to resolve frequency variations. Time resolution is determined by a sensor's ability to distinguish time variations, while the radial resolution gives rise to the possibility of a sensor to distinguish variation in the distance.

Image depth or dynamic range. It shows how a sensor can represent differences of intensity usually expressed as the number of gray levels or bits. As an example, 8 bits or 256 gray levels is a typical dynamic range of fingerprint image. Typical of the face image are 24 bits or 256 red, green, and blue (RGB) levels.

Modulation transfer function (MTF). Spatial-frequency response gives the relationship between the input and the output signal of a sensor and measures the ability of sensor to resolve detail in the input image. For the more extended response, the image will be sharper.

Geometric image accuracy. Let X be the distance measured between any two points on the input image and Y the distance measured between those same two points on the output image. Then geometric image accuracy denotes the absolute value of the difference $D = X - Y$.

Field of view. This parameter represents the capacity of a sensor to capture the whole biometrics in a single image. For example, in the case of digital camera, this is angular extent of the observable object that is seen at any given moment.

Depth of field (DoF). The range of distance in which the object remains focused is presented by DoF. It shows the location in which biometrics must be placed to remain always focused.

Intensity linearity. This parameter refers to the capacity of a device to reproduce correctly the intensity level. In order to measure the accuracy, the grayscale levels on the output image are compared with the grayscale levels on the input. Large variations lead to the conclusion that the sensor is calibrated.

Signal-to-noise ratio. This ratio is a measure that compares the level of the background noise introduced by the sensor during the biometric capture.

Frame rate. This parameter must be taken into account when the object's movements are involved during the biometric capture (e.g., in touchless devices, face recognition devices). In fact, this is the number of frames per time unit that a sensor can generate and is measured in frames/s.

Optics. Here, f-number (focal ratio or relative aperture) is the dimensionless number that represents the quantitative measure of lens speed, which is a concept used in photography. In fact, in an optical system, it represents the focal length divided by the aperture diameter.

To conclude, the choice of a sensor will depend on many characteristics: electrical, ergonomic, mechanical, optical (Clark and Yulle, 2009).

10.3 Internet of Biometric Things

While invoking IoT services, integration of cloud computing with the IoT architecture is a necessary process. In the open literature, this is referred to as cloud-centric IoT (Ren, Gong, Hao, Cai, and Wu, 2016). The application of biometrics for access control in cloud computing has been proposed as a solution in the literature (Huang, Shi, Xian, and Liu, 2013; Sharma and Balasubramanian, 2014). With cloud-centric IoT to manage access control to the object via biometrics, it is possible to introduce the so-called Internet of Biometric Things (IoBT) (Katarci, Kantarci, and Schuckers, 2015). The conception is demonstrated in Figure 10.1.

Biometric data are gathered using IoT devices and then transmitted to the cloud where liveness detection runs for identification and authentication (Johnson, Hua, and Schuckers, 2011). The integration of cloud computing leads to web-based interfaces, scalable on-demand and cost-efficient access to computing and store capacity, inter-operability between the sensing objects, and finally security assurance. This presents a basis for deployment, development, and management for the IoBT applications over the cloud (Zhou et al., 2013). Biometric context-aware identification components in IoBT are shown in Figure 10.2.

FIGURE 10.1
Internet of Biometric Things concept.

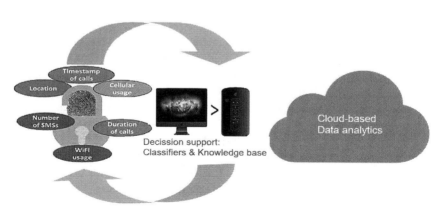

FIGURE 10.2
Cloud-centric IoBT architecture.

This architecture relies on biometric and behaviometric identification together and is called cloud-centric IoBT. As identification and authentication are in the cloud, this fact will provide cloud-based data analytics as a service. In that way, data analytics will offer distributed components for the case of biometrics classification and context analysis. The platform consists of biometrics classifiers, context classifiers, and knowledge bases. The output of the back-end computer undertakes the decision-making process which authenticates the user who carries the IoBT object.

Security of biometric information is another concern to deal with. Biometric treats are leveraging the cloud not only for storage, but also for performance and scalability features. Thus, it is important to transfer them securely from the client to the remote server. Biometric security treat is an approach of active attack against vulnerability in a biometric system. Threats can be classified as presentation attacks, biometric processing attacks, and social and presentation attacks.

During presentation attacks, the appearance of the biometric sample is changed or replaced. Biometry processing attacks are characterized by the fact that an algorithm is used to cause incorrect processing and decisions. Finally, in social and presentation attacks, the authorities using the systems are fooled.

10.3.1 IoBT Devices

Each of the IoBT devices has a subset of multi-modal biometric authentication techniques. It should be noted that the context of each device can be different. On the other hand, the framework has to be expanded to take into account the needs of all applied devices that can be identified using biometrics and behaviometrics. As for location, movement, incoming and outgoing data rate, traffic characteristics, they are defined by a more generalized feature set. For their context analysis, device-specific features can be included. Most systems using liveness detection can be strengthened against physical attacks. Starting from the point that local identification via biometrics cannot provide continuous authentication, cloud-based data analytics is obliged in the case of seamlessly connected IoBT devices.

10.3.2 Examples of Cloud-Centric Benefits

Cloud-centric secure IoBT takes into account the multi-modal biometrics and liveness detection. Improvement is provided by introducing behaviometrics. Behaviometric identification and authentication have been studied and were proposed a few years ago. For example in Shabtai, Kanonov, and Elovici (2010) the knowledge-based temporal abstraction is proposed in order to keep track of SMSs and software updates sent to date targeting mobile devices. As another example of a benefit, authentication through learning user behavior is proposed, too (Shi, Niu, Jakobsson, and Chow, 2010). False acceptance ratios (FAR) indicates the consequences of combining biometric and behaviometric systems. While FAR quantifies zero effort, imposter-case attacks, such as physical attacks and/or replay attacks, lead to *non-zero* fall. In Sousedik and Busch (2013), liveness detection techniques have been proposed for mitigating attacks. Local identification, on the other hand, cannot provide continuous authentication. Therefore, cloud-based data analytics is welcome in case of seamlessly connected IoBT devices (Figure 10.3). It means that as the context is determined by a large feature set, spatio-temporal knowledge as well as context classification should be accelerated by offloading the mobile device and, at the same time, requesting data analysis service using cloud platform.

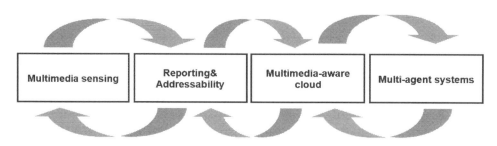

FIGURE 10.3
Block scheme of an IoMT architecture.

10.3.3 Behavioral Analytics Based on Multimedia Content

Behavioral analytics have many applications. Usually, behavior in the video streams is automatically detected, and at the same time, warnings are issued. The same happens in the case of disallowed dimensions or traveling in the wrong detection. Also, videos are integrated with other sensors in wireless network (Akpakwu, Silva, Hancke, and Abu-Mahfouz, 2017) for the processes of tracing and tracking, traffic management, and evaluation of traffic patterns.

IoT has got a strong influence on the surveillance and security systems. Some of the features significant for the superior security are object detection, face recognition, theft detection and are important for a high level of security. All the events are identified when streaming multimedia-video content in a control center. On the other hand, in multimedia networks, the behavioral analytics is carried out at camera nodes in order to show the event (Alvi, Afzal, and Shah, 2015).

10.4 Internet of Multimedia Things

Internet of Multimedia Things (IoMT) belongs to a special part of IoT integrating and cooperating with heterogeneous multimedia devices, and it is characterized by distinct sensing and computational and communication capabilities including resources. A multimedia cloud can help in designing multimedia services and applications. Multimedia content has evident characteristics when compared to the scalar data obtained by IoT devices. However, multimedia transmission requires much more bandwidth compared to the conventional scalar data traffic in IoT. Introducing multimedia applications, such as real-time security/monitoring systems in smart homes, intelligent surveillance systems deployed in smart cities, transportation management using smart video cameras, remote monitoring of an ecological system, etc., requires higher-processing and higher-memory resources. Introducing IoT systems with multimedia devices and content requires additional functionalities as well as the revision of existing ones. In this way, it is possible to have a special set of IoT called IoMT. Starting from the point of vision, it can be concluded that IoMT is the network of interconnected multimedia things which are identifiable and addressable to acquire sensed multimedia data. At the same time, there exist multimedia devices and services, with and without direct human intervention.

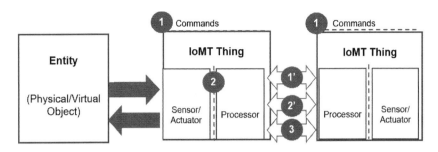

FIGURE 10.4
Scope of MPEG–IoMT standardization.

10.4.1 Multimedia Communication in IoMT

The function of IoMT-based service can be presented using IoMT architecture shown in Figure 10.4. It is composed of the following blocks: multimedia sensing, reporting and addressability, multimedia-aware cloud, and multiagent systems. Each of these blocks is characterized by the fact that it includes processes which have to be executed at each step of a service including various aspects, constraints, and challenges.

Multimedia sensing demands processing as well as continuous data acquisition and results in higher energy consumption. It is expected to operate IoMT devices on batteries. These may not last longer due to the nature of multimedia data. That is the reason why efficient energy procedures need to be applied in energy sensor prolonging the network lifetime.

The next step is to report the multimedia content to the cloud. This is realized by incorporating efficient communication and addressing techniques. That is why the multimedia content is stored, processed, and disseminated as per end user demand at the cloud.

Multimedia-aware cloud provides services for huge data storage and distributed processing services to the end user. This is the case when the cloud is a computer providing access to the multimedia content from sensor networks (Zhu, Luo, Wang, and Li, 2011; He, Wu, Zeng, Hei, and Wen, 2013). Multimedia-aware cloud incorporates heterogeneity at various levels including devices, networks, applications, and services. The users can be positioned to access sensor data from remote devices in sensing as a service. Also, rule engines can be implemented to control the actuator's operation automatically from the cloud in sensing and actuation as a service providing control to identity and policy management systems.

A multiagent system is composed of many independent cooperating and communicating agents responding to some events in accordance with the user demands. The role of the agents in such a system is to organize, optimize, and configure themselves by interacting with each other. Multiagents are suitable software components for handling the cloud functionalities.

10.4.2 IoMT Standardization Process

The MPEG (Moving Picture Experts Group) is an ISO/IEC standardization working group for audio/video (A/V) compression and transmission. MPEG IoMT standard specifies interfaces, protocols, and associated formats that enable advanced services based on human-to-device and device-to-device interaction (MPEG, 2016). The heterogeneity of devices as part of complex distributed systems makes it difficult to standardize descriptions, data formats, and API (Application Programming Interface) for discovering devices and enabling communication between them. However, in well-established ecosystems like IoBT, this can be done.

IoMT information refers to data sensed and processed by a device and/or communicated to a human or another device. *Thing* is anything that can communicate with other things; in addition, it may sense and/or act on any physical or virtual object. The *Media Thing* is a thing with at least one of the A/V sensing and actuating capabilities.

The objective of IoMT standard is to define interfaces 1, 1′, 2, 2′, and 3:

Interface 1 User commands (setup info.) between a system manager and an MThing

Interface 1′ User commands (setup info.) forwarded by an MThing to another MThing, possibly in a modified form (e.g., subset of 1)

Interface 2 Sensed data (raw or processed data) (compressed or semantic extraction) and actuation information

Interface 2′ Wrapped interface 2 (e.g., for transmission)

Interface 3 MThing characteristics, discovery.

Within this large field of IoTs, the IoMT applications and services that offer the provision, interpretation, representation, and analysis of multimedia content collected from IoMT devices (cameras, microphones) are challenging yet appealing to users. IoMT is a particular case of IoT (that by definition has the communication capability, and it may sense or act on a physical or virtual object), with the specificity that an IoMT has A/V capabilities. IoMT applications and services can be designed and implemented if a variety of technologies are available such as information-centric networking, media analytics, cloud technologies, big data for media content, content streaming, and caching. Orchestration and synchronization between these components is an absolute requirement, and this can be done in a sustainable manner only by developing open standards. The objective of the MPEG standard is to contribute with major technologies in the field of data representation and APIs for IoMTs.

The MPEG IoMT standard covers five aspects addressed in the following individual parts:

Part 1. Global architecture, use cases, and common requirements

Part 2. System aspects (discovery, communication)

Part 3. Individual IoMTs, data formats, and APIs (sensors & actuators, processing, & storage)

Part 4. Wearable IoMTs, data formats, and APIs (smart glasses, camera, gesture recognizer, microphone, display, smart watch, fitness tracker, smart textile, sleep monitoring)

Part 5. IoMT aggregates combining individual IoMTs, eventually with other IoTs (video surveillance systems, car sensing and communication system, smart cities).

MPEG produced working documents of use cases (June 2016) and requirements (October 2016) on IoMT. The IP surveillance camera (MThing Camera) captures A/V data and sends them to both a storage (MThing Storage) and a human tracker unit (MThing Processing Unit). When the human tracker detects a person in the visible area, it traces the person and extracts the moving trajectory. If the person gets out of the visual scope of the first IP camera, another IP camera in the vicinity takes over the control and keeps capturing A/V data of the corresponding person. The specific descriptors (e.g., moving trajectory information, appearance information, media locations of detected moments) can be extracted and sent to a storage. In this use case, the human tracker module can control the activation and deactivation of cameras in the area. The user (e.g., a system designer) can set up either all the MThings in the system (centralized manner) or only an MThing camera that can exchange necessary information with other MThings to achieve the mission (in a distributed manner).

As a first stage of IoMT standardization, MPEG issued a call for proposals in compliance with the requirements defined in the working documents (October 2016). Based on responses and cooperative work, MPEG issued in January 2018 the committee draft (CD) for the IoMT. This standard specifies APIs providing IoMT (cameras/displays and microphones/loudspeakers, possibly capable of significant processing power) with the capability of being discovered, setting up ad-hoc communication protocols, exposing usage conditions, and providing media and metadata as well as services processing them. IoMT APIs cover a large variety of sophisticated devices such as smart glasses, image/speech analyzers, gesture recognizers, and biometric devices.

10.5 Conclusions

Biometric technology reveals a new significant period based on IoT. The benefits of incorporating IoT are low cost, low space occupancy, low power consumption as well as portability of the system. IoT is a dynamic and distributed network system with a number of smart objects to produce information. The related technologies are cloud computing, big data, and so on. The IoT allows objects to be sensed and controlled remotely across existing and future network infrastructure.

Research in the field of IoBT is quite interdisciplinary. The final goal is to provide a new class of advanced applications and services. The success of the IoBT depends strongly on the existence and effective operation of global standards. The CD of IoMT technical specification was a great achievement for the MPEG working group. With this framework, MPEG facilitate the large-scale deployment of distributed biometric systems with interoperable A/V data and metadata exchange.

References

Akpakwu, G.A., Silva, B.J., Hancke, G.P., Abu-Mahfouz, A.M. (2017). A survey on 5G networks for the Internet of Things: Communication technologies and challenges. *IEEE Access, vol. 6*, 3619–3647.

Alvi, S., Afzal, B., Shah, G. (2015). Internet of multimedia things: Vision and challenges. *Ad Hoc Networking, vol. 33*, 87–111.

Bhattacharyya, D. et al. (2009). Biometric authentication: A review, *International Journal of eService, Science and Technology, vol. 2, no. 3*, 13–28.

Bojkovic, Z., Bakmaz, B., Bakmaz, M. (2017). The IoT vision from an opportunistic networking perspective. In *The Internet of Things: Foundation for Smart Cities, eHealth, and Ubiquitous Computing* (Eds. R. Armentano, R.S. Bhadoria, P. Chatterjee, G.C. Deka), pp. 1–18, CRC Press, Boca Raton, FL.

Clark, J.J., Yulle, A.L. (2009). *Data Fusion for Sensory Information Processing Systems*, Kluwer Academic, Boston, MA.

Fisher, B.A.J., Fisher, D.R. (2012). *Techniques of Crime Scene Investigation*, CRC Press, Boca Raton, FL.

He, J., Wu, D., Zeng, Y., Hei, X., Wen, Y. (2013). Toward optimal deployment of cloud-assisted video distribution services. *IEEE Transactions on Circuits and Systems for Video Technology, vol. 23, no. 10*, 1717–1728.

Huang, K., Shi, J., Xian, M., Liu, J. (2013). Achieving robust biometric based access control mechanisms for cloud computing. In *Proceedings of the International Conference on Information and Network Security ICINS*, Beijing, China.

Huynh, C. et al. (2015). Forensic identification of gender from fingerprints. *Analytical Chemistry, vol. 87, no. 22*, 11531–11536.

Jain, A.K. (2007). Biometrics recognition. *Nature, vol. 449*, 38–40.

Jain, A.K., Nandakumar, K., Nagar, A. (2007). Biometric template security, *EURASIP Journal of Advances in Signal Processing, vol. 2008*, Article ID 579416, 1–17.

Johnson, P.A., Hua, F., Schuckers, S. (2011). Comparison of quality-based fusion of face and iris biometrics. In *Proceedings of the IEEE International Joint conference on Biometrics IJCB*, Washington, DC.

Katarci, B., Kantarci, M.E, Schuckers, S. (2015). Towards secure cloud-centric Internet of Biometric Things. In *Proceedings of the IEEE International conference Cloud Networking*, Niagara Falls, ON.

MPEG (2016). Use cases for Internet of Media-Things and Wearables, Doc. N16345, Geneva.

Peer, P., Bule, J. (2013). Building cloud-based biometric services. *Informatica, vol. 37*, 115–122.

Raghavendra, R., Busch, C., Yang, B. (2013). Scaling-robust fingerprint verification with smartphone camera in real-life scenarios. In *Proceedings of the IEEE International Conference on Biometrics: Theory, Applications and Systems BTAS*, Washington, DC.

Ren, C., Gong, Y., Hao, F., Cai, X., Wu, Y. (2016). When biometrics meet IoT: A survey. In *Proceedings of the International Asia Conference on Industrial Engineering and Management Innovation*, 635–643, Bali, Indonesia.

Sayeemuddin, S.M., Pithadia, P.V., Vandra, D. (2014). A simple and novel fingerprint image segmentation algorithm. In *Proceedings of the IEEE International conference on Issues and Challenges in Intelligent Computing Techniques*, 761–764, Ghaziabad, India.

Shabtai, A., Kanonov, U., Elovici, Y. (2010). Intrusion detection for mobile devices using the knowledge-based temporal abstraction method, *Journal of Systems and Software, vol. 83, no. 8*, 1524–1537.

Shah, C.M., Sangoi, V.B., Visharia, R.M. (2014). Smart security solutions based on Internet of Things. *International Journal of Current Engineering Technology, vol. 4, no. 5*, 3401–3404.

Sharma, S., Balasubramanian, V. (2014). A biometric based authentication and encryption framework for sensor health data in cloud. In *Proceedings of the International Conference on Technology and Multimedia*, 49–54, Putrajaya, Malaysia.

Shi, E., Niu, Y., Jakobsson, M., Chow, R. (2010). Implicit authentication through learning user behavior. In *International Conference on Information Security ISC*, 99–113, Boca Raton, FL.

Sousedik, C., Busch, C. (2013). Presentation attack detection methods for fingerprint recognition systems: A survey. *IET Biometrics, vol. 3, no. 4*, 219–233.

Zhang, D. (2014). *Palmprint Authentication*, Kluwer Academic Publishers, Norwell, MA.

Zhou, J. et al. (2013). Cloud things: A common architecture for integrating the Internet of Things with cloud computing. In *Proceedings of the IEEE International Conference on Computer Supported Cooperative Work in Design CSWD*, 651–657, Whistler, BC.

Zhu, W., Luo, C., Wang, J., Li, S. (2011). Multimedia cloud computing. *IEEE Signal Processing Magazine, vol. 28, no. 3*, 59–69.

Part III

The Biometric Computing – Futuristic Research & Case Studies

11

Deep Neural Networks for Biometric Identification Based on Non-Intrusive ECG Acquisitions

João Ribeiro Pinto and Jaime S. Cardoso

University of Porto

INESC TEC

André Lourenço

CardioID Technologies, Instituto Superior de Engenharia de Lisboa (ISEL)

Instituto de Telecomunicações (IT)

CONTENTS

11.1 Introduction

The benefits of biometric systems are well known: they avoid the risk of loss, copying, or theft of credentials by using the user itself as the credential and make the user authenticable at all times, without the need to carry physical credentials or remember codes or passwords (Jain et al. 2011, Kaur et al. 2014). The merit of biometrics has been increasingly recognized in all fields of industry, as biometrics quickly replace traditional authentication methods in smartphones, computers, building entrances, and airports.

Biometric recognition is largely dominated by four traits: face, iris, fingerprints, and voice. These, in 2015, represented 75% of the biometrics market (Mani and Nadeski 2015), and their state of development is significantly more advanced than other traits such as retinal scans, signature, gait, or keystroke (Adeoye 2010, Chauhan et al. 2010). Nevertheless, the techniques for circumvention of such systems, based on recordings of the traits, are constantly evolving (Belgacem et al. 2012, Fratini et al. 2015) and require the system to become more sophisticated to ensure the liveness of the acquired trait.

Medical or physiological biometrics, such as the electrocardiogram (ECG), have inherent liveness information. The ECG is the most promising, on par with the most common traits (Kaur et al. 2014), due to its combination of universality, permanence, and measurability (Li and Narayanan 2010, Agrafioti et al. 2012), along with increased acceptability and comfort when acquired on non-intrusive off-the-person settings, on the hands of the subjects using ungelled electrodes. Furthermore, there are very few and recent circumvention techniques for ECG biometrics (Eberz et al. 2017), and this presents many constraints (like continued contact with both hands of the subject) that reduce the likelihood of success.

The algorithms for biometric recognition based on electrocardiographic signals have the goal to output an identity or match score, based on a given short segment of ECG and the data stored by the biometric system (Bolle et al. 2004, Jain et al. 2011). The existing algorithms can generally be divided into four stages, according to their purpose, as illustrated in Figure 11.1.

First, we have the denoizing stage that generally consists of Butterworth filters, line-fitting smoothing procedures, and other preprocessing methods that aim toward the attenuation of noise in the received signal. Then, we have the preparation stage, where minor preparations of the denoised signal are applied to ease the extraction of meaningful features. These preparations usually include fiducial detection (the most frequent is R-peak localization), heartbeat segmentation, time and amplitude normalization, and outlier removal (Odinaka et al. 2010, Lourenço et al. 2011, Pinto et al. 2017). The processes included in denoizing and signal preparation have also usually been grouped in one larger stage, designated as signal preprocessing stage (Bolle et al. 2004, Jain et al. 2011). However, as the field evolves toward more comfortable and acceptable acquisitions, the importance and complexity of such processes increase to face the growing influence of noise and variability.

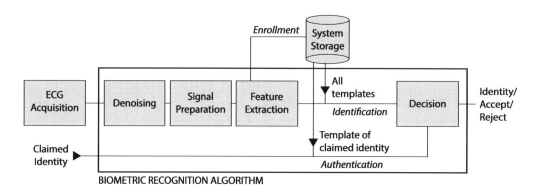

FIGURE 11.1
Schema of the typical structure of an ECG-based biometric system, with emphasis on the four stages of an ECG biometric recognition algorithm.

The third stage is feature extraction that aims to enhance the information that is linked with the identity and individual characteristics of each individual and remove useless information that can harm the process of recognition. The fourth and last stage is decision, where a classifier can be used to compare all stored data templates to the received signal and return an identity (identification) or to compare the received signal with the claimed identity's template and, based on a measure of similarity or dissimilarity, return a decision of acceptance or rejection (authentication).

Deep learning has many times shown its potential to integrate different functionalities into a single model (LeCun et al. 2015). The flexibility of the convolutional and fully connected layers, together with the techniques of regularization and data augmentation, enables it to autonomously learn the most fitting features for the classification task at hand, while keeping the ability to generalize and be robust against high variability and noise dominance over the signal (LeCun et al. 2015, Zhang et al. 2017). The integration of all the often separate stages into a single, unified model means that the optimization is not individual and partitioned but, instead, holistic and synergic: all stages are optimized as a whole toward the best model possible.

Here, we propose a novel algorithm for biometric recognition based on convolutional neural networks (CNNs). The proposed method was evaluated on the most complete and challenging off-the-person collection currently available and replaces all the traditional pipeline stages from denoizing to decision. Furthermore, we studied how the progressive integration of traditional stages on a CNN can benefit the process of biometric identification using ECG signals acquired in non-intrusive settings, and the algorithm previously proposed in Pinto et al. (2017) was improved and adapted to be used as the baseline algorithm for direct comparison.

Besides this introduction, this chapter presents a brief but comprehensive overview of the state of the art in ECG-based biometrics and some recent applications of deep learning in biometrics with signal acquisitions, in Section 11.2; a description of the proposed CNN architecture and its constituent parts and regularization and data augmentation techniques, in Section 11.3; the baseline algorithm adapted from Pinto et al. (2017), in Section 11.4; the results of the evaluation of the proposed architecture, the study of progressive integration of pipeline stages, and the comparison with the baseline and state-of-the-art algorithms, in Section 11.5; and final remarks and conclusions on the work performed and results obtained, in Section 11.6.

11.2 State-of-the-Art Overview

The research topic of ECG-based biometrics is fairly recent. Despite being prophesied by Forsen et al. (1977), the first studies specifically aiming to use the ECG for individual recognition were only published by the start of the millennium (Biel et al. 1999, Kyoso et al. 2000), after important studies on the inter- and intrasubject variability of the signal (Hoekema et al. 1999, Schijvenaars 2000). On the other hand, deep learning methodologies, specifically CNNs, were first used by LeCun et al. (1998) but have become trendy only in the last few years, due to exceptional computational requirements that could only be efficiently met with recent technology. Below, we present a more detailed overview of both the state of the art in ECG biometrics and the use of deep learning for biometrics and signal analysis.

11.2.1 ECG-Based Biometrics

One of the most successful algorithms is that of Plataniotis et al. (2006) and Agrafioti and Hatzinakos (2008) that extracted normalized autocorrelation features from fixed-length ECG segments and applied dimensionality reduction with linear discriminant analysis. Without requiring fiducial detection, they used a nearest-neighbor classifier for classification, achieving 100% identification rate (IDR) in a dataset with 27 subjects. The algorithm was then adapted for continuous identification by Matta et al. (2011). Odinaka et al. (2010) used Log-STFT (Short-Time Fourier Transform) spectrograms of heartbeat segments to train Gaussian Mixture Models, and log-likelihood maximization was used to output an identity, attaining 99% IDR (or accuracy) in a private dataset with 269 subjects.

Wang et al. (2013) used representation elements from max pooling of sparse coding coefficients of ECG segments without denoizing, fed to a nearest-neighbor classifier, obtaining 99.5% IDR with 100 subjects of the public Physicalisch-Technische Bundesanstalt (PTB) database. Brás and Pinho (2015) also evaluated their algorithm on the PTB database. They proposed feature extraction using Kolmogorov-based normalized relative compression, from unsegmented signals denoised using a combination of moving average, notch, and low-pass filters, and reported 99.9% IDR with 52 subjects.

More recently, Carreiras et al. (2016) used heartbeats segmented after denoizing with a 5–20 Hz bandpass filter and obtained 84.4% IDR with 618 subjects using a nearest-neighbor classifier. Tan and Perkowski (2017) fed temporal, amplitude, and angle fiducials into a Random Forest (RF) classifier, used Discrete Wavelet Transform (DWT) coefficients as features on a nearest-neighbor classifier using wavelet distance, and fused the two classifiers at the score level. It resulted in 99.5% IDR with recordings of 184 subjects gathered from several public datasets.

The performance results reported by these works is quite encouraging, with identification rates equal or very near 100%. Nevertheless, it is important to recall the nature of the signal acquisitions used: these early initiatives focused mainly on on-the-person ECG acquisitions: intrusive recordings using several gelled electrodes, characterized by high signal quality and low influence of noise.

The performance degrades when we consider more acceptable and comfortable acquisitions, on non-intrusive off-the-person settings. These are characterized by the use of few ungelled electrodes placed on the hands or fingers, with increased influence of noise due to free movements and variability due to unstable electrode placement and contact with the skin.

Using off-the-person signals, Lourenço et al. (2011) proposed the use of averaged heartbeats, passed through processes of filtering and time and amplitude normalization. Using a nearest-neighbor classifier in identification tasks, the method attained 94.3% IDR with 16 subjects. More recently, Pinto et al. (2017) selected Discrete Cosine Transform (DCT) and Haar coefficients from an ensemble of average heartbeats, extracted from 5-s segments denoised by a combination of Savitzky–Golay filter with a moving average filter. With data acquired from six drivers, during normal unconstrained driving activity using a conductive leather steering wheel cover, the method attained 94.9% IDR.

Wieclaw et al (2017) proposed the use of a multilayer perceptron (MLP) on individual heartbeats extracted from segments denoised using a bandpass filter, including methods for the rejection of noisier heartbeats, achieving 88.97% IDR with ECG signals acquired in off-the-person settings from 18 subjects. These last three works illustrate the significantly negative effect of higher noise and variability of off-the-person acquisitions on the performance of ECG biometric algorithms, despite the relatively small number of subjects in the datasets used.

As for the trending deep learning methodologies, few have yet ventured into their application on ECG biometrics, and they are yet to be explored to the fullest extent of their potential. Zhang et al. (2017) aimed to replace only the classification stage with a CNN that separately receives and processes (on the convolutional layers) the autocorrelation of the approximation and detail coefficient sets of the wavelet transform of 2-s ECG segments, after component selection. The information on those levels was unified only at the classification level (on the fully connected layers) and resulted in an average 93.5% accuracy for separate datasets with 18–47 subjects. Eduardo et al. (2017) used autoencoders to replace only the denoising and feature extraction stages of the biometric system. The autoencoders learned lower-dimensional representations of heartbeats, used on a k-nearest-neighbors (kNN) classifier, which, for a dataset with 706 subjects in medical settings, rendered 0.91% identification error.

Salloum and Kuo (2017) used Recurrent Neural Networks (RNNs) with Long Short-Term Memory (LSTM) and Gated Recurrent Units (GRUs) with a sequence of segmented heartbeats and obtained 100% IDR with 90 subjects of the public ECG-ID database. At last, Luz et al. (2019) proposed the substitution of the feature extraction and classification stages, through the fusion of two separate convolutional networks at the score level, one receiving segmented heartbeats and the other receiving the heartbeats' spectrograms. Tested for authentication tasks, their method achieved 14.27% Equal Error Rate (EER) on the University of Toronto ECG Database (UofTDB).

Unlike these recent works, our aim is to more completely take advantage of the potential of deep learning: to integrate all stages (denoising, preparation, feature extraction, and classification) of processing into a CNN, to use it for biometric identification purposes, and to achieve competitive performance on the largest and most complete public dataset of off-the-person signals acquired from 1,019 individuals.

11.2.2 Deep Learning for Signals and Biometrics

Besides the recent efforts toward deep ECG-based biometric systems, deep learning has been extensively explored for biometric applications with other traits and for various other purposes. In the realm of biometric recognition, Taigman et al. (2014) have proposed the well-known DeepFace algorithm for biometric face authentication. Trained with a large dataset with more than 4,000 identities, the algorithm was able to learn representations that proved robust to unconstrained settings, even when transferred to different datasets. Raja et al. (2015) explored the extraction of deep sparse filtering features that proved robust for biometric recognition based on unconstrained smartphone iris photographs. Nogueira et al. (2016) proposed the use of a CNN for liveness detection (the important process of determining if the measured trait is real or fake) in biometric fingerprint recognition and were awarded for the high accuracy achieved.

Deep learning has also been used with signal acquisitions for diverse purposes, resulting in significant performance benefits. Rajpurkar et al. (2017) proposed a 34-layer CNN for detection of several types of arrhythmias on single-lead ECG acquisitions and reported performance superior to that of professional cardiologists. Hannun et al. (2014) aimed to discard overly engineered processing pipelines for speech recognition and instead use a much simpler RNN that proved robust against common problems like speaker variation and background noise, without needing to be personally tuned for that purpose. Um et al. (2017) explored CNNs for classification of the motor state of Parkinson's disease patients, along with several unidimensional data augmentation techniques to work around dataset size limitations, predominant noise, and high variability, which resulted in significant performance improvement.

With these and many other examples, deep learning has impressed, not only by the outstanding results it achieved, but also by its robustness against noise and variability and its adaptability to a very broad range of tasks and their specificities. Thus, deep learning presents itself as a very promising alternative to common pattern recognition, image and signal processing, and computer vision techniques applied in biometric recognition. For ECG-based biometrics, it could be the way to overcome current limitations and increase competitiveness between ECG and more advanced traits like the face and fingerprints.

11.3 A CNN for ECG Biometrics

11.3.1 General Structure Overview

The proposed methodology follows the typical structure of a CNN: the first layers are convolutional, along with max pooling, to allow the network to learn the most advantageous representation of the input signal segment for classification, quickly reduce the dimensionality to keep the number of parameters low, and thus control the computational cost and training time.

The proposed CNN (cf. Figure 11.2) was developed to integrate all common pipeline stages into a single, end-to-end model, receiving raw 5-s ECG segments and delivering the corresponding identity. In the following subsections, each part of the convolutional networks and its purpose is presented, along with the baseline algorithm used for comparison and to complement them in some settings.

11.3.2 Convolutional and Pooling Layers

The convolutional and pooling layers compose the first part of a CNN (LeCun et al. 2015). Each convolutional layer (cf. Figure 11.3) is composed of feature maps that are connected to the previous layer or input by sets of weights called filter banks. The filter banks are used for discrete convolution to obtain higher representations of the inputs

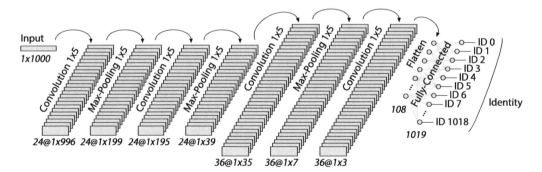

FIGURE 11.2

Architecture of the proposed CNN model. (The number of neurons on the fully connected layer refer to the entire dataset with 1,019 possible identities.)

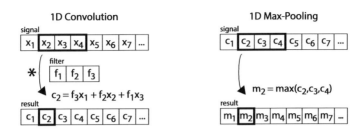

FIGURE 11.3
Illustration of the behavior of convolutional and max-pooling layers in unidimensional CNNs.

at each convolutional layer (Zhang et al. 2017). Each unit of the feature map of a convolutional layer will correspond to a local patch of the previous layer or input, and these local weighted sums are passed through an activation function (LeCun et al. 2015). The dimensions of the local patches and the feature maps can be controlled by tuning the size of the filters and the stride of the convolution and by using pooling layers. In the proposed architecture, the activation function used was the Rectified Linear Unit (ReLU) (Krizhevsky et al. 2012); the size of the filters was set at 5, and the stride was set at 1, through empirical tuning.

Pooling layers (cf. Figure 11.3) have the main goal of reducing the dimension of the feature maps (and thus avoid unnecessary or harmful information and control the computational costs), by merging similar features into one (LeCun et al. 2015). In the proposed CNN, max pooling was used, and it works by scanning along each feature map and keeping only the maximum value in each local neighborhood considered. Here, the pooling stride was set at 1, and the pool size was set at 5.

11.3.3 Fully Connected Layers

The feature maps output by the last convolutional layer are concatenated (flattened) into a single unidimensional vector of features that serve as input to the fully connected layers, the classification structure of the CNN. The fully connected layers act as a multilayer perceptron, where each neuron is a function of the outputs of the neurons of the previous layer and the weights of the respective connections (Rumelhart et al. 1986), which means the features will be appropriately weighted and combined at each neuron, for the classifier to output the expected scores.

In the proposed architecture, a single fully connected layer is used, composed of N neurons (where N is the number of subjects), with softmax activations to output a normalized distribution over the identity labels (Taigman et al. 2014). The neuron that outputs the highest value will correspond to the predicted identity.

11.3.4 Optimization and Regularization

11.3.4.1 Optimizer and Loss

At each moment of the training phase, based on a batch of train samples fed to the network, a measure of loss is computed by comparing the output of the network with the true labels of the batch. Based on that loss, the weights/parameters that compose the neural network are adjusted to reduce the loss, through error backpropagation (Rumelhart et al. 1986, LeCun et al. 1998).

An optimizer is a function that controls the way the weights are updated. In this work, the optimizer Adam was used. Adam is a first-order gradient-based optimization method for stochastic functions and has been widely used for the simplicity of implementation, hyper-parameter tuning, and effectiveness in diverse tasks (Kingma and Ba 2015). The learning rate was empirically adjusted to each situation, inside the range [0.01, 0.001], without decay, in order to allow for a quick and stable optimization. Being a biometric identification task, a classification task with several labels, the loss measure selected was the sparse categorical cross-entropy to more effectively work with large quantity of labels and a single identity for each object.

11.3.4.2 Dropout

In order to reach the loss minimum on the train set, neural networks will tend to memorize the train samples, i.e., overfit by learning overly specific patterns on the train data that do not correspond with the validation data. Dropouts are used to avoid this situation (Krizhevsky et al. 2012, Srivastava et al. 2014). They are placed between two layers on the convolutional network and act upon the connections between them, setting the corresponding input to zero. In the proposed method, dropouts are used on the connections between the flattened vector of features and the fully connected layer, effectively blocking the access of the classifier to a part of the features and requiring it to become less specific to the training set and more robust to unexpected variability and noise.

11.3.4.3 Data Augmentation

Dropout is effective in keeping the balance between train and test performance and avoiding an overly specific neural network. However, it sometimes falls short, and data augmentation is used to avoid overfitting and attain a more robust classifier. Data augmentation works by applying small transformations/changes to the train samples, while protecting the integrity underlying label of each sample, to simulate larger datasets and ensure the network is robust to such variabilities (Krizhevsky et al. 2012, Chatfield et al. 2014).

Like deep learning in general, data augmentation techniques are significantly more frequent in 2D networks (for images) than in 1D (signals). Nevertheless, based on the recent work of Um et al. (2017), and taking into account the unique characteristics of the electrocardiographic signals, we propose and explore seven different types of data augmentation for 1D CNNs, presented below and illustrated in Figure 11.4.

- *Baseline wander* – Based on the typical ECG noise with the same name, this technique of data augmentation simulates a periodic undulation on the signal, by adding a sinusoidal wave with frequency near 1 Hz.
- *Cropping* – This technique mutates an original train sample by taking a smaller, contiguous subsegment out of it and resampling to match the original length. In the case of ECG signals, this technique effectively simulates slower cardiac frequencies.
- *Flip* – This consists of the inversion of the signal along the time axis: the first sample becomes the last, and similarly for all samples of the segment. This technique causes the inversion of the waveforms P, Q, R, S, and T of the heartbeats and the inversion of their relative locations.

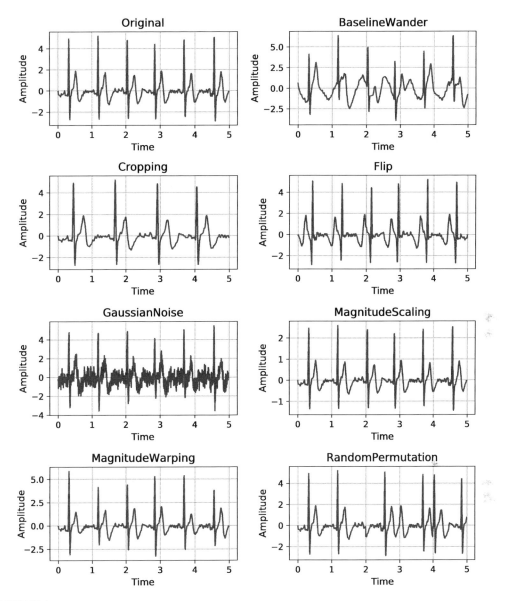

FIGURE 11.4
Illustration of the effects of the different data augmentation techniques on a sample 5-s ECG segment. (For easier visualization, the original segment was denoised with a bandpass filter of 1–30 Hz and had its amplitude z-score normalized.)

- *Gaussian noise* – Gaussian noise (with mean zero and standard deviation about ten times lower than the signal amplitude) is introduced to the signal to cause high-frequency distortions, similar to movement artifacts and powerline interference noise on the ECG signal.

- *Magnitude scaling* – This technique consists of the rescaling of the original train sample by the multiplication of the amplitude by a factor inferior or superior (but close) to 1.

- *Magnitude warping* – Similar to the previous technique, this type of data augmentation rescales the signal in a non-uniform fashion, using a sinusoidal wave instead of a fixed factor, so that certain parts of the signal will have their amplitude shrunk, and others will see their amplitude expanded.

- *Random permutations* – The signal is divided into N contiguous subsegments with similar length, and their order is randomly changed. This may cause discontinuities in the heartbeats and their constituent waveforms, simulating sensor faults or abrupt segment terminations.

These data augmentation techniques were implemented on an online data generator for unidimensional data that applied them, randomly, to the samples of each batch before feeding them to the neural network.

11.4 Baseline Algorithm

Throughout the development of the proposed CNN, an algorithm, adapted and improved from the prior work of Pinto et al. (2017) and Pinto (2017), was used for direct comparison. As a pipeline algorithm, it is composed of stages of denoizing, preparation, feature extraction, and decision that are presented below and in Figure 11.5.

- *Denoizing stage* – A bandpass filter with cutoff frequencies 1 and 30 Hz is used to clean the 5-s ECG segments from both low- and high-frequency noise such as baseline wander (>1 Hz) and powerline interference (60 Hz for the Canadian UofTDB signals).

- *Preparation stage* – The Engelse–Zeelenberg algorithm is used to locate the R-peaks in the 5-s ECG segments. Heartbeat segments are extracted by cropping the original segment 0.25 s before and 0.4 s after the R-peak locations. The heartbeats are then z-score normalized, subtracting the mean amplitude and dividing by the standard deviation. Among the heartbeats extracted from a 5-s segment, the less

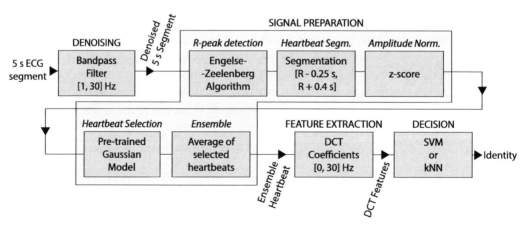

FIGURE 11.5
Schema of the baseline algorithm adapted from Pinto et al. (2017) and Pinto (2017).

noisy are selected using a Gaussian model, previously trained with a selection of cleaner heartbeats from train data of 50 subjects. The selected heartbeats are then averaged to build a single ensemble heartbeat that represents the 5-s segment.

- *Feature extraction stage* – The DCT is applied to each ensemble heartbeat. As most useful information on the ECG signal is included in the frequency range [1, 30], the DCT coefficients that correspond to this range are selected and used as features in the decision stage.

- *Decision stage* – The features are fed to a Support Vector Machine (SVM) classifier, previously deemed the best option, by Pinto et al. (2017), that will output a predicted identity. The second-best classifier, kNN, was also explored.

11.5 Results and Benchmarking

The performance of the proposed CNN architecture, as previously described, was evaluated on electrocardiographic recordings of the UofTDB (Wahabi et al. 2014). This signal collection includes signals acquired from a total of 1,019 subjects, using ungelled electrodes placed on their fingertips. The subjects were measured in five different postures (supine, tripod, exercise, standing, and sitting) on up to six occasions over a time period of 6 months, with a sampling frequency of 200 Hz.

Besides the entire database of 1,019 subjects, two subsets were also used, with 25 and 100 subjects, to evaluate the performance in smaller datasets. The datasets were divided, 70% of the data for training and 30% for testing. Also, besides the proposed method, the baseline algorithm as described here, the state-of-the-art algorithm based on autoencoders proposed by Eduardo et al. (2017), and the algorithm based on AC/LDA features by Matta et al. (2011) were also evaluated in the same conditions. For these, it was occasionally necessary to perform adaptations in order to attend to different sampling frequencies.

The proposed method also suffered slight adaptations to allow the study of the progressive integration of the traditional pipeline stages into the CNN model (cf. Figure 11.6).

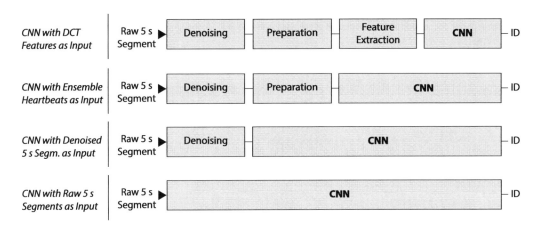

FIGURE 11.6
Illustration of the progressive phases of integration of the traditional pipeline stages into the CNN architecture, as explored in this chapter. For further details on each pipeline stage, cf. Figure 11.5.

Thus, besides the proposed end-to-end version that receives raw 5-s ECG segments, three other variants were evaluated: the first receives denoised 5-s segments (integrating all stages but denoizing in the CNN); the second receives average ensemble heartbeats (integrating the stages of feature extraction and decision); and the third receives DCT features (the CNN replaces only the decision stage). In these cases, the stages not integrated on the CNN correspond to the baseline algorithm (cf. Section 11.4). The pool size of max pooling that was set at 5 for 5-s segments as input was changed to 3 for ensemble heartbeats or 2 for DCT features.

Analyzing the results of the proposed algorithm with DCT features as input (cf. Figure 11.7), it is possible to verify that its performance is similar to that of the baseline algorithm for 25 subjects. However, with the increase of subjects on the dataset (with 100 and 1,019 subjects), the proposed algorithm falls behind. We hypothesize this may be caused by the very concise information that the input carries, which is fitted for typical pipeline algorithms as the baseline but not for the proposed deep learning method. The results of the evaluation without a feature extraction stage, using ensemble heartbeats as input (cf. Figure 11.8), support this hypothesis, as the performance increases and approaches that of the baseline methods, and even surpasses that of the kNN classifier on the two smaller datasets.

Integrating the stages of decision, feature extraction, signal preparation, and even denoizing into the deep learning model allows us to simplify the traditional pipeline structure and use longer signal segments as inputs (in this case, 5 s). This means not only an increase of complexity of the input, which can harm the performance of the CNN, but also an increase of available information and variability, which can allow for a more robust model.

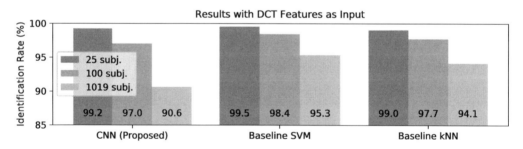

FIGURE 11.7
Results of the proposed and baseline algorithms when using DCT features as input.

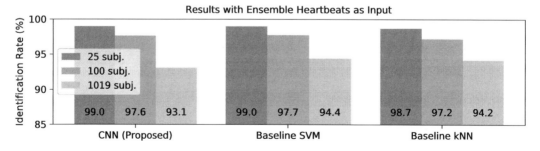

FIGURE 11.8
Results of the proposed and baseline algorithms when using ensemble heartbeats as input.

The results of the use of denoised or raw 5-s segments (cf. Figure 11.9) show the tradeoff between signal complexity and the increase of robustness due to extra information and variability, and the results are similar to those of the CNN receiving ensemble heartbeats. One other interesting observation is that, in general, the results of the CNN with raw segments surpassed those of the CNN with denoised segments, which likely result from the benefit of increased variability.

Increased variability is, in turn, the goal of data augmentation (DA). The aforementioned techniques were separately tested on the datasets of 25 and 100 subjects, and the results are presented in Figure 11.10. As presented on the figure, most of the techniques of data augmentation bring improvements to the algorithm in the form of increased identification rates. The exceptions were magnitude scaling in the 25-subject dataset, cropping and magnitude warping in the 100-subject dataset, and Gaussian noise in both datasets. Knowing that the highest risk of data augmentation is the corruption of the underlying labels, especially sensitive in biometric tasks; this is the likely cause of the performance decay with these techniques.

Analyzing the remaining techniques, the most promising were random permutations (that excelled in both datasets), baseline wander, and flip. These were evaluated in groups, by order of performance, to assess if the combination of two or three techniques would be beneficial to the performance of the algorithm. The results (cf. Figure 11.11) show that the sole use of random permutations is the best option, despite the fact the combinations also caused the improvement of identification performance.

We compared the proposed and baseline algorithms with state-of-the-art algorithms. As they were implemented and tested in the same context, the algorithms of

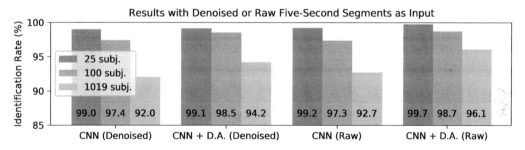

FIGURE 11.9
Results of the proposed and baseline algorithms when using 5-s ECG segments as input raw or after the denoizing stage.

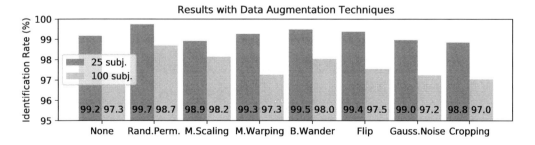

FIGURE 11.10
Results of the proposed algorithm receiving raw 5-s segments as input with each technique of data augmentation on the datasets of 25 and 100 subjects.

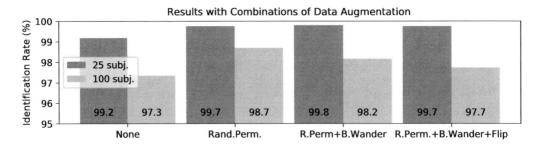

FIGURE 11.11
Results of the proposed algorithm receiving raw 5-s segments as input with combinations of the proposed data augmentation techniques.

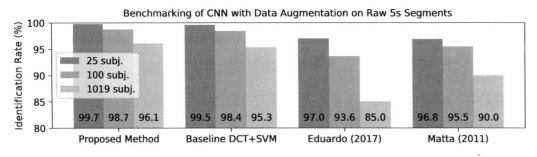

FIGURE 11.12
Direct benchmarking between the proposed architecture (CNN with random permutations as data augmentation and raw 5-s segments as input), the best baseline algorithm, and the two implemented state-of-the-art algorithms.

Eduardo et al. (2017) and Matta et al. (2011) can be used for a direct benchmarking (cf. Figure 11.12). The proposed method presents better results than all alternatives and a slightly slower decay with the increase of the number of subjects, which can denote better scalability to larger populations. The state-of-the-art algorithms likely suffer from using nearest-neighbor classifiers, prone to overfitting, as the results of the baseline algorithm with kNN were consistently worse than with SVM. The method of Eduardo et al. (2017), despite showing remarkably good results in the denoizing of signals during our experiments (using the entire encoder–decoder), falls short in these settings.

Finally, recalling the state-of-the-art works previously described, we can also compare the results of the proposed and baseline algorithms with the results reported by the most recent prior art works. The results of this comparison can be analyzed in Table 11.1. The IDR of the proposed and baseline algorithms may pale in comparison with some results reported in some of the considered prior works, but it is important to consider the evaluation settings. Only Wieclaw et al. (2017) used an off-the-person database as well, as opposed to the much cleaner signals of on-the-person databases still used by most researchers. Also, the UofTDB collection allowed us to evaluate our algorithm with a much larger set of subjects than any other identification method.

However, it is important to recall that deep learning both requires and benefits greatly from large datasets, where each class is represented by a large amount of samples. While, as visible on the results presented here, data augmentation attenuates the prejudicial effects of scarce data, it is difficult to acquire sufficient ECG signals from each subject to

TABLE 11.1

Comparison of the Proposed and Baseline Algorithms with Recent State-of-the-Art Methods

Authors	Brief Description	Dataset	IDR
Proposed method	Raw segments + CNN with data augment	UofTDB (off-the-person) – 1,019 subjects	96.1%
Baseline	DCT features + SVM	UofTDB (off-the-person) – 1,019 subjects	95.3%
Salloum and Kuo (2017)	LSTM-GRU RNN	ECG-ID (on-the-person) – 90 subjects	100%
Zhang et al. (2017)	Multiscale CNN	Several (on-the-person) – 18–47 subjects	93.5%
Wieclaw et al. (2017)	Heartbeats + Multilayer Perceptron	Private (off-the-person) – 18 subjects	89.0%
Tan and Perkowski (2017)	Fiducials + Random Forestsfused with DWT + Wavelet DistancekNN	Several (on-the-person) – 184 subjects	99.5%
Carreiras et al. (2016)	Heartbeats + kNN	Private (on-the-person) – 618 subjects	84.4%
Brás and Pinho (2015)	Kolmogorov-based compression	PTB (on-the-person) – 52 subjects	99.9%
Wang et al. (2013)	Max pooling of sparse coding coefficients	PTB (on-the-person) – 100 subjects	99.5%

compensate for the increased noise and variability of off-the-person settings. In the data-sets used, each subject was represented, in average, by just 170 5-s ECG segments, which is arguably too few to train a CNN to robustly discriminate between 1,019 individuals. Considering this, with future efforts devoted to adequately deal with scarce data, deep learning methodologies could see their potential for ECG biometrics better harnessed and place themselves as clearly better alternatives to traditional pipeline algorithms.

11.6 Conclusion

A method for biometric identification based on non-intrusive ECG acquisitions was proposed in this chapter. The method is based on the common structure of CNNs that have proven to have the ability to completely and adequately replace traditional pipeline algorithms. With this goal in mind, the proposed method was evaluated when integrating more and more stages of the traditional ECG biometric pipeline, including a complete substitution by the CNN architecture, that received raw 5-s ECG segments and output a decision on the corresponding identity.

Besides this study on the progressive integration of stages on the traditional pipeline, seven data augmentation techniques for unidimensional signals were explored, and their individual and collective impact on the algorithm's performance was assessed. Furthermore, a previously proposed algorithm was adapted, improved, and evaluated as a baseline algorithm, along with some promising state-of-the-art methods.

The proposed CNN, the baseline, and the implemented state-of-the-art algorithms were evaluated using the UofTDB, currently the most complete and challenging ECG database for biometrics, using the entire population of 1,019 subjects and subsets of

25 and 100 individuals. The results have shown that the total integration of traditional pipeline processes in the CNN architecture was successful, as the results of the proposed CNN with data augmentation and receipt of raw 5-s segments surpassed, in all settings, those of the baseline and state-of-the-art algorithms in direct benchmarking. Among other recent state-of-the-art methods, considering the diverse dataset characteristics, the proposed method has also shown success as an accurate and robust biometric identification algorithm.

Acknowledgments

This work was partially funded by the Project "NanoSTIMA: Macro-to-Nano Human Sensing: Towards Integrated Multimodal Health Monitoring and Analytics/NORTE-01-0145-FEDER-00001" financed by the North Portugal Regional Operational Programme (NORTE 2020), under the PORTUGAL 2020 Partnership Agreement, and through the European Regional Development Fund (ERDF). The authors also wish to acknowledge the access to the University of Toronto ECG Database, gracefully granted by Professor Dimitris Hatzinakos and the team of Biometrics Security Laboratory that provided a challenging environment that significantly improved the quality of our work.

References

Adeoye, O. S. 2010. A survey of emerging biometric technologies. *International Journal of Computer Applications*, 9 (10): 8887–8899.

Agrafioti, F., Bui, F. M., and Hatzinakos, D. 2012. Secure telemedicine: Biometrics for remote and continuous patient verification. *Journal of Computer Networks and Communications*, 2012: 924791.

Agrafioti, F., and Hatzinakos, D. 2008. ECG based recognition using second order statistics. *CNSR 2008: 6th Annual Communication Networks and Services Research Conference*, Halifax, NS, pages 82–87.

Belgacem, N., Nait-Ali, A., Fournier, R., and Bereksi-Reguig, F. 2012. ECG based human authentication using wavelets and random forests. *International Journal on Cryptography and Information Security (IJCIS)*, 2 (2): 1–11.

Biel, L., Pettersson, O., Philipson, L., and Wide, P. 1999. ECG analysis: A new approach in human identification. *IMTC/99: Proceedings of the 16th IEEE Instrumentation and Measurement Technology Conference*, Venice, Italy, pages 557–561.

Bolle, R. M., Connell, J. H., Pankanti, S., Ratha, N., and Senior, A. W. 2004. *Guide to Biometrics*. New York: Springer Science + Business Media, 1st edition.

Brás, S., and Pinho, A. J. 2015. ECG biometric identification: A compression based approach. *37th Annual International Conference of the IEEE Engineering in Medicine and Biology Society (EMBC)*, Milan, Italy, pages 5838–5841.

Carreiras, C., Lourenço, A., Silva, H., Fred, A., and Ferreira, R. 2016. Evaluating template uniqueness in ECG biometrics. *Informatics in Control, Automation and Robotics: 11th International Conference, ICINCO 2014*, Vienna, Austria, pages 111–123.

Chatfield, K., Simonyan, K., Vedaldi, A., and Zisserman, A. 2014. Return of the devil in the details: Delving deep into convolutional nets. *Proceedings of the British Machine Vision Conference 2014 (BMVC)*, Nottingham, UK.

Chauhan, S., Arora, A. S., and Kaul, A. 2010. A survey of emerging biometric modalities. *Procedia Computer Science*, 2: 213–218.

Eberz, S., Paoletti, N., Roeschlin, M., Patané, A., Kwiatkowska, M., and Martinovic, I. 2017. Broken hearted: How to attack ECG biometrics. *Network and Distributed System Security Symposium 2017*, San Diego, CA.

Eduardo, A., Aidos, H., and Fred, A. 2017. ECG-based biometrics using a deep auto encoder for feature learning: An empirical study on transferability. *Proceedings of the INSTICC International Conference on Pattern Recognition Applications and Methods—ICPRAM*, Porto, Portugal, pages 463–470.

Forsen, G. E., Nelson, M. R., and Staron, Jr., R. J. 1977. Personal attributes authentication techniques. Pattern Analysis and Recognition Corporation, Rome Air Development Center.

Fratini, A., Sansone, M., Bifulco, P., and Cesarelli, M. 2015. Individual identification via electrocardiogram analysis. *BioMedical Engineering OnLine*, 14 (1): 78.

Hannun, A., Case, C., Casper, J., et al. 2014. Deep speech: Scaling up end-to-end speech recognition. arXiv preprint arXiv:1412.5567.

Hoekema, R., Uijen, G. J. H., and van Oosterom, A. 1999. Geometrical aspects of the inter-individual variability of multilead ECG recordings. *Computers in Cardiology*, 1999: 499–502.

Jain, A. K., Ross, A. A., and Nandakumar, K. 2011. *Introduction to Biometrics*. New York: Springer Publishing Company, Incorporated, 1st edition.

Kaur, G., Singh, G., and Kumar, V. 2014. A review on biometric recognition. *International Journal of Bio-Science and Bio-Technology*, 6: 69–76.

Kingma, D. P., and Ba, J. L. 2015. Adam: A method for stochastic optimization. *ICLR 2015- International Conference on Learning Representations*, San Diego, CA.

Krizhevsky, A., Sutskever, I., and Hinton, G. E. 2012. ImageNet classification with deep convolutional neural networks. *Advances in Neural Information Processing Systems*, 1: 1097–1105.

Kyoso, M., Ohishi, K., and Uchiyama, A. 2000. Development of ECG identification system. *Japanese Journal of Medical Electronics and Biological Engineering*, 38 (Supplement): 392.

LeCun, Y., Bengio, Y., and Hinton, G. 2015. Deep learning. *Nature*, 521: 436.

LeCun, Y., Bottou, L., Bengio, Y., and Haffner, P. 1998. Gradient-based learning applied to document recognition. *Proceedings of the IEEE*, 86 (11): 2278–2324.

Li, M., and Narayanan, S. 2010. Robust ECG biometrics by fusing temporal and cepstral information. *20th International Conference on Pattern Recognition (ICPR), 2010*, Istanbul, Turkey, pages 1326–1329.

Lourenço, A., Silva, H., and Fred, A. 2011. Unveiling the biometric potential of finger-based ECG signals. *Computational Intelligence and Neuroscience*, 2011: 720971.

Luz, E., Moreira, G., Oliveira, L. S., Schwartz, W. R., and Menotti, D. 2019. Learning deep off-the-person heart biometrics representations. *IEEE Transactions on Information Forensics and Security*, 13 (5): 1258–1270.

Mani, A., and Nadeski, M. 2015. Processing solutions for biometric systems. Texas Instruments Incorporated.

Matta, R., Lau, J. K. H., Agrafioti, F., and Hatzinakos, D. 2011. Real-time continuous identification system using ECG signals. *24th Canadian Conference on Electrical and Computer Engineering (CCECE)*, Niagara Falls, ON, pages 001313–001316.

Nogueira, R. F., Lotufo, R. A., and Machado, R. C. 2016. Fingerprint liveness detection using convolutional neural networks. *IEEE Transactions on Information Forensics and Security*, 11 (6): 1206–1213.

Odinaka, I., Lai, P. H., Kaplan, A. D., et al. 2010. ECG biometrics: A robust short-time frequency analysis. *2010 IEEE International Workshop on Information Forensics and Security*, Seattle, WA, pages 1–6.

Pinto, J. R. 2017. Continuous biometric identification on the steering wheel. Master's Thesis. Faculdade de Engenharia da Universidade do Porto.

Pinto, J. R., Cardoso, J. S., Lourenço, A., and Carreiras, C. 2017. Towards a continuous biometric system based on ECG signals acquired on the steering wheel. *Sensors*, 17 (10): 2228.

Plataniotis, K. N., Hatzinakos, D., and Lee, J. K. M. 2006. ECG biometric recognition without fiducial detection. *2006 Biometrics Symposium: Special Session on Research at the Biometric Consortium Conference*, Baltimore, MD, pages 1–6.

Raja, K. B, Raghavendra, R., Vemuri, V. K., and Busch C. 2015. Smartphone based visible iris recognition using deep sparse filtering. *Pattern Recognition Letters*, 57: 33–42.

Rajpurkar, P., Hannun, A. Y., Haghpanahi, M., Bourn, C., and Ng, A. Y. 2017. Cardiologist-level arrhythmia detection with convolutional neural networks. arXiv preprint arXiv:1707.01836.

Rumelhart, D. E., Hinton, G. E., and Williams, R. J. 1986. Learning representations by back-propagating errors. *Nature*, 323: 533–536.

Salloum, R., and Kuo, C. -C. J. 2017. ECG-based biometrics using recurrent neural networks. *2017 IEEE International Conference on Acoustics, Speech and Signal Processing (ICASSP)*, New Orleans, LA, pages 2062–2066.

Schijvenaars, R. J. A. 2000. Intra-individual variability of the electrocardiogram: Assessment and exploitation in computerized ECG analysis. Erasmus University Rotterdam.

Srivastava, N., Hinton, G., Krizhevsky, A., Sutskever, I., and Salakhutdinov, R. 2014. Dropout: A simple way to prevent neural networks from overfitting. *Journal of Machine Learning Research*, 15: 1929–1958.

Taigman, Y., Yang, M., Ranzato, M., and Wolf, L. 2014. DeepFace: Closing the gap to human-level performance in face verification. *2014 IEEE Conference on Computer Vision and Pattern Recognition (CVPR)*, Columbus, OH, pages 1701–1708.

Tan, R., and Perkowski, M. 2017. Toward improving electrocardiogram (ECG) biometric verification using mobile sensors: A two-stage classifier approach. *Sensors*, 17 (2): 410.

Um, T. T., Pfister, F. M. J., Pischler, D., et al. 2017. Data augmentation of wearable sensor data for Parkinson's disease monitoring using convolutional neural networks. *ICMI 2017—Proceedings of the 19th ACM International Conference on Multimodal Interaction*, Glasgow, Scotland, pages 216–220.

Wahabi, S., Pouryayevali, S., Hari, S., and Hatzinakos, D. 2014. On evaluating ECG biometric systems: Session-dependence and body posture. *IEEE Transactions on Information Forensics and Security*, 9 (11): 2002–2013.

Wang, J., She, M., Nahavandi, S., and Kouzani, A. 2013. Human identification from ECG signals via sparse representation of local segments. *IEEE Signal Processing Letters*, 20 (10): 937–940.

Wieclaw, L., Khoma, Y., Fałat, P., Sabodashko, D., and Herasymenko, V. 2017. Biometric identification from raw ECG signal using deep learning techniques. *2017 9th IEEE International Conference on Intelligent Data Acquisition and Advanced Computing Systems: Technology and Applications (IDAACS)*, Bucharest, Romania, pages 129–133.

Zhang, Q., Zhou, D., and Zeng, X. 2017. HeartID: A multiresolution convolutional neural network for ECG-based biometric human identification in smart health applications. *IEEE Access*, 5: 11805–11816.

12

Recent Advances in Biometric Recognition for Newborns

Abhinav Kumar and Sanjay Kumar Singh
IIT (BHU)

CONTENTS

12.1 Introduction

Abduction, mixing, and swapping of newborns are some of the major problems faced by neonatological societies. According to "World Population Prospects," the world's population reached 7.6 billion as on December 2017. Thus, around 1 billion people have been added to the world population over the last 12 years. In addition, approximately 80–90 million newborns are added worldwide each year, and the world population is estimated to reach 9.8 billion by 2050 (World Population Prospects, 2017). Also, due to better health management services and vaccination programme, a 70% reduction in under-five deaths since 2000 has been reported worldwide. Thus, with the increased inflow of newborns, a major challenge for neonatological units is the recognition of babies. Throughout the world, a number of incidences related to the delivery of newborns to wrong parents have been reported (Burgess et al., 2008; National Center for Missing and Exploited Children, 2018; BBC News 2011; ABC News, 2003; HuffPost, 2014). A recent study in the USA (Gray et al.,

2006) claimed that the probability of incorrect recognition of babies is as high as 50%. Further, it was reported that around 100,000–500,000 newborns were mistakenly switched in the USA (www.loricarangelo.com/StolenBabies.html). In 2005, Dalton et al. reported that approximately 23 million times a year, babies are shifted to and from the mother for medical care, and 10% of these transfers had been estimated to be erroneous. However, most cases are corrected before discharge.

Another social challenge, which needs attention is the abduction of newborns. The "National Center for Missing and Exploited Children" (NCMEC) has accounted for 325 child abductions in the United States since 1965, including the one in 2017. Majority of these abductions have been reported from medical care units, with greater than 50% from the room of the mother (National Center for Missing and Exploited Children, 2018). There are occurrences of deceased newborns swapped with conscious ones (Akola hospital swaps, 2016), babies held for ransom (Daily News, 1932), and newborn girls exchanged for newborn boys (Bhatti, 2016). In developing countries including India, the problems of newborn abduction and swapping have become a menace for neonatological societies (Indian Express, 2015; Times of India, 2013; Tribune, 2015; Telegraph, 2015). Currently the techniques used for recognition of newborns include usage of color-coded tags, radio frequency identification (RFID) bracelets, and footprinting; however, these have limited reliability. With advances in medical science, highly sensitive and accurate technologies such as human leukocyte antigen (HLA) and deoxyribonucleic acid (DNA) typing are now available, but these are time-consuming, expensive, and thus difficult to implement for use in developing countries. Thus, the development of a reliable and user-friendly biometric system for automatic recognition of newborns is the need of the hour. However, automatic recognition of newborns is a challenging task as the face of newborn changes day by day. Apart from this, the behavior and unique nature of newborns are other challenges for developing a biometric system for recognition of newborns. However, with recent advances in image recognition and processing methodologies, these challenges can be overcome. Inferable from the improvement of gadgets and computer technology innovation, the digital image processing has entered the era of rapid advancement, and now we can obtain massive pictures from the imaging equipment, the image database, or the Internet. Researchers have recently explored the possibilities of using baby's fingerprints, footprints, face, ear, headprint, etc. for developing biometric and other soft biometric data (Tiwari et al., 2012a–e; Tiwari and Singh, 2012a,b, 2013; Singh and Om, 2016). If such a recognition system is developed, it may prove quite useful for various applications, including efficient vaccine delivery, apart from upholding children's right to be with their biological parents. All the recent developments in this direction are discussed in detail in this chapter.

12.2 State-of-the-Art

Biometric infant authentication is the emerging domain in biometrics research that can be attributed to important applications such as tracking vaccination scheme, preventing newborn swap, and civil ID programs. The early recommendation on using infant biometrics is discussed by Galton (1899). The practical application of the biometrics was first investigated by Weingaertner et al. (2008) wherein contact-based sensing of footprints and palmprints are compared for the infant verification. They noticed that the palmprint quality was better than that of footprints. Biometric recognition has been shown to be acceptable in terms of face and fingerprint (Jain et al., 2017). They concentrated on data collection using specialized fingerprint sensors designed to capture the fingerprint images of young children at 1,270 ppi. Face biometrics (Best-Rowden et al., 2016; Bharadwaj et al., 2010) has been widely explored for infant identification, as the face can be captured without difficulty in a non-intrusive manner. However, due to the non-cooperative nature of the infants, infant face recognition fails to be as accurate as adult face recognition using commercial systems. Ear biometrics was explored by Tiwari and Singh (2014) who indicated the reliability of ears for verification of the infants. Based on the evolving studies on newborn biometric verification, it is noted that only commercial sensors are evaluated, especially in face and fingerprint biometrics. Furthermore, all the biometric characteristics (fingerprint, footprint, and palmprint) are captured using touch-based sensors which are extremely inconvenient for capturing newborn's images.

12.3 Current Technological Response and Its Limitations

Recent advances in the newborn-baby identification are gaining interest. Since the biometric characteristics of an infant are still evolving as compared to those of the adult, it is challenging to adopt the existing techniques for infant biometrics. Most of the existing work focuses on fingerprint and face biometrics, and the results disclosed in the literature are limited to commercial solutions. The newborn-baby identification framework presents a unique challenge of identifying a baby from the day it is born (day 1) until day 3 or its discharge from the hospital. This imposes a challenge for biometric verification because of the non-cooperative nature of the newborn baby, hygiene issues, and non-cooperative nature of parents to use conventional biometric sensors. In addition, no publicly available database for newborn babies is available for benchmarking and tailoring techniques for this unique application. To effectively address applications such as vaccination tracking and child swapping, a biometric database of newborn babies collected from day 1 until 9 months is needed. Further, the need for "Presentation Attack Detection" is also an essential component to address the misuse of social benefit schemes in India, which was not addressed by the community, especially for the newborn babies. A recent ruling from the Supreme Court of India has stressed the importance of the privacy that was not addressed especially for infant biometrics.

A set of limited works have investigated biometrics for establishing the identity of infants through the use of biometric characteristics, especially face, ear, palmprint, and fingerprint (Bharadwaj et al., 2016; Jain et al., 2014, 2016, 2017; Koda et al., 2016; Kotzerke et al., 2014, 2017). However, many of the key problems associated with infant recognition are not yet addressed. These include the following.

- The biometric identity of the infants is explored using face, ear, palmprint, and fingerprint biometrics. As they are bound to change physiologically from the time of birth until a certain age, a reliable biometric modality has not been established yet, but a recent study advocates the use of fingerprint biometrics (Jain et al., 2017). However, the sample size in this study is very small to determine the reliability of the biometric modality especially for infants from day 1 to day 3 (when discharged from hospital).

- A major problem in using fingerprint and palmprint is that the sensors are contact based which can result in unhygienic practices that can compromise the health of infants. Thus, contactless biometrics should be investigated to identify the infants.

- Further, as these studies are independent and have disjoint subjects, multi-modal approaches for identifying the infants have not been explored.

- The imaging of infants needs to be carried out in low-light conditions as capturing the images with ash light is not advocated.

- The biometric data of infants need to be secured and communicated to a server in a privacy-preserved manner.

12.4 Newborn Recognition Techniques

The main advantage of biometric technology over other conventional methods is automation in identification or verification of persons or newborns based on physiological or behavioral traits. The traits which have been regularly used for biometric identification include fingerprints, footprints, palmprints, face, the geometry of the hand, voice, iris, etc. (Jain et al., 2004). Today, there are several occasions in which personal validation is required. Interestingly, in recent years, scientists have used newer features like finger knuckle print, hand vein, etc. to develop biometric systems. However, to date, majority of the biometric systems discussed above have focused on adults (Kumar and Prathyusha, 2009; Kumar and Ravikanth, 2009; Zhang et al., 2009, 2010). Various techniques used in the past for recognition of newborns are mentioned below.

12.4.1 RFID Bracelets

The problem of identification of newborn babies arises in the hospitals where a number of babies are born and put in the same ward, i.e., baby ward. One of the prominent methods used in hospitals is the use of bracelets (with ID) wrapped around the legs or hands of the baby immediately after birth. A sample RFID bracelet is shown in Figure 12.1. However, this method lacks reliability, provides insufficient security for newborns, and is not very effective in the information and network society.

12.4.2 Medical Techniques

HLA typing and DNA typing techniques have shown accuracy and high efficiency for identification of infants. However, these techniques have not come into routine use due to the involvement of high costs, labor, and long time periods in processing HLA or DNA samples. Thus such techniques are not feasible for routine identification of all individuals. Further, being invasive, these methods cannot be used repeatedly for recognition of newborns. Although Rapid DNA technology has recently evolved as a cost-efficient and faster way of using DNA for newborn identification, this technology may not be available in developing countries and is still quite early to be implemented for newborn verification.

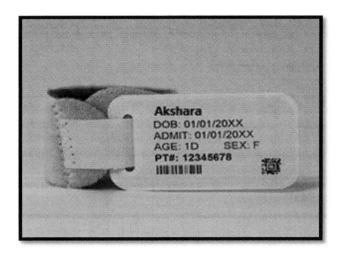

FIGURE 12.1
Samples of RFID bracelets used in hospitals.

12.4.3 Footprint

According to Wierschem (1965), the analysis of newborn footprints collected from Chicago's hospitals in the USA revealed that 98% of footprints are not suitable for recognition. However, a new analysis of the collected footprints, after giving the right equipment and proper training to the staff, showed that newborns could be recognized in 99% of cases using this method. However, this method had a disadvantage of using flexion creases of the foot that are likely to change in the first month of life. Shepard et al. (1966) also explored the usage of footprints for recognition; however, only 20% footprints could be identified accurately, and it was also observed that the larger part of these prints could not be presented for legal analysis in courts. The use of footprints captured at birth has been supported by Stapleton (1994) as a reliable, fast, and cost-effective technique for newborn identification, and poor performances in identification were attributed to poor image capturing and untrained staff. Further, poor performance was also found to be associated with improper ridge quality in newborn footprints. Blake (1959) provided more accurate results based upon use of ridge line structure or flexion creases.

The Federal Bureau of Investigation (FBI), USA, recommended footprinting and fingerprinting of the infant and mother as a method for newborn recognition (Stapleton, 1994). As per the survey, the majority (90%) of the hospitals in the USA take footprints of newborns immediately after their birth. In a study by Pelá et al. (1976) in Brazil, 1,917 footprints were collected by skilled hospital personnel; however, the majority of the images could not provide sufficient information for infant recognition. Therefore, the American Academy of Pediatrics and others concluded that although hospitals could continue to take fingerprints and footprints of the newborn's mother, the universal application of this method is not recommended. Thompson et al. (1981) collected 100 footprints from 20 newly born babies and found that 11% could be technically accepted and only 1% had all the necessary elements for legal verification. They also concluded that footprints of babies born with less than 1,500 g weight (premature babies) are not suitable for recognition.

Jia et al. (2010) found that the time while the child is asleep or calm, and not crying or moving its hands/feet, is the most convenient time for capturing the images. Capturing images this way, they could achieve an equal error rate (EER) of 3.82%. Further, Balameenakshi and Sumathi (2013) collected 240 images from 40 newborns within 2 days of their birth. Jia et al. (2010) adopted a two-person approach and confirmed the importance of the newborn's mood. The majority of their pictures were taken from calm subjects, and they could correctly match 130 of 200 images. Kotzerke et al. (2013) proposed an algorithm for newborn footprint crease feature extraction, but no matching performance has been reported. Footprints are larger than fingerprints and palmprints, so the number of their minutiae is also larger. Since such a large friction ridge or minutiae area is not necessary for recognition, Kotzerke et al. (2014) used only the ridge structure characteristics of the ballprint, i.e., the hallucal area under the big toe for the process of recognition. The left and right ballprints' images were collected from 54 newborns at different time intervals: 2 days (first session), 2 months (second session), and 6 months (third session) age. Out of these, the ballprints captured in the first session were removed owing to poor quality; only 192 ballprints were manually selected from the other two sessions; and the EER of matching ballprint images between session 2 and session 3 was observed to be 7.28%. Though a newborn's footprints captured offline have been used in many countries, the effectiveness of image quality of offline footprints is still debated.

Table 12.1 provides a compilation of all the prior studies in this direction for infant footprint recognition. Further, during the past 100 years, no innovation has been reported for offline footprint acquisition, and most of the offline footprint images are not suitable because of the usage of inadequate materials like paper, ink, cylinder and employment of untrained staff for image acquisition. The skin of the newborn is also covered with an oily substance and, due to the low thickness of the epidermis of the newborn, the ridges easily deform on contact, and the valleys between the ridges are filled with ink. Moreover, the newborn's ridges are 3–5 times smaller than those of adults (Weingaertner et al., 2008).

It can therefore be concluded from past studies that matching baby footprints is very difficult because of poorly captured images, minute and fragile ridges, and images/prints captured at the wrong time after birth. After a few days of birth, the skin begins to dry off, and cracks begin to appear. Thus the ridge pattern which helps in identification is obscured (Blake, 1959; FBI and US Department of Justice, 1971). However, the flexure creases are not much affected.

12.4.4 Face Identification

Face recognition is becoming more popular, even for infants and newborns, negating the notion that all the newborns look alike. For passive recognition of newborns, headprint is an important source of data. Tiwari and Singh (2012a) have shown the feasibility of using headprint for newborn recognition. They generated a biometric database of headprints of newborns and then developed a texture-based algorithm for verification of 200 newborns. Using a database of 34 newborns, Bolle et al. (2013) reported 86.9% accuracy when using headprints for their recognition. Nevertheless, the method is limited by facial expression of the newborns, as getting a neutral face image of babies is a challenging task. Tiwari et al. (2012b) proposed a newborn face recognition method to mitigate the influence of newborn face covariates and to provide an experimental as well as analytical underpinning consequence of different facial expressions shown by the newborns. Sample neutral, crying, and sleeping face images from the newborn database of the Indian Institute of

TABLE 12.1

Prior Works on Newborn Footprint Recognition

Author	Sensor Used	Database Information	Comment
Liu (2017)	Watson Mini (500 ppi)	756 images from 42 infants, captured in three sessions (age: 1 month (session 1), 3 months (session 2), and 6 months (session 3)).	This method claims GAR of 61%, 55%, and 83%, when matching session 1 vs. session 2, session 1 vs. session 3, and session 2 vs. session 3, and fusing left and right footprint images, respectively, at a false acceptance rate (FAR) of 0.01.
Kotzerke et al. (2014)	NEC PU900-10 (1,000 ppi)	54 newborn footprints have been collected in three sessions (age: 2 days, 2 months, and 6 months).	An EER of 7.28% was observed by matching of the footprints acquired at 6 months and 2 months age.
Kotzerke et al. (2013)	Nekoosa Printed Products Identifier and HP Scanjet G4010	Dataset includes 54 footprints collected within 3 days after birth, 41 footprints collected at 8 weeks, 4 footprints at 6 months age.	In this study, no matching experiments were conducted.
Balameenakshi and Sumathi (2013)	Cannon EOS 7D camera	Images (240 nos.) collected from 40 babies within 2 days after birth.	This method reveals 65% rank-1 identification accuracy with background size of 40 images.
Jia et al. (2012)	Cannon power shot SX110 IS camera	1,968 footprint images from 101 infants acquired within 2 days of their birth.	An EER of 1.34% was obtained with this strategy.
Thompson et al. (1981)	Inked	Footprints of 100 babies (full-term) and 20 premature babies.	Accuracies of 11% and 0% were obtained for full-term and premature babies, respectively.
Pelá et al. (1976)	Inked	Footprints collected from 1,917 infants.	Ridge information was not sufficient for manual recognition.
Shepard et al. (1966)	Inked	Footprints of 51 babies taken immediately after birth and 5–6 weeks later.	Only 19.6% of the newborns matched correctly as the images were acquired using incorrect capture practices.
Blake (1959)	Inked	1,388 newborn footprints collected immediately.	About 79% of the footprints were recognized accurately by using only flexure creases.

Technology (BHU) are shown in Figure 12.2. The method has been shown to be 87.04% accurate for newborn face recognition. An unconstrained database comprising 280 subjects and ten images per individual with slight variations in pose, facial expression, and illumination were used for the study.

Bharadwaj et al. (2010) introduced a method for identifying the newborns – an automatic face recognition system with an unconstrained setup. The local binary patterns and robust features of different levels of the Gaussian pyramid were extracted in this algorithm, and using the weighted sum rule, feature descriptors gathered at each Gaussian level were combined. In two sessions, 374 images of 34 individuals were collected. The first was held within 2 h of birth, and the second at discharge. The method gave rank-1 identification accuracy of 86.9% using a 34-infant database. It has been demonstrated based on a database of 210 newborns that integration of soft-biometric traits with face recognition improves accuracy by 5.6% over the face recognition system (Tiwari et al., 2012c).

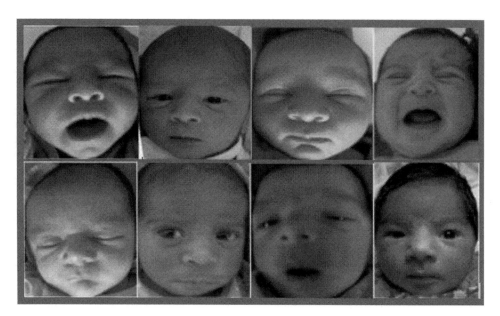

FIGURE 12.2
Samples of neutral, crying, and sleeping face images of a newborn.

12.4.5 Fingerprints

Though recognition based on fingerprints is used widely and successfully for adults, for newborns, it is not feasible as the fingers of newborns are very small, and thus ridges or minutiae points are not very clear (Holt, 1973; Cummins and Midlo, 1961; Kücken and Newell, 2005). Recognition based on palmprints and fingerprints is a well-established technique for children above 5 years of age; however, for newborns, it isn't good enough in the first days of life. A study using newborn fingerprints with ink and paper found that fingerprints taken before the age of 17 months were not useful for recognition (Galton, 1899). Therefore fingerprint studies (Schneider, 2010; Gottschlich et al., 2011) do not focus on newborns and deal mainly with infants and teenagers, studying infant growth and its effect on the pattern of ridge lines. Castellanos (1953) states that newborns' ridge line structures are 3–5 times smaller than those of adults. Weingaertner et al. (2008) have supported the earlier claim by showing an adult as well as a newborn fingerprint next to each other. They stressed that the use of this feature forces a high burden on sensor technology and the newborn, as the skin ridges are very fragile and easily deformed on contact. Further, the newborns usually keep their hands clenched, requiring them to be opened for capturing fingerprints. Thus, fingerprints have utility as an infant biometric modality rather than a newborn biometric modality, and researchers are now trying to study how growth affects the physical biometric features. However, in one study, Schneider (2010), working with digital scans taken 1 year apart from 186 subjects initially 1–18 years old, fingerprint growth could not be approximated with any linear affine transformation. On the other hand, Gottschlich et al. (2011), after studying the criminal records of 48 juveniles (6–15 years of age) taken at varying time gaps, proposed a gender-specific ratio of median population height taken at different ages as an appropriate isotropic scaling factor for approximating growth. They observed that the usage of the scaling factor for a test set of 462 fingers reduces the error rates by nearly 50%, leading to an EER of 1.8% in a verification scenario.

Recently, Jain et al. (2014) have also investigated the suitability of fingerprints for toddler and infant recognition. They collected two separate databases: the East Lansing dataset, comprising 1,600 images of 20 infants aged 0–4 years (mostly older than 6 months) collected at intervals of about 1 week, and the Benin dataset, comprising 420 images of 70 newborns collected in a single session and without a specified age group, but on observation, most subjects were younger than 6 months. Jain et al. (2014) reported three major challenges for data acquisition in case of infants: first being their non-cooperative nature; second, their skin condition (oily or wet fingers); and last, the small size of their fingers. The researchers found that, owing to the condition of skin of newborns, the image quality may be influenced by their habit of sucking fingers. They emphasized the use of a compact and comfortable sensor device for making the capture process as pleasant as possible. They also suggested the need for fast capture speed to reduce the burden forced upon the infant while holding its finger onto the sensor and also to avoid smudged images due to movement. The studies resulted in 83.8% and 98.97% accuracy of rank-1 identification with a commercial fingerprint software development kit (SDK) and a latent fingerprint SDK for East Lansing data, respectively. For the Benin data, accuracies of 40.0% and 67.14% were reported, respectively. In addition, fingerprint, palmprint, and hand geometry image acquisition are not easy for newborns, as they usually keep their hands closed, making it difficult to take images with all fingers in the right position. Thus there are very few reports related to the use of newborns' fingerprints for their verification.

12.4.6 Palmprint

Palmprint-based recognition is inappropriate for newborns as they often keep their hands closed and it is difficult to make a newborn open his/her hand. Morgan and Pauls (1939) presented a method for collection of a newborn's palmprints and reported that collected images could be used for recognition. However, authors did not provide any objective analysis of the images nor did they support their studies with any matching test. Weingaertner et al. (2008) have reported a new high-resolution sensor to acquire newborn palmprints and footprints. Based on two sets of images of 106 infants (one before 24 h of birth and another at 48 h), newborns could be identified using palmprints and footprints, with accuracies of 83% and 67.7%, respectively. Lemes et al. (2011) applied a high-resolution scanner to the 250-infant database and found that the method is limited due to poor quality images, which are highly intrusive, apart from the high cost involved.

12.4.7 Ear

Ear shape is another biometric modality that has been proposed recently for infant identification. Ear shape as a biometric modality has been shown to be distinctive and can be used to identify newborns, identical twins or triplets, and adults. Samples of ear images of newborns are shown in Figure 12.3.

The applicability of recognizing newborns using ear shape has been studied by Fields et al. (1960) who observed that two juveniles could be differentiated by their ear images. Tiwari et al. (2011) acquired five images per ear of 210 newborns with minor variations, such as lighting or partial occlusion changes. They stressed that this modality is useful as well as hygienic because the capture process can be carried out from a distance without a contact sensor. Moreover, the image taken with a standard camera is also sufficient because of the size of the ear. The authors achieved 83.67% identification accuracy. In 2016, Kumar et al. (2016) used an unspecified database consisting of 750 photographs of

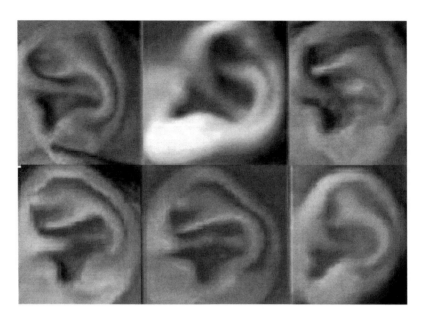

FIGURE 12.3
Samples of ear images of newborns.

infant/children's ears, aged between 0 and 6 years. Up to five mobile phone images of the same individual were included in the database. For further application of this technique in future, the longitudinal behavior of ear shape needs to be investigated to determine whether it changes over time as the baby grows.

12.4.8 Iris Recognition

In adults, a biometric system based upon iris recognition is highly accurate; however, for newborns, biometric capturing of this trait is not feasible. Daugman (2007) also reported that capturing iris patterns for newborns is a challenging task. Newborns keep their eyes closed and also lack the ability to look into the scanner, and touching their eyelids to take images might harm them. Thus utilizing iris scan for recognizing babies is not possible, especially for the premature ones (Bolle et al., 2013). Moreover, the iris pattern of newborns keeps on changing and stabilizes 2 years after birth (Jain et al., 2004).

12.4.9 Soft Biometrics

The biometric traits like age, height, gender, weight, blood group, eye color, ethnicity, etc., which provide information about the particular individual, however, are not unique and reliable for individual recognition and are called as soft biometric traits. These traits help in providing complementary information for biometric traits. Soft biometric traits are easy to understand and record, require low-cost equipment, and have no issues related to identity theft. These traits are also helpful in filtering large databases. The concept of complementing soft biometrics with primary biometrics (ear) for recognition of newborns has been demonstrated by Tiwari et al. (2012c). The researchers designed and implemented soft biometrics and ear fusion to recognize 210 newborns and showed a 5.59% improvement over the primary biometric system, i.e., ear.

Very recently, Akhtar et al. (2017) have reported a novel newborn biometric recognition method combining the face and soft biometrics. In this study, face and soft biometric databases of 210 newborns were prepared and evaluated for identification of 221 infants. The authors reported an improvement of approximately 6% over the primary biometric system (face) resulting in the combination of all four soft biometric traits (height, weight, gender, and blood group).

12.5 Multimodal Biometrics for Newborn Recognition

Though unimodal biometric approaches based upon recognition using a single biometric trait have shown considerable performance, these systems are still facing various limitations owing to the type of data and methodology used. The performance and security of these unimodal systems are compromised mainly because of noisy input data, intraclass variability, non-universality, limited degrees of freedom, etc. (Monwar and Gavrilova, 2009). These problems can be managed by using numerous biometric traits, and such a biometric system that uses the knowledge from several sources is referred to as a multimodal biometric system. This new approach of multimodal biometrics can provide an improvement in the security level, recognition performance, apart from improving population coverage, increasing the degrees of freedom, preventing spoof attacks, etc. These multimodal systems require large processing time, storage, and sophisticated computational inputs; however, their advantages fascinate users to employ them in real-world authentication systems (Down and Sands, 2004). Figure 12.4 shows the sample block diagram of a multimodal biometric system.

The development of an effective multimodal biometric system requires a sound fusion scheme for combining the information from various biometric sources. In a biometric system, after each level of fusion, the extent of the information available for fusion declines. An integration scheme is necessary to fuse the information from the individual biometric modalities. The fusion could be carried out at four different levels of information (Kumar and Imran, 2010), corresponding to four important modules of a biometric system (i.e., Sensor module, Decision-Making module, Matching module, and Feature

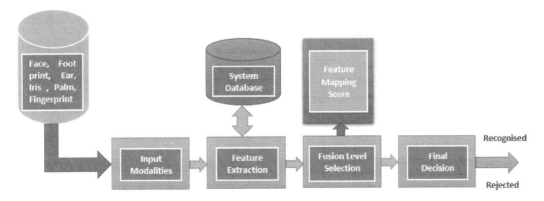

FIGURE 12.4
Broad multimodal biometric system (block diagram).

TABLE 12.2

Comparative Evaluation of Different Biometric Modalities for Newborn Recognition

Biometric/Non-Biometric Methods	Universality	Uniqueness	Permanence	Measurability	Performance	Acceptability	Circumvention
				Biometric Trait Performance Parameters			
ID-Band	Low	Low	Low	High	High	Low	Low
Barcode	Low	Low	Low	High	High	Low	Low
DNA	High	High	High	Low	High	Low	Low
Palmprint	Low	Low	Low	Low	Low	Low	Low
Ear	Moderate	Moderate	Moderate	Moderate	High	Moderate	Moderate
Face	Moderate	Moderate	Low	Moderate	High	Moderate	Moderate
Footprint	Moderate	Moderate	Moderate	Low	High	High	Moderate
Iris	Not Available	Not Available	Not Available	Not Available	Not Available	Not Available	Not Available
Fingerprint	Not Available	Not Available	Not Available	Not Available	Not Available	Not Available	Not Available

Extraction module). Despite its many advantages, little work has been done to develop a multimodal recognition system for newborns. A limited availability of benchmark databases for young children is one of the main reasons for limited research in this direction. A multimodal database comprising of ear shapes, face images, and headprints of 210 newborns has been described by Tiwari et al. (2013). Another multimodal biometric database, comprising face, iris, and fingerprint images of more than 100 young children has been developed by Basak et al. (2017). The researchers found that with SVM fusion of images of iris, all ten fingerprints, and face, a general acceptance rate (GAR) of 100% at 0.1% FAR was observed. The high performance was mainly because of high accuracy of iris matching. In cases where the matching of iris fails, face and fingerprint scores can match correctly. This research also demonstrated that iris provides the highest GAR, particularly upon combining iris images of both right and left eyes. The applicability of different biometric traits to recognize newborns is summarized in Table 12.2.

12.6 Proposed Framework and Future Work

Future studies should concentrate on providing an identity management solution through the use of a smartphone for unconstrained infant biometrics wherein the images/videos of the biometric characteristics are captured in non-invasive and contactless manner. Figure 12.5 presents a new concept of identifying infants using smartphones by capturing the images and videos of biometric characteristics.

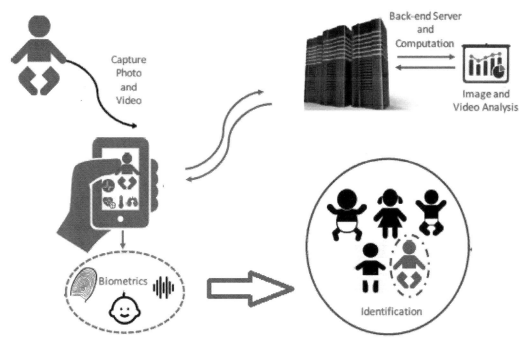

FIGURE 12.5
A proposed framework for smartphone-based biometric system for newborn recognition.

12.7 Conclusions

Biometric recognition of newborns has several applications including prevention of newborn baby swapping, identification of missing children, tracking of vaccination schedules, and other civil ID programs. However, the development of an efficient biometric system for newborn recognition is a challenging task as the biometric characteristics of the infants are still evolving as compared to the adult biometrics. Further, there are very few publicly available databases for the newborn babies to benchmark and tailor techniques for this unique application. Various biometric traits like footprints, fingerprints, headprints, face, palmprints, ear, iris, etc. have been explored with variable performances for their ability to accurately recognize the infants. All biometric features have their own unique challenges, but they are essentially all based on the will of the newborn to cooperate, the difficulties of capturing its small and fragile physical features, and the need to please parents. The performance and security of these unimodal systems can be substantially enhanced by developing multimodal biometric systems combining information from multiple biometric traits. This new approach of multibiometrics can provide an improvement in the security level and infant recognition performance.

References

ABC News 2003. Hospital Newborn Swap Outrages Parents. 2003, December. *ABC News*. Retrieved from http://abcnews.go.com/GMA/story?id=128144. Accessed on January 10, 2018.

Akhtar, R., Siddiqui, J. and Sharma, D. 2017. Soft-biometric in recognition of new born? *International Journal of Computer Science and Mobile Computing*, 6(5), 316–322.

Akola hospital swaps dead infant with a live one. 2016, January. *The Times of India*. Retrieved from https://timesofindia.indiatimes.com/city/nagpur/Akola-hospital-swaps-dead-infant-with-a-live-one/articleshow/50597475.cms. Accessed 2 January, 2018.

Balameenakshi, S. and Sumathi, S. 2013, April. Biometric recognition of newborns: Identification using footprints. In *IEEE Conference on Information & Communication Technologies (ICT), 2013*, pp. 496–501, Thuckalay, Tamil Nadu, India.

Basak, P., De, S., Agarwal, M., Malhotra, A., Vatsa, M. and Singh, R. 2017, October. Multimodal biometric recognition for toddlers and pre-school children. In *2017 IEEE International Joint Conference on Biometrics (IJCB)*, pp. 627–633. IEEE, Denver, CO.

BBC News 2011. Switched at birth, then meeting aged 12. 2011, October. *BBC News*. Retrieved from www.bbc.com/news/magazine-15432846. Accessed on January 25, 2018.

Best-Rowden, L., Hoole, Y. and Jain, A. 2016, September. Automatic face recognition of newborns, infants, and toddlers: A longitudinal evaluation. In *2016 International Conference of the Biometrics Special Interest Group (BIOSIG)*, pp. 1–8. IEEE, Darmstadt, Germany.

Bharadwaj, S., Bhatt, H.S., Singh, R., Vatsa, M. and Singh, S.K. 2010, September. Face recognition for newborns: A preliminary study. In *2010 Fourth IEEE International Conference on Biometrics: Theory Applications and Systems (BTAS)*, pp. 1–6. IEEE, Washington, DC.

Bharadwaj, S., Bhatt, H.S., Vatsa, M. and Singh, R. 2016. Domain specific learning for newborn face recognition. *IEEE Transactions on Information Forensics and Security*, 11(7), 1630–1641.

Bhatti, M. 2016. Parents protest at CHK after baby boy 'swapped' with girl's body. *The News*. Retrieved from www.thenews.com.pk/print/93706-Parents-protest-at-CHK-after-baby-boy-swapped-with-girls-body. Accessed on 5 January 2018.

Blake, J.W. 1959. Identification of the new-born by flexure creases. *Identification News*, 9(9), 3–5.

Bolle, R.M., Connell, J.H., Pankanti, S., Ratha, N.K. and Senior, A.W. 2013. *Guide to Biometrics*. Springer Science & Business Media, New York; London.

Burgess, A.W., Carr, K.E., Nahirny, C. and Rabun Jr, J.B. 2008. Nonfamily infant abductions, 1983–2006. *The American Journal of Nursing*, 108(9), 32–38.

Castellanos, I. 1953. Dermopapiloscopia clínica. Imp. P. Fernández y Cia.

Cummins, H., & Midlo, C. 1961. *Finger Prints, Palms and Soles: An Introduction to Dermatoglyphics*, Volume 319. Dover Publications, New York.

Daily News 1932. The day Charles Lindbergh's baby was kidnapped. 1932, March. *Daily News*. Retrieved from www.nydailynews.com/news/crime/charles-lindbergh-baby-kidnapped-1932-article-1.2127396. Accessed on January 15, 2018.

Dalton, J., Kim, I. and Lim, B. 2005. RFID technologies in neonatal care. White Paper by Intel Corporation, LG CNS, ECO Inc and Wonju Christian Hospital, p. 122.

Daugman, J. 2007. New methods in iris recognition. *IEEE Transactions on Systems, Man, and Cybernetics, Part B (Cybernetics)*, 37(5), 1167–1175.

Down, M.P. and Sands, R.J. 2004. Biometrics: An overview of the technology, challenges and control considerations. *Information Systems Control Journal*, 4, 53–56.

Federal Bureau of Investigation and US Department of Justice. 1971. Classification of footprints. *FBI Law Enforcement Bulletin*, 40, 18–22.

Fields, C., Falls, H.C., Warren, C.P. and Zimberoff, M. 1960. The ear of the newborn as an identification constant. *Obstetrics & Gynecology*, 16(1), 98–102.

Galton, F. 1899. Finger prints of young children. *British Association for the Advancement of Science*, 69, 868–869.

Gottschlich, C., Hotz, T., Lorenz, R., Bernhardt, S., Hantschel, M. and Munk, A. 2011. Modeling the growth of fingerprints improves matching for adolescents. *IEEE Transactions on Information Forensics and Security*, 6(3), 1165–1169.

Gray, J.E., Suresh, G., Ursprung, R., Edwards, W.H., Nickerson, J., Shiono, P.H., Plsek, P., Goldmann, D.A. and Horbar, J. 2006. Patient misidentification in the neonatal intensive care unit: Quantification of risk. *Pediatrics*, 117(1), e43–e47.

Holt, S.B. 1973. Clinical review: The significance of dermatoglyphics in medicine: A short survey and summary. *Clinical Pediatrics*, 12(9), 471–484.

HuffPost 2014. Parents horrified after hospital mixup allows their baby to be breastfed by stranger. 2014, January. HuffPost. Retrieved from www.huffingtonpost.com/2014/01/11/baby-breastfed-by-stranger_n_4578108.html. Accessed on January 9, 2018.

Indian Express, The 2015. Newborn's Death: Parents Accuse Hospital of Changing Their Child With Somebody Else's. 2015, July. *The Indian Express*. Retrieved from http://indianexpress.com/article/cities/chandigarh/newborns-death-parents-accuse-hospital-of-changing-their-child-with-somebody-elses. Accessed on January 10, 2018.

Jain, A.K., Arora, S.S., Best-Rowden, L., Cao, K., Sudhish, P.S., Bhatnagar, A. and Koda, Y. 2016, June. Giving infants an identity: Fingerprint sensing and recognition. In *Proceedings of the Eighth International Conference on Information and Communication Technologies and Development*, pp. 29–32. ACM, Ann Arbor, MI.

Jain, A.K., Arora, S.S., Cao, K., Best-Rowden, L. and Bhatnagar, A. 2017. Fingerprint recognition of young children. *IEEE Transactions on Information Forensics and Security*, 12(7), 1501–1514.

Jain, A.K., Cao, K. and Arora, S.S. 2014, September. Recognizing infants and toddlers using fingerprints: Increasing the vaccination coverage. In *2014 IEEE International Joint Conference on Biometrics (IJCB)*, pp. 1–8. IEEE, Clearwater, FL.

Jain, A.K., Ross, A. and Prabhakar, S. 2004. An introduction to biometric recognition. *IEEE Transactions on Circuits and Systems for Video Technology*, 14(1), 4–20.

Jia, W., Cai, H. Y., Gui, J., Hu, R. X., Lei, Y. K. and Wang, X. F. 2012. Newborn footprint recognition using orientation feature. *Neural Computing and Applications*, 21(8), 1855–1863.

Jia, W., Hu, R.X., Gui, J. and Lei, Y.K. 2010, June. Newborn footprint recognition using band-limited phase-only correlation. In *International Conference on Medical Biometrics*, pp. 83–93. Springer, Berlin, Heidelberg.

Koda, Y., Higuchi, T. and Jain, A.K. 2016. Advances in capturing child fingerprints: A high resolution CMOS image sensor with SLDR method. BIOSIG 2016.

Kotzerke, J., Arakala, A., Davis, S., Horadam, K. and McVernon, J. 2014, October. Ballprints as an infant biometric: A first approach. In *Proceedings, 2014 IEEE Workshop on Biometric Measurements and Systems for Security and Medical Applications (BIOMS)*, pp. 36–43. IEEE, Rome, Italy.

Kotzerke, J., Davis, S., Horadam, K. and McVernon, J. 2013, September. Newborn and infant footprint crease pattern extraction. In *Proceedings, 2013 20th IEEE International Conference on Image Processing (ICIP)*, pp. 4181–4185. IEEE, Melbourne, Australia.

Kotzerke, J., Davis, S.A., Hayes, R. and Horadam, K.J. 2017. Newborn and infant discrimination: Revisiting footprints. *Australian Journal of Forensic Sciences*, 51, 1–14.

Kücken, M. and Newell, A.C. 2005. Fingerprint formation. *Journal of Theoretical Biology*, 235(1), 71–83.

Kumar, A. and Prathyusha, K.V. 2009. Personal authentication using hand vein triangulation and knuckle shape. *IEEE Transactions on Image Processing*, 18(9), 2127–2136.

Kumar, A. and Ravikanth, C. 2009. Personal authentication using finger knuckle surface. *IEEE Transactions on Information Forensics and Security*, 4(1), 98–110.

Kumar, G.H. and Imran, M. 2010. Research avenues in multimodal biometrics. *IJCA Special Issue on "Recent Trends in Image Processing and Pattern Recognition", RTIPPR*, (3), pp. 1–8.

Kumar, M., Insan, A., Stoll, N., Thurow, K. and Stoll, R. 2016. Stochastic fuzzy modeling for ear imaging based child identification. *IEEE Transactions on Systems, Man, and Cybernetics: Systems*, 46(9), 1265–1278.

Lemes, R.P., Bellon, O.R., Silva, L. and Jain, A.K. 2011, October. Biometric recognition of newborns: Identification using palmprints. In *2011 International Joint Conference on Biometrics (IJCB)*, pp. 1–6. IEEE, Washington, DC.

Liu, E. 2017. Infant footprint recognition. In *Proceedings of 2017 IEEE International Conference on Computer Vision (ICCV)*, pp. 1662–1669. IEEE, Venice.

Monwar, M.M. and Gavrilova, M.L. 2009. Multimodal biometric system using rank-level fusion approach. *IEEE Transactions on Systems, Man, and Cybernetics, Part B (Cybernetics)*, 39(4), 867–878.

Morgan, L.E. and Pauls, F. 1939. Palm prints for infant identification. *The American Journal of Nursing*, 39(8), 866–868.

National Center for Missing and Exploited Children 2018. Infant Abduction. Retrieved from www.missingkids.com/InfantAbduction. Accessed on January 10, 2018.

Pelá, N.T.R., Mamede, M.V. and Tavares, M.S.G. 1976. Analise critica de impressoes plantares de recem-nascidos. *Revista Brasileira de Enfermagem*, 29(4), 100–105.

Schneider, J.K. 2010. Document title: Quantifying the dermatoglyphic growth patterns in children through adolescence.

Shepard, K.S., Erickson, T. And Fromm, H. 1966. Limitations of foot printing as a means of infant identification. *Pediatrics*, 37(1), 107–108.

Singh, R. and Om, H. 2016. Pose invariant face recognition for new born: Machine learning approach. In *Computational Intelligence in Data Mining*, Volume 1, H.S. Behera and D.P. Mohapatra, 29–37. Springer, New Delhi.

Stapleton, M. E. 1994, November. Best foot forward: Infant footprints for personal identification. The Free Library. Retrieved from www.thefreelibrary.com/Best foot forward: infant footprints for personal identification.-a016473798. Accessed on January 9, 2018.

Telegraph, The 2015. Couple Alleges Newborn Swap. 2015, December. *The Telegraph*. Retrieved from www.telegraphindia.com/1151212/jsp/northeast/story_58007.jsp#.VnZk0GSGSko. Accessed on 6 January, 2018.

Thompson, J.E., Clark, D.A., Salisbury, B. and Cahill, J. 1981. Foot printing the newborn infant: Not cost effective. *The Journal of Pediatrics*, 99(5), 797–798.

Times of India, The 2013. Family Alleges Swapping of Newborn at Nursing Home. 2013, October. *The Times of India*. Retrieved from http://timesofindia.indiatimes.com/city/chandigarh/Family-alleges-swapping-of-newborn-at-nursing-home/articleshow/23756717.cms. Accessed on January 6, 2018.

Tiwari, S., Singh, A. and Singh, S.K. 2011, November. Newborn's ear recognition: Can it be done?. In *Proceedings, 2011 International Conference on Image Information Processing (ICIIP)*, pp. 1–6. IEEE, Shimla, India

Tiwari, S., Singh, A. and Singh, S.K. 2012a. Fusion of ear and soft-biometrics for recognition of newborn. *Signal and Image Processing: An International Journal*, 3(3), 103–116.

Tiwari, S., Singh, A. and Singh, S.K. 2012b. Intelligent method for face recognition of infant. *International Journal of Computer Applications*, 52(4), 46–50.

Tiwari, S., Singh, A. and Singh, S.K. 2012c. Integrating faces and soft-biometrics for newborn recognition. *International Journal of Advanced Computer Engineering and Architecture*, 2(2), 201–209.

Tiwari, S., Singh, A. and Singh, S.K. 2012d, March. Can face and soft-biometric traits assist in recognition of newborn? In *Proceedings, 2012 1st International Conference on Recent Advances in Information Technology (RAIT)*, pp. 74–79. IEEE, Dhanbad, India.

Tiwari, S., Singh, A. and Singh, S.K. 2012e. Can ear and soft-biometric traits assist in recognition of newborn? In *Advances in Computer Science, Engineering & Applications*, 179–192. Springer, Berlin, Heidelberg.

Tiwari, S., Singh, A. and Singh, S.K. 2013. Multimodal database of newborns for biometric recognition. *International Journal of Bio-Science and Bio-Technology*, 5(2), 89–100.

Tiwari, S. and Singh, S.K. 2012a. Face recognition for newborns. *IET Biometrics*, 1(4), 200–208.

Tiwari, S. and Singh, S.K. 2012b. Newborn verification using headprint. *Journal of Information Technology Research (JITR)*, 5(2), 15–30.

Tiwari, S. and Singh, S.K. 2014. Multimodal biometrics recognition for newborns. In *Research Developments in Biometrics and Video Processing Techniques*, R. Srivastava, S.K. Singh and K.K. Shukla, 25–51. IGI Global, Hershey, PA.

Tribune, The 2015. Get Newborn Clicked With Family to Avoid Swapping Controversies: NCSC Member. 2015, September. *The Tribune*. Retrieved from www.tribuneindia.com/news/haryana/community/get-newborn-clicked-with-family-to-avoid-swapping-controversies-ncsc-member/131875.html. Accessed on January 7, 2018.

Weingaertner, D., Bellon, O.R.P., Silva, L. and Cat, M.N. 2008. Newborn's biometric identification: Can it be done?. In *VISAPP*, 1, pp. 200–205.

Wierschem, J. 1965. Know them by their feet. *Journal of Medical Record News*, 168, 158–160.

World Population Prospects: The 2017 Revision. 2017, June. Department of economic and social affairs, United Nations. Retrieved from www.un.org/development/desa/publications/world-population-prospects-the-2017-revision.html. Accessed on January 15, 2018.

Zhang, L., Zhang, L. and Zhang, D. 2009, November. Finger-knuckle-print: A new biometric identifier. In *Proceedings, 2009 16th IEEE International Conference on Image Processing (ICIP)*, pp. 1981–1984. IEEE, Cairo, Egypt.

Zhang, L., Zhang, L., Zhang, D. and Zhu, H. 2010. Online finger-knuckle-print verification for personal authentication. *Pattern Recognition*, 43(7), 2560–2571.

13

Paradigms of Artificial Intelligence in Biometric Computing

Pradeep Kundu

Indian Institute of Technology (IIT) Delhi

Manish Rawat

Manipal University Jaipur

CONTENTS

13.1 Introduction

Biometric system allows natural interaction between human and machines, and it measures the unique human characteristics such as physical and behavioral. These interactions can be related to fingerprint, image, touch, speech, body language, etc. The biometric technology identifies or verifies the individual based on these individual traits or combination of these traits. Due to excellent system performance, decreasing costs and increasing security requirements, the biometric systems are used in many organizations such as military, government, education, business, etc. (Xiao, 2007).

For example, India has successfully deployed the biometric technologies for the purpose of national identification and entitlements distribution through the program Aadhaar launched in 2010. According to this program, a 12-digit unique identification number is given to each Indian citizen based on biometric identifying data such as iris and

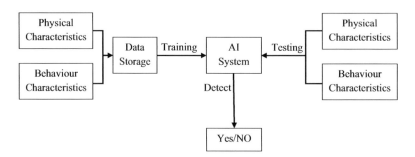

FIGURE 13.1
Biometric computing framework.

fingerprint patterns. The unique number generates based on the individual's unique eye, and fingerprint traits enable fair distribution to government benefits, services and subsidies (Daugman, 2014). Similarly, over the globe, banks and other financial outfits are increasingly using the biometrics to authenticate customers accessing their services to reduce the financial fraud, identify theft and threats from the cyberspace.

Biometrics application has gained a lot of popularity for solving the wide range of complex problems. Still many issues need to be resolved for successful implementation of the biometric systems on a large scale. Hence, nowadays, most of the biometric industries are focusing on integrating the biometric system with the artificial intelligence (AI) based algorithms for solving the biometric computational problems such as sample nonidealities and difficulty in recognition.

Figure 13.1 shows how an AI system can be integrated with the biometric system to solve computational problem. The technology uses distinctive and measurable characteristics of the particular part or behavior of the human body. All these information are converted into a database understandable by the AI technology. For authentication and verification, the information from this database is digitally sampled and used for security purpose. For example, the surveillance team of a country has the criminal information, and it is very difficult for them to do visual check of individual ones at the airport, railway station, etc. In such cases, the best way is to install the security camera at the entry and exit points of these locations. Using the video surveillance with facial recognition technology and artificial intelligence, suspicious can be detected.

The remaining chapter is organized in the following manner: Section 13.2 provides a general overview of the types of biometric technologies. Section 13.3 discusses few AI approaches used for biometric applications. Section 13.4 details the metrics which can be used for testing the performance of the biometric system. Finally, Section 13.5 sets out the conclusions.

13.2 Types of Biometric Technologies

There are many parameters to distinguish the users for using biometric solutions. These parameters vary from physical selection to behaviour selection. However, selection of those depends on the types of security and access control requirements. In general, the biometric technologies which can work with the AI-based approaches can be divided into the following six categories:

13.2.1 Fingerprint Recognition

The fingerprint-based biometric solutions check for ridge line patterns, the valley between the ridges, etc., for fingerprint match for verification and authorization.

13.2.2 Face Recognition

Face recognition-based biometric system select the facial features such as eye distance, nose width, nose shape, cheekbone shape, jaw line length, etc., from a digital image or a video frame (Ashok, 2010). Combination of these features creates the framework which is natural and hence is more acceptable than other biometrics. The biometric solution based on the face recognition is little computationally expansive and costly compared to the fingerprint recognition system.

13.2.3 Voice Recognition

The voice recognition system characterizes the parameters such as pitch, cadence, tone, etc., of the voice.

13.2.4 Iris Recognition

Iris is the colored part between pupil and sclera. Iris has a complex structure and is unique from person to person. It is stable throughout the life and forms a good biometric (Ashok, 2010).

13.2.5 Handwriting and Signature Recognition

The signature recognition is a behavioral biometric, it helps in recognizing the kind of artistic handwriting which may be unreadable and may have special characters and attracts attention. The handwriting and signature recognition-based biometric system for verification works in two ways: (i) online signature recognition and verification systems (SRVS) and (ii) offline SRVS. In the offline SRVS, the signatures are handled as an image which are based on the image processing and through AI techniques they can be recognized. Whereas the online SRVS measures the hand speed and pressure on the human hand when it creates the signature for recognition.

13.2.6 Behavioral Recognition

Most of the physical biometric security systems do the authentication at the beginning of an action, for example, device login or access control. However, the original user may provide their credentials to another person after initial authentication. In that scenario, the behavioral biometric system provides an additional layer of security along with the physical biometric solution. The behavioral biometric system works on the human activities such as keystroke dynamics, voice print, signature analysis, error pattern, etc. The behavioral biometric system analyze users interactions with their devices and record the activities that vary from normal usage patterns.

13.3 Artificial Intelligence in Biometric Applications

The matching of a biometric is a "fuzzy comparison" as the biometric trait of the same person cannot be captured in precisely the same way twice. This feature makes use of AI-based approaches for solving different biometric computing problems. AI focuses on the creation of intelligent machines that work and react like or better than humans. The AI approaches work similar to the human brain and hence has wide scope of research nowadays. The major issue with the biometric recognition system is how to extract discriminative corresponding biometric trait-based features. When AI integrates with the biometric computing system, it reduces such problems in addition to the computational issues. Figure 13.2 shows the outline of the AI-based biometric recognition system. The main components of the AI-based biometric computing system are sensor, segmentation, quality enhancement, features extraction and matching or classification (Labati et al., 2016).

The purpose of the sensor is to collect the biometric traits in digital format. These traits are converted by the sensor into the format understandable by the computer language or convert the biometric traits of a person into an electric signal. The selection of the sensor depends on the type of biometric recognition system it requires. For example, for face recognition a good quality camera is required, and for voice recognition a good quality microphone is required. Several times while collecting the biometric traits, some other information such as surrounding noise, background area in the image, etc., are also recorded which does not have any relevance with the biometric traits. This extra information may cause a computational problem while analyzing the biometric traits and hence needs to be removed.

The segmentation step is used to extract the region that only contains biometric information. For example, for fingerprint-based biometric process the ridge lines are important, so the segmentation step removes the extra information of the finger nonrelated to the ridge lines. In addition, for face recognition this step removes the background area. In this context, Tikoo and Malik (2017) used the artificial neural network (ANN)-based classifier for segmenting the face and its major features. The captured image is converted into red, green and blue (RGB) format and the histogram of the same is calculated. The obtained histogram data are used as an input for face recognition and segmentation.

Hilado et al. (2011) also segments the faces in images using the ANN-based classifier. Using the edge detection approach, skin colors are used as an input in the classifier model for detecting the possible faces in the image. The output of the network in these models is binary as faces or no faces. Similarly, Da Costa et al. (2015) used the kernel method such as a support vector machine (SVM) classifier for facial recognition. Zhao et al. (2008) used the similar SVM classification approach for fingerprint image segmentation. The fingerprint image was portioned into 12×12 blocks, and low grey variance background blocks were segmented by the contrast for fingerprint segmentation. The iris-based biometric system also faces the noise in the eye image such as blur, off-axis, specular reflections and

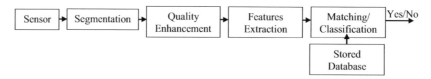

FIGURE 13.2
Outline of the artificial intelligence based biometric recognition system.

occlusions which affects the accuracy of the iris-based recognition system. To overcome that, Sinha et al. (2017) proposed the deep neural network approach for semantic pixel-wise iris segmentation.

Many times, the quality of the image captured, fingerprints collected, voice recorded, etc., may not be of good quality due to various reasons such as the exerted pressure of the finger on the sensor, poor skin conditions, off-angle gaze, environmental and camera effects, etc. (Xiao, 2007). Hence, the quality enhancement step is required to increase the biometric sample quality. In this direction, many AI-based approaches are used. For example, a convolution neural network approach was proposed by Grundhofer and Rothlin (2018) for camera specific image quality enhancement. The proposed approach reduces the image artefacts and enhances the image quality.

The feature extraction step involves extraction of the most distinctive parameters from the collected biometric traits after segmentation and quality enhancement so that the identity comparison can be easily performed. Various features which can be used in the different kinds of biometric recognition systems are found in Xiao (2007). At last, the matching process compares the live biometric traits with the traits stored in the database for acceptance or rejection.

Many AI approaches are used in the field of biometrics. However, among them, ANN and SVM are quite popular because of their excellent performance. In the present chapter, the working procedure of only these two is discussed.

13.3.1 Artificial Neural Network

Artificial neural networks (ANNs) are nonlinear data-driven self-adaptive approach. The ANN is the electronic model of neural structure of the brain. These models learn from the training in a similar way as biological neural network learns from the experience (Anderson and Mcneill, 1992). Like the biological neural network, it contains a large number of highly interconnected processing elements (neurons) and works as a single language to solve a specific problem. Figure 13.3a,b shows the pictorial view of the biological and artificial neural network, respectively.

In biological neurons, the electric signals are received by the dendrite from the axons of other neurons. In the ANN, the numerical values of the data are a representation of these electrical signals. In biological neurons, the electrical signals are modulated by synapses in various amounts. In a similar way, each input value in ANN is multiplied by a value called the weight. If input signal total strength crosses a certain threshold, the biological neuron fires an output signal. In the ANN, the weighted sum of the inputs represents the input signal total strength. The step function applies on the weighted sum determine its output.

During training of the ANN, the error between the desired and the actual output is reduced by adjusting the weights of each unit. Deviation in error is calculated with a slight increase or decrease in the weight value. After a certain point, validation mean square error (MSE) starts increasing while the training MSE starts decreasing. The training should stop at this point, as the ANN starts to model the noise in the training set.

Characteristics of ANN:

1. It is very difficult to determine the optimum network structure and a number of nodes, and hence trial and error search method is the only way to obtain optimum network topology (Jardine, Lin, and Banjevic, 2006; Mahamad, Saon, and Hiyama, 2010).

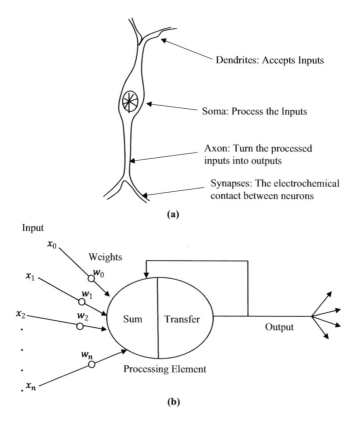

FIGURE 13.3
(a) Biological neural network (b) Artificial neural network.

2. ANN with fewer input nodes has better generalization capability and gives more accurate results (Tian, 2012).

3. The ANN training algorithms is uncertain and hence will not give the same results for different numbers of run during training with the same input, output and model structure (Tian, 2012).

4. Without a validation mechanism, one cannot say that the network is best even though training MSE is low (Tian, 2012).

13.3.2 Support Vector Machine

In AI techniques after ANN, SVM is the most popular method used for pattern recognition and classification problems. The SVM does not depend on the feature space dimensionality and less prone to the over fitting problems as observed with the neural networks (Ramesh Babu and Jagan Mohan, 2017). It maps the input data x in a nonlinear fashion into a higher-dimensional feature space via kernel functions (Roulias, 2014; Siegel, Ly, and Lee, 2011). Each instance in the input feature space is assigned with a class label of +1 or −1. For example, if x_i is the input vector and d_i is the desired value, then SVM consider the data set in ways, i.e., $(x_1, d_1), (x_2, d_2), \dots , (x_n, d_n)$, where, $d_i \in \{-1, +1\}$. The SVM classifies the dataset by constructing a hyper plane as shown in Figure 13.4.

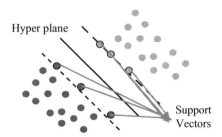

FIGURE 13.4
Support vector machine example.

13.4 Performance Metrics for Testing the Biometric System

The implementation of the biometric system has several advantages such as uniqueness, universality, user-friendly, etc. However, they have several disadvantages also such as costliness, facial imaging, false acceptances and rejections. Hence, the performance testing of a biometric system is very crucial for the implementation of these technologies in real applications. The biometric system performance mainly can be evaluated using the metrics such as false rejection rate, false acceptance rate and receiving operating characteristic curve.

a. **False Rejection Rate:** It is the rate at which the biometric security system incorrectly rejects the attempt access by an authorized user.

b. **False Acceptance Rate:** It is the rate at which the biometric security system incorrectly accepts the attempt access by an unauthorized user.

c. **Receiving Operating Characteristic Curve:** The ROC curve is a graphical plot which represents the diagnosis capability of a binary classifier. This curve is obtained by plotting the true positive/acceptance rate versus the false positive/acceptance rate as shown in Figure 13.5. The area under the ROC curve represents

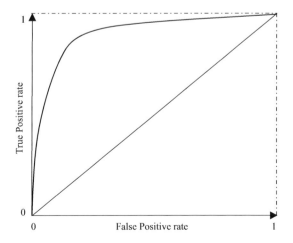

FIGURE 13.5
ROC curve example.

the accuracy of the classification model. For example, the curve closer to LHS border and top border of the ROC curve shows the more accurate classification results. Contrary, the curve closer to 45° diagonal to the ROC curve shows the less accurate classification results.

13.5 Conclusion

Biometric computing system uses the measurable physical or behaviour characteristics to enhance the security and control access. Because of processing ability, the AI techniques dominate over humans. The adoption of a biometric system with AI has proved to be profitable in terms of operational efficiency and customer satisfaction. Based on the results obtained from the various researchers this technology looks promising and won customer satisfaction largely. However, the issue such as privacy has to be sorted out for completely successful implementation of these technologies. In addition, the biometric computing system time efficiency needs to be improved by reducing the false rejection and the false acceptance rate.

References

Anderson, D., & Mcneill, G. (1992). Artificial neural networks technology. Data & Analysis Center for Software, 82, 87. doi:10.1093/rpd/ncm084.

Ashok, J. (2010). An overview of biometrics. *International Journal on Computer Science and Engineering*, 2. Retrieved from http://citeseerx.ist.psu.edu/viewdoc/download?. doi:10.1.1.301.2569&rep=repl&type=pdf.

Da Costa, D. M. M., Peres, S. M., Lima, C. A. M., & Mustaro, P. (2015). Face recognition using Support Vector Machine and multiscale directional image representation methods: A comparative study. In *Proceedings of the International Joint Conference on Neural Networks* (Vol. 2015 September), Killarney. doi:10.1109/IJCNN.2015.7280699.

Daugman, J. (2014). 600 million citizens of India are now enrolled with biometric ID. In *SPIE Newsroom*. doi:10.1117/2.1201405.005449.

Grundhofer, A., & Rothlin, G. (2018). Camera-specific image quality enhancement using a convolutional neural network. In *Proceedings of the International Conference on Image Processing, ICIP* (Vol. 2017 September), Beijing. doi:10.1109/ICIP.2017.8296510.

Hilado, S. D. F., Dadios, E. P., & Gustilo, R. C. (2011). Face detection using neural networks with skin segmentation. In *2011 IEEE 5th International Conference on Cybernetics and Intelligent Systems (CIS)*, Qingdao, pp. 261–265. doi:10.1109/ICCIS.2011.6070338.

Jardine, A. K. S., Lin, D., & Banjevic, D. (2006). A review on machinery diagnostics and prognostics implementing condition-based maintenance. *Mechanical Systems and Signal Processing*, 20(7), 1483–1510. doi:10.1016/j.ymssp.2005.09.012.

Labati, R. D., Genovese, A., Muñoz, E., Piuri, V., Scotti, F., & Sforza, G. (2016). Computational intelligence for biometric applications: A survey. *International Journal of Computing*, 15(1), 40–49. Retrieved from www.computingonline.net/index.php/computing/article/view/829.

Mahamad, A. K., Saon, S., & Hiyama, T. (2010). Predicting remaining useful life of rotating machinery based artificial neural network. *Computers and Mathematics with Applications*, 60(4), 1078–1087. doi:10.1016/j.camwa.2010.03.065.

Ramesh Babu, N., & Jagan Mohan, B. (2017). Fault classification in power systems using EMD and SVM. *Ain Shams Engineering Journal, 8*(2), 103–111. doi:10.1016/J.ASEJ.2015.08.005.

Roulias, A. D. (2014). Methodologies for remaining useful life estimation with multiple sensors in rotating machinery (June).

Siegel, D., Ly, C., & Lee, J. (2011). Evaluation of Vibration-Based Health Assessment and Diagnostic Techniques for Helicopter Bearing Components. In *The 14th Australian International Aerospace Congress* (Hums).

Sinha, N., Joshi, A., Gangwar, A., Bhise, A., & Saquib, Z. (2017). Iris segmentation using deep neural networks. In *2017 2nd International Conference for Convergence in Technology, I2CT 2017* (Vol. 2017 January), Mumbai. doi:10.1109/I2CT.2017.8226190.

Tian, Z. (2012). An artificial neural network method for remaining useful life prediction of equipment subject to condition monitoring. *Journal of Intelligent Manufacturing, 23*(2), 227–237. doi:10.1007/s10845-009-0356-9.

Tikoo, S., & Malik, N. (2017). Detection, segmentation and recognition of Face and its features using neural network. Retrieved from https://arxiv.org/ftp/arxiv/papers/1701/1701.08259.pdf.

Xiao, Q. (2007). Biometrics-technology, application, challenge, and computational intelligence solutions. *IEEE Computational Intelligence Magazine, 2*(2). doi:10.1109/MCI.2007.353415.

Zhao, S., Hao, X., & Li, X. (2008). Segmentation of Fingerprint Images Using Support Vector Machines. In *2008 Second International Symposium on Intelligent Information Technology Application*, Shanghai, pp. 423–427. doi:10.1109/IITA.2008.323.

14

Face Recognition in Low-Resolution Space

Shyam Singh Rajput

Bennett University

Karm Veer Arya

ABV-Indian Institute of Information Technology & Management

Poonam Sharma

Visvesvaraya National Institute of Technology

CONTENTS

14.1 Introduction

Face recognition is a task that people perform regularly and effortlessly in everyday lives. In recent years, computational power and embedded technologies have rapidly improved which create an immense impact in the automated processing of digital images in numerous applications such as biometric authentication, multimedia management, human-computer interaction and surveillance. Among all applications, research and advancement for automatic face recognition is growing naturally.

Biometric using facial images has many advantages over other biometric modalities, e.g., iris and fingerprint. Apart from reasonable and non-intrusive, the most valuable benefit of the facial images based biometric is that the faces can be captured at a distance and in a secret manner. According to literature [1], the face-based biometric approach is most appropriate as compared to other biometric techniques.

As the face-based biometric models utilize the features of the input facial images, the captured faces should have adequate information or characteristic for distinguishing one person from other. Accuracy of recognition system depends on the input features. The most of the current biometric systems require high-quality input faces means faces with sufficient features known as high-resolution (HR) faces. However, in many practical and surveillance scenario the recorded pictures often lose some features due to various reasons called as low-resolution (LR) images which make the recognition task challenging. The major challenges in face-based biometric techniques are as follows:

- **Low-resolution capturing devices:** At most of the public places, the installed imaging systems are cheap, LR and weak circuitry. The images recorded through these types of imaging systems are often blurry, noisy and in low-resolution. Faces extracted from such images do not have adequate information for performing the biometric task. To overcome this, either we replace low-cost systems with expensive HR capturing systems or develop more robust and efficient face recognition techniques that can adequately recognize the low-quality probe images. Hence, it is always a big issue performing face recognition in an uncontrolled environment.

- **Object distance:** Location of an object also plays the vital role in the performance of face recognition systems. The far distance of a subject from the camera may result in very small size captured faces which cause the significant problem in performing their matching with gallery images. Because by performing matching between different dimensions images (i.e., larger size gallery images and smaller size probe faces) is not possible directly. This is the biggest issue for modern recognition systems.

- **Lightning:** The performance of face recognition systems also varies with the presence of different lighting and illumination conditions. Varying illumination conditions rapidly change the features of captured images. Hence, the recognition system should be capable of doing recognition task under diverse illumination conditions.

- **Environmental factors:** Various environmental factors like dust, shadow, atmospheric noise, atmospheric blur, etc., also affect the recognition performance. These factors make biometric challenging under un-controlled environment.

- **Object view:** The performance of the face recognition systems also varies with the deferent poses of the performance. Maximum features are available in frontal images rather than other poses. Hence, frontal view faces are more appropriate for performing recognition as compared to other poses. But in practical scenario it is not necessary that the captured images are always frontal. Hence, the recognition model should pose robust.

- **Person expressions:** The expression of the human at the time of photo capturing also affects the performance of the systems. In general, face recognition systems give the best performance for normal expression faces as compared to those of faces with other expressions like the surprise, smiley, etc.

- **Other factors:** Apart from the above challenges, there are some other factors which are also challenging for face recognition. These factors include occlusion, blocking, partial detection of an object, different types of image processing noise, etc.

The above factors create the significant problem in performing face recognition task. However, to handle factors like pose, illumination, expression, and occlusion numerous methods have been introduced in the literature, and they have achieved the reasonable performance. But performing face recognition for small size, noisy and LR images are still challenging in the literature. The small size of captured faces create the massive problem in face recognition because performing matching operation between the smaller size LR probe and larger size gallery faces using the method like the nearest neighbour is not possible. Nowadays, most of the research works are focusing to solve the problem of recognition of small and LR faces. It is known as LR face recognition problem [2]. In this chapter, we are discussing the problems which arise due to LR images in performing face recognition and discussing the possible solutions to this problem.

14.2 Low-Resolution Face Recognition Problem

In numerous real-world and surveillance scenario, faces extracted from the images captured by the camera are noisy, small in size and blur and are called LR faces due to many reasons as discussed above. These types of LR faces produce the massive problems in many computer vision applications like face recognition. This problem is termed as LR face recognition problem in the literature.

14.3 Solutions of Low-Resolution Face Recognition Problem

There are three possible solutions to the problem which arises due to LR captured images in face recognition. These solutions are as follows:

- *Up-sampling*: The first approach used to solve the dimensionality mismatch problem in the LR probe image and HR gallery images is up-sampling. In this approach, the size of the LR probe images is made equal to gallery images by performing up-sampling. The popular techniques used for up-sampling are known with the name of super-resolution. Apart from making the size of probe image compatible with gallery images, super-resolution techniques also reconstruct the missing features of input LR probe images.

- *Down-sampling*: The second approach used to solve the dimensionality mismatch problem in LR probe image and HR gallery images is down-sampling. In this approach, instead of improving the size of the probe as performing in the up-sampling procedure, the gallery images are down-scaled to make their size compatible with LR probe images.

- *Unified feature space*: The latest approach to solve the LR face recognition problem is performing matching in a unified or common feature space (UFS). In UFS-based methods, the feature of the LR probe and HR gallery images are projected in common or unified space where matching or classification operation is performed for obtaining recognition accuracy.

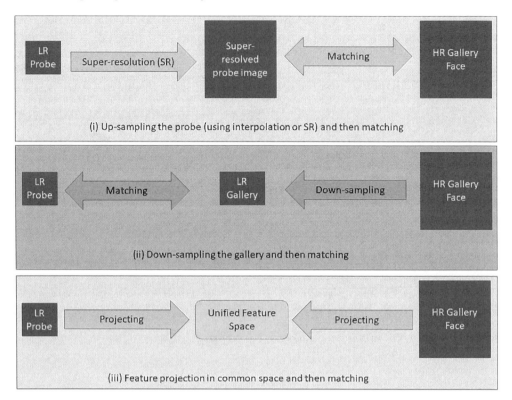

Out of three popular solutions above, the UFS-based [3,4] and super-resolution (SR)-based techniques [5–7] are widely used in the literature because of their specific characteristics. Apart from several advantages, the UFS and SR-based techniques also have few disadvantages. The advantages and disadvantages of both the approaches are discussed below.

- The UFS-based techniques insist only on resolving the size mismatch problem in LR probe and HR gallery images. Whereas, the SR-based methods resolve the size mismatch problem along with reconstructing the corrupted features of captured probe images for better recognition performance. It allows the SR-based approach to provide robust performance in the presence of noise, blur, etc. as compared to UFS-based techniques.

- As SR-based LR face recognition models first perform the up-sampling and then matching, the SR-based approaches are computationally inefficient as compared to UFS-based approaches. Hence, SR-based techniques may be less appropriate for real-time applications.

- The SR techniques are universal as compared to UFS-based techniques. Apart from face super-resolution, SR techniques can also be used for other types of biometric models like iris-based and fingerprint-based biometric.

14.4 Literature Review

As discussed above the UFS-based and SR-based approaches are most popular in the literature to solve the LR face recognition problem, hence the research papers published related to these two approaches are discussed separately in Sections 14.4.1 and 14.4.2.

14.4.1 SR-Based Methods

Here we are providing the discussion about the works available in the literature for image super-resolution (SR).

The SR is the class of techniques which generate HR version of captured LR images. As the SR techniques can be applied to any kind of input images like iris, fingerprint, face, etc., the SR techniques are appropriate for performing biometric under uncontrolled environment. In this chapter, there will be main focus on face super-resolution techniques also known as face hallucination.

The face hallucination approaches are basically of three types: (i) reconstruction-based [8–10], (ii) interpolation-based and (iii) learning-based. Due to superiority of learning-based SR methods over interpolation and reconstruction-based methods under uncontrolled environment (presence of blur, noise, etc.), the learning-based SR methods are much popular in the literature. The learning-based methods are basically of two types namely patch-based and global-face–based. The learning-based models require well aligned faces. However, recently one more category is also getting much attention that is deep-learning–based SR techniques. The deep-learning–based techniques avoid the requirement of aligning of input and training faces. But the limitation with these techniques is that these require huge computation power for training the network. The overview of popular and widely used face hallucination model is presented in Table 14.1.

TABLE 14.1

Overview of Several Global-Face–Based, Pixel-Based, Neighbor-Patch–Based, Position-Patch–Based and Deep-Learning–Based Face Hallucination Models

Year	Authors	Category	Methodology	Remark
2000	Baker and Kanade [5]	Global face based model	• Learn the resolution enhancement function. • It used pyramid-based algorithm.	First paper in the field of face SR also known as face hallucination model
2005	Liu et al. [11]	Neighbour-patch based model	• This model assumed that patches in LR and HR space form manifold with the same local structure that is described by the weights of neighbouring patches. • Based on this, it synthesizes the HR patch by utilizing the weights obtained according to input low-resolution patches.	This is the first neighbor-patch–based face hallucination model

(Continued)

TABLE 14.1 (*Continued*)

Overview of Several Global-Face–Based, Pixel-Based, Neighbor-Patch–Based, Position-Patch–Based and Deep-Learning–Based Face Hallucination Models

Year	Authors	Category	Methodology	Remark
2005	Wang and Tang [12]	Global face-based model	• This model is distinct from most of the previous probabilistic-based models. • It observes hallucination process as a transformation between distinct image styles. • It utilizes principal component analysis (PCA) to represent the test LR face as linear combination of LR faces in the training set. • The HR face is furnished by substituting the LR training faces by HR ones with same coefficients.	By choosing the number of Eigen faces, it secures the maximum facial information from input LR faces also removes the noise.
2010	Ma et al. [13]	Position-patch–based model	• The concept of least square representation (LSR) is introduced here. • The reconstruction weights for each input LR patch is obtained from its weighted linear representation in term of same position patches in LR training faces. • Obtained weights are applied to counterpart patches of HR training set to reconstruct the output HR patch.	This is the state-of-the-art face hallucination model. Its limitation is that when the dimensions of input patch is smaller than the size of training set then its solution is not unique.
2014	Jiang et al. [14]	Position-patch–based model	The concept of locality constrained representation (LcR) is introduced with LSR to provided robustness against the Gaussian noise.	This is the first noise robust model in the literature.
2011, 2014	Hu et al. [15]	Pixel-based models	To generate more discriminable face these tow paper proposed the pixel-based reconstruction instead of global face or patch based reconstruction.	Not appropriate for blurry and noisy input images.
2016	Jiang et al. [16]	Position-patch–based model	In this the work of LcR is further improved by introducing sparsity constrained. This model is known as Tikhonov regularized neighbor representation (TRNNR)	The LcR and TRNR model work well only for Gaussian noisy input faces while their performance degrade drastically for other type of noise e.g., Impulse noise.
2016	Zhu et al. [17]	Deep-learning-based model	This model is capable for hallucinating very low-resolution and un-constrained poses faces.	Unlike to the previous models, this work avoids the requirement of pre-alignment of faces.
2017	Jiang et al. [18]	Position-patch–based model	It uses the characteristic of reconstruction-based, interpolation-based, and learning based approaches for generating more discriminable faces.	This model fails for noisy input faces.
2017	Liu et al. [19]	Position-patch–based model	In this model, the robustness of LcR is improved against the impulse noise by introducing the model called robust LcR with by-layer representation (RLcBR).	First impulse noise robust face hallucination model.

(*Continued*)

TABLE 14.1 (*Continued*)

Overview of Several Global-Face–Based, Pixel-Based, Neighbor-Patch–Based, Position-Patch–Based and Deep-Learning–Based Face Hallucination Models

Year	Authors	Category	Methodology	Remark
2018	Rajput et al. [7]	Position-patch-based model	Error shrunk nearest neighbor representation model (ESNNR) is introduced to make hallucination process robust to mixed Gaussian-impulse noise.	First mixed noise robust face hallucination algorithm
2018	Rajput et al. [6]	Position-patch-based model	In this, the robustness of RLcBR is further improved by introducing iterative sparsity and locality constrained representation (ISLcR) model.	This is the best impulse noise robust model among all methods available in the literature till date.
2017	Chen et al. [20]	Deep learning	• In this, deep end-to-end trainable Face SR Network (FSRNet) is presented. This model fully utilizes geometry prior for hallucinating very LR faces without the well-aligned requirement. • It utilizes different prior related to human face geometry, e.g., parsing and facial landmark heat maps.	This is the latest and popular deep-learning-based model.

14.4.2 UFS-Based Methods

The unified feature-space-based methods first learn two projection matrix (separate for LR and HR images) from training faces and then used these matrix to project the features of LR probe and HR gallery images in the common (or unified) feature space where recognition operation performs. The overview of several good works is provided in Table 14.2.

TABLE 14.2

Overview of Several Global-Face–Based, Pixel-Based, Neighbor-Patch–Based, Position-Patch–Based and Deep-Learning–Based Face Hallucination Models

Year	Authors	Methodology	Remark
2005	Jia and Gong [21]	• A Bayesian framework is proposed here to compute multi-modal (such like changes in illumination and viewpoint) face SR for recognition in tensor space. • Unlike doing pixel-domain SR and recognition separately, in this paper, the SR and recognition tasks are integrated by calculating the vector of maximum likelihood identity parameters in HR tensor space.	This approach overcomes the time complexity of SR-based low-resolution face recognition models.
2011	Huang and He [22]	• Similar to Ref. [21], this paper also focuses on simultaneous super-resolution and recognition task. • Canonical correlation analysis (CCA) is used to build the coherent subspaces between the principal component analysis (PCA)-based features of HR and LR faces. • The radial basis functions (RBFs) is used to build a nonlinear mapping between HR/LR features in the coherent feature space. • Then the trained RBF model is used to produce super-resolved coherent features for an input LR face. Further, the identity of the input face is accomplished by supplying the obtained super-resolved features to the nearest neighbor classifier.	This approach is computationally fast as compared to two step SR-based approaches, also robust to variation in pose and expressions.

(Continued)

TABLE 14.2 (*Continued*)

Overview of Several Global-Face–Based, Pixel-Based, Neighbor-Patch–Based, Position-Patch–Based and Deep-Learning–Based Face Hallucination Models

Year	Authors	Methodology	Remark
2012	Biswas et al. [23]	• In this paper, the concept of multidimensional scaling (MDS) is used to project the LR probe and HR gallery images in a common space. • An iterative majorization algorithm is used to learn the two mapping functions simultaneously from HR training images. • The distance between projected images in the transformed space is approximately equal to the distance between the HR versions of both the images. It used pyramid-based algorithm.	This model is capable to work with non-frontal images also.
2010	Li et al. [4]	• The concept of coupled mappings (CMs) is introduced in this paper to project the faces of varying resolutions into a UFS, supports the responsibility of classification. • The CM functions or matrix are calculated by simplifying the objective function in such a way that the difference between the LR image and its corresponding HR image is minimized. • Further, locality preserving constraint is incorporated with CMs (called CLPM) to improve the recognition accuracy further.	This is the first model that performs recognition of low-resolution images directly without performing the SR process. As this model does not use level information while projecting the features in the single manifold, there is the possibility to easily connect samples belonging to different manifolds (means wrong recognition).
2016	Jiang et al. [3]	• A concept of discriminative analysis is introduced with CLPM [4] to avoid the chances of wrong recognition. • It first explores the neighbourhood knowledge and the local geometric structure of the multi-manifold space created from the samples. • Then it learns two projection matrices to project the LR and HR features in the unified discriminative feature space (UDFS) in a supervised manner, where the discriminative knowledge is maximized for better classification.	This model gives excellent performance for clean images. While the performance of this and other previous models degrades drastically in the presence of noise.
2017	Chu et al. [24]	• This is the first model that investigate low-resolution face recognition under single sample per person. • It uses cluster-based regularization approach. • It regularizes the within-class, and between-class scatter matrices with intra-cluster and inter-cluster scatter matrices. • The cluster-based scatter matrices help in overcoming the singular matrix and over-fitting problems.	This model is appropriate for the application like law-enforcement where only a single HR image of an individual subject is available in the database.

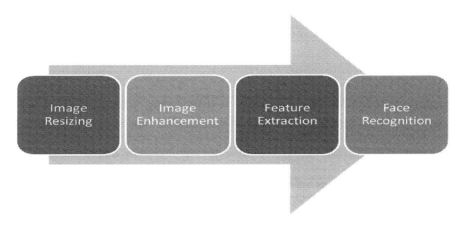

FIGURE 14.1
The stages of proposed methodology for face recognition on low resolution images.

14.5 Low-Resolution Face Recognition System via SR Technique

In this work, the FSRNet [20] model is utilized to develop a face recognition system for LR facial images, which can be called as LR-FR system. The detailed description of proposed methodology is given below (see Figure 14.1).

The three stages are image resizing, image super-resolution, feature extraction and face recognition. Each stage increases resistance to low resolution and makes the information needed for recognition more manifested.

14.5.1 Image Super-Resolution

Since our focus is on low resolution images, the images of high resolution have been resized down to low resolution. The MATLAB® function, imresize() which uses nearest neighbour interpolation has been used for resizing. The algorithm of image resizing is presented in Algorithm 14.1 below.

Algorithm 14.1: Image Super-Resolution Algorithm

1. for each folder, N in the LFW dataset
2.   for each file, j in the N
3.   im = imread(path[j]);   //read the image in the specified path of the file
4.   imr = imresize(im, [15 15]);   //image resize to 15×15 image size
5.   end for
6. end for

In order to enhance the quality of the input images for further analysis, there is a need for performing image enhancement on the images. The goal of this stage is to avoid the unwanted noise in the input images. For this purpose, the multi-layer neural network made up of three layers: input layer, hidden layer and output layer has been built. The error back propagation algorithm which is presented in Algorithm 14.2 was used for learning and updating the weights by applying error correcting learning rule and stochastic gradient descent optimizer.

In the case of 15 × 15 resolution images, the network consists of 225 nodes in both the input and output layers each. And the hidden layer constitutes 25 nodes after performing trial and error. Only the subjects whose number of images is more than eight have been considered. Each pixel value of an image corresponding to each node in the input layer has been fed to the network. The parameters like weights and biases of the layers were initialized with random values (between −1 and 1). The initialization of the learning rate with 0.95 and the momentum with 0.95 has been done. And these two parameters vary during training by using step decay function after each epoch. Here the weights are learnt for each subject separately and while testing if noise image is given then it gets inclined to any of the trained images of the corresponding subject.

Algorithm 14.2: Error Back Propagation Algorithm for Super-Resolution Image

1. Let the learning rate be η, momentum be μ
2. Initialize the weights to some random numbers.
3. repeat
4.   for each training sample, do
5.   Input the training sample to the network and compute the outputs.
6.   for each output unit k
7. $\delta_k = (t_k - O_k) O_k (1 - O_k)$
8.   end for
9.   for each hidden unit h
10. $\delta_h = O_h (1 - O_h) \Sigma_{k_{output}} W_{kh} \sigma_{kh}$
11.   end for
12.   Update each network weight wji:
13.   wji = wji + μ Δwji where Δwji = η δj xji
14.   end for
15. until termination condition (no. of epochs completes) is met

The step decay of the parameters helped in convergence of the algorithm without encountering the saturation problem. It drops the learning rate by a factor every few epochs. We have used the step decay function that drops the learning rate by half every ten epochs. And the same is applicable for momentum.

The number of epochs used for training was 2,500 since the error values obtained around this epoch number were less than 0.01 which is approximately 0 (convergence). During the network training phase, the weights updating takes place when all the images of a subject were fed once (epoch). This goes on for 2,500 times and then the resulted network with updated weights has been used for testing purpose. In the testing phase, each image of the subject was fed and the corresponding predicted (enhanced) image was obtained from the output layer. These two phases were performed on each subject of the dataset and then these enhanced images of the entire dataset were given as input to the following module.

14.6 Feature Extraction

After performing the image enhancement, local sampling has been performed on the resulted images. Since the components of the face extend horizontally or vertically and converge diagonally to the ends, the sampling is done in 0, $\pi/4$, $\pi/2$, $3\pi/4$, π, $5\pi/4$, $3\pi/2$ and $7\pi/4$ directions. In local sampling, for each pixel O, i.e., $M(x, y)$ where x and y are the coordinates of the image say M, the sampling has been done on the local 16 neighbourhood pixels: $\{M(x-2, y-2), M(x-1, y-1)\}$; $\{M(x-2, y), M(x-1, y)\}$; $\{M(x-2, y+2), M(x-1, y+1)\}$; $\{M(x, y+2), M(x, y+1)\}$; $\{M(x+2, y+2), M(x+1, y+1)\}$; $\{M(x+2, y), M(x+1, y)\}$; $\{M(x+2, y-2), M(x+1, y-1)\}$ and $\{M(x, y-2), M(x, y-1)\}$ which covers eight directions that corresponds to the extension directions of major facial textures. The resulted sample points are denoted as $\{A0, B0; A1, B1; \ldots ; A7, B7\}$

The nearest patterns in each of the eight directions have been encoded and these were combined to form the NP codes as shown in the equation below.

$$NP_i = T\left(L\left(I_{B_i}, I_{A_i}\right), L\left(I_{A_i}, I_{O_i}\right)\right), \quad 0 \le i \le 7$$

where

$$L(x,y) = \begin{cases} 1, x \ge y \\ 0, x < y \end{cases} \quad \text{and} \quad T(x,y) = \begin{cases} 3, x = 1 \wedge y = 1 \\ 2, x = 1 \wedge y = 0 \\ 1, x = 0 \wedge y = 1 \\ 0, x = 0 \wedge y = 0 \end{cases} \quad \text{and IO, IAi and IBi are the grey scale}$$

values of pixels O, Ai and Bi, respectively.

From the above diagrams, it is clear that the first four direction numpy (NP) codes are reflecting the same result as remaining four direction NP codes. Hence, the nearest four direction NP codes are considered for further process in order to reduce the total number of NP codes. It also reduces the time complexity of NP to the same level as LBP. This results in the subset $\{NP0, NP1, NP2 \text{ and } NP3\}$. The NP descriptor for each pixel O in an image is

$$NP = \sum_{i=0}^{3} NP_i \times 4^i$$

After encoding each pixel in the face image using the nearest encoders, the image produced is divided into a grid of non-overlapping regions. Histograms of NP codes are computed in each region (block size of 9) by using histogram() method of numpy package in python. The following formula is for building histogram:

$$H(k) = \sum_{x=1}^{N} \sum_{y=1}^{M} f\left(E_L(x,y), k\right), \quad k \in [0, K]$$

$$f(u, v) = \begin{cases} 1, & u = v \\ 0, & \text{otherwise} \end{cases}$$

where K is the maximal NP pattern value, $N \times M$ is image size, and $E_L(x, y)$ is an encoded image

Thus all the histograms of each region are concatenated and given as features to the deep learning framework, Facenet for face recognition purpose. And the NP encoded images are fed as inputs to Facenet.

14.7 Face Recognition

The concatenated histograms of DCP resulted above codes from the images which were given as embeddings (features) to Facenet architecture, i.e. inception resnet v1 model. From the inception-resnet-v1 architecture the reduction-A, inception-resnet-B and reduction-B layers are removed in order to make the training and testing image sizes compatible. This resulted in fast convergence of network.

After embeddings are learnt, we have evaluated our model on LFW dataset. Nine training splits are used to learn the optimal L2-distance threshold. Classification (same or different) is then performed on the tenth test split. If the Euclidean distance between two embeddings is less than the threshold value, then the embeddings are recognized as of the same person. Otherwise, the embeddings are recognized as of different persons.

Thus the classification accuracy is calculated for the given number of samples that are correctly classified (true positives (TP) and true negatives (TN)) and is evaluated by the formula:

$$\text{Accuracy} = \frac{\text{TP} + \text{TN}}{\text{Total no. of samples}}$$

14.8 Results and Discussions

In this work, we conducted experiments on LFW, ORL, AR and EYB databases to evaluate the proposed method, local binary pattern (LBP), local ternary pattern (LTP), and dual-cross pattern (DCP). In the experiments, the radius of neighbours for LBP and LTP (with threshold, 1.5) is taken as 1 and for both DCP and NP inner radius is 1 and outer radius is 2. To evaluate that our proposed method is not sensitive to noise, we added Gaussian noise on the face databases (using the 'imnoise' function in MATLAB with mean as 0, standard deviation as 0.05).

14.8.1 Experimental Results on the LFW Database

From Table 14.3, it is clear that the NP descriptor produces overall best results when compared to other descriptors upon evaluating on LFW database. However, the effect of NP is not much on high resolution images as they have already been recognized well enough by Facenet. More than 15% accuracy increase has been achieved by NP with Facenet approach than Facenet without any descriptor approach.

14.8.2 Experimental Results on the ORL Database

The ORL face database contains 40 subjects and for each subject there are ten images that differ in illumination, pose and expression from each other. Since the images size is 112 × 92, they are resized to 100 × 100 image size in order to make them appeal as square matrix. The training is performed on eight images of each individual and then the

TABLE 14.3

Facenet Performance on LFW Database

Resolution	Descriptor Facenet	LBP	LTP	DCP	NP
160×160	99.2	99.2	99.2	99.2	99.2
100×100	97.4	98.7	98.7	99.1	99.1
80×80	84.6	95.3	96.4	98.4	99.2
60×60	80.9	83.2	83.9	84.6	98.5
40×40	78.5	78.8	79.3	80.5	98.5
20×20	64.6	72.1	72.3	77.7	81.3
10×10	58.2	66.3	66.9	68.1	71.6

TABLE 14.4

Facenet Performance on ORL Database

Resolution	Descriptor LBP	LTP	DCP	NP
100×100	98.3	98.3	98.3	98.3
80×80	94.3	95.4	98.1	98.2
60×60	80.6	82.8	83.6	83.9
40×40	76.9	77.3	81.5	82.4
20×20	71.9	73.3	75.5	79.8
10×10	64.8	65.9	67.1	70.2

performance is tested on remaining two images of the same individuals. For histogram representation each encoded image has been divided into regions of 10×10 size. Due to the variation effects, LTP showed better performance than LBP. DCP performed better than LTP. Furthermore, the NP improved the accuracy rate by nearly 8% compared to LBP as observed from Table 14.4.

14.8.3 Experimental Results on the AR Database

For this experiment, we have used AR face database. The database consists of 76 men images and 60 women images with difference in illumination, expressions and occlusions. These images are also resized from 92×92 to 80×80 image sizes. From Table 14.5, it has been observed that the accuracy obtained for low resolution images of AR database is higher when compared with that of LFW database. This is becuase its training set is much

TABLE 14.5

Facenet Performance on AR Database

Resolution	Descriptor LBP	LTP	DCP	NP
80×80	98.7	98.7	98.7	98.7
60×60	87.1	88.9	90.6	93.5
40×40	80.3	82.6	84.6	87.9
20×20	76.5	78.3	79.1	82.7
10×10	70.3	72.2	77.0	80.6

less than LFW and slightly more than ORL. The obtained accurate rate of the proposed approach is 10% more than that of LBP.

14.8.4 Experimental Results on the EYB Database

The EYB database has 38 individuals with pose and illumination differences. First, these images are resized to 80×80 images and then training has been performed. In this experiment, out of 73 images of an individual 53 images are taken as training set and remaining 20 images as testing set. Due to less variation of the images, the accuracy obtained is the highest compared to the other databases. The proposed NP result increased by more than 7% compared to that of LBP as observed from Table 14.6.

The results of these Facenet-based experiments with local binary pattern (LBP), local ternary pattern (LTP), dual-cross pattern (DCP), and numpy (NP) on the four face database are summarized in Tables 14.3–14.6. Here only the 10×10 image size has been considered since the results obtained are worth noting for least resolution images. The accuracy list of the conventional Facenet on 10×10 resolution images on each of LFW, ORL, AR and EYB databases is 58.2%, 61.5%, 66.9% and 70.2% whereas the corresponding accuracy list of the proposed system is 71.6%, 70.2%, 80.6% and 82.3%. It has been observed that EYB database obtained highest accuracy while ORL databases obtained lowest accuracy compared to all other databases. This is due to the number of training samples and variation range of illumination, pose, expression and occlusion. Furthermore, the accuracy list comparison of NP-based Facenet on 10×10, 20×20, 40×40, 60×60 and 80×80 images has been depicted in Figures 14.2–14.6.

TABLE 14.6

Facenet Performance on EYB Database

Resolution	Descriptor LBP	LTP	DCP	NP
80×80	92.4	92.4	95.3	95.8
60×60	87.1	88.9	90.6	93.5
40×40	84.4	85.0	87.5	89.3
20×20	81.3	82.2	84.1	84.7
10×10	75.3	76.2	78.0	82.3

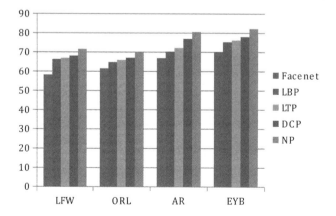

FIGURE 14.2

The comparison of NP-based Facenet performance on 10×10 with that of LBP, LTP and DCP and Facenet on LFW, ORL, AR and EYB databases in terms of accuracy (percentage).

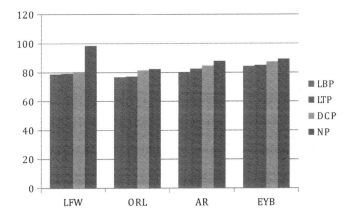

FIGURE 14.3
The comparison of NP-based Facenet performance on 20 × 20 with that of LBP, LTP and DCP on LFW, ORL, AR and EYB databases in terms of accuracy (percentage).

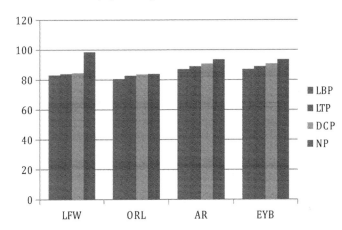

FIGURE 14.4
The comparison of NP-based Facenet performance on 40 × 40 with that of LBP, LTP and DCP on LFW, ORL, AR and EYB databases in terms of accuracy (percentage).

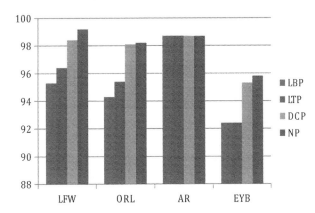

FIGURE 14.5
The comparison of NP-based Facenet performance on 60 × 60 with that of LBP, LTP and DCP on LFW, ORL, AR and EYB databases in terms of accuracy (percentage).

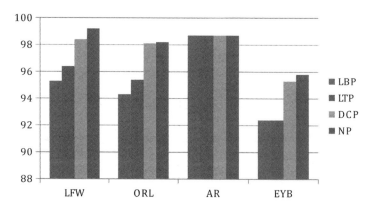

FIGURE 14.6
The comparison of NP-based Facenet performance on 80 × 80 with that of LBP, LTP and DCP on LFW, ORL, AR and EYB databases in terms of accuracy (percentage).

14.9 Conclusion

This chapter presented the study of different existing methods for LR images having different resolution. Experimental results performed on LR images from different popular datasets show that the performance of face recognition systems degrades with its reduction in the resolution of input images. Results show that as compared to other existing methods, the deep-learning–based Facenet method performing well for LR images.

References

1. Shakhnarovich, G., Moghaddam, B.: *Face Recognition in Subspaces*. Springer, London (2011) 19–49.
2. Zou, W.W.W., Yuen, P.C.: Very low resolution face recognition problem. *IEEE Transactions on Image Processing* 21(1) (2012) 327–340.
3. Jiang, J., Hu, R., Wang, Z., Cai, Z.: CDMMA: Coupled discriminant multi-manifold analysis for matching low-resolution face images. *Signal Processing* 124 (2016) 162–172.
4. Li, B., Chang, H., Shan, S., Chen, X.: Low-resolution face recognition via coupled locality preserving mappings. *IEEE Signal Processing Letters* 17(1) (2010) 20–23.
5. Baker, S., Kanade, T.: Hallucinating faces. In: *Proceedings Fourth IEEE International Conference on Automatic Face and Gesture Recognition* (2000) 83–88.
6. Rajput, S.S., Arya, K., Singh, V.: Robust face super-resolution via iterative sparsity and locality-constrained representation. *Information Sciences* 463–464 (2018) 227–244.
7. Rajput, S.S., Singh, A., Arya, K., Jiang, J.: Noise robust face hallucination algorithm using local content prior based error shrunk nearest neighbors representation. *Signal Processing* 147 (2018) 233–246.
8. Zhu, Y., Li, K., Jiang, J.: Video super-resolution based on automatic key-frame selection and feature-guided variation optic low. *Signal Processing: Image Communication* 29(8) (2014) 875–886.
9. Li, K., Zhu, Y., Yang, J., Jiang, J.: Video super-resolution using an adaptive super pixel-guided auto-regressive model. *Pattern Recognition* 51 (2016) 59–71.

10. Lin, Z., Shum, H.Y.: Fundamental limits of reconstruction-based super-resolution algorithms under local translation. *IEEE Transactions on Pattern Analysis and Machine Intelligence* 26(1) (January 2004) 83–97.
11. Liu, W., Lin, D., Tang, X.: Neighbor combination and transformation for hallucinating faces. In: *2005 IEEE International Conference on Multimedia and Expo* (July 2005) 1–4.
12. Wang, X., Tang, X.: Hallucinating face by eigen transformation. *IEEE Transactions on Systems, Man, and Cybernetics, Part C (Applications and Reviews)* 35(3) (2005) 425–434.
13. Ma, X., Zhang, J., Qi, C.: Hallucinating face by position-patch. *Pattern Recognition* 43(6) (2010) 2224–2236.
14. Jiang, J., Hu, R., Wang, Z., Han, Z.: Noise robust face hallucination via locality-constrained representation. *IEEE Transactions on Multimedia* 16(5) (2014) 1268–1281.
15. Hu, Y., Lam, K.M., Qiu, G., Shen, T.: From local pixel structure to global image super-resolution: A new face hallucination framework. *IEEE Transactions on Image Processing* 20(2) (2011) 433–445.
16. Jiang, J., Chen, C., Huang, K., Cai, Z., Hu, R.: Noise robust position-patch based face super-resolution via tikhonov regularized neighbor representation. *Information Sciences* 367–368 (2016) 354–372.
17. Zhu, S., Liu, S., Loy, C.C., Tang, X.: Deep cascaded bi-network for face hallucination. CoRR abs/1607.05046 (2016).
18. Jiang, J., Chen, C., Ma, J., Wang, Z., Wang, Z., Hu, R.: SRLSP: A face image super-resolution algorithm using smooth regression with local structure prior. *IEEE Transactions on Multimedia* 19(1) (2017) 27–40.
19. Liu, L., Chen, C.L.P., Li, S., Tang, Y.Y., Chen, L.: Robust face hallucination via locality-constrained bi-layer representation. *IEEE Transactions on Cybernetics* (99) (2017) 1–13.
20. Chen, Y., Tai, Y., Liu, X., Shen, C., Yang, J.: Fsrnet: End-to-end learning face super-resolution with facial priors. CoRR abs/1711.10703 (2017).
21. Jia, K., Gong, S.: Multi-modal tensor face for simultaneous super-resolution and recognition. In: *Tenth IEEE International Conference on Computer Vision (ICCV'05)* Volumes 1 and 2 (2005) 1683–1690.
22. Huang, H., He, H.: Super-resolution method for face recognition using nonlinear mappings on coherent features. *IEEE Transactions on Neural Networks* 22(1) (2011) 121–130.
23. Biswas, S., Bowyer, K.W., Flynn, P.J.: Multidimensional scaling for matching low-resolution face images. *IEEE Transactions on Pattern Analysis and Machine Intelligence* 34(10) (2012) 2019–2030.
24. Chu, Y., Ahmad, T., Bebis, G., Zhao, L.: Low-resolution face recognition with single sample per person. *Signal Processing* 141 (2017) 144–157.

Index

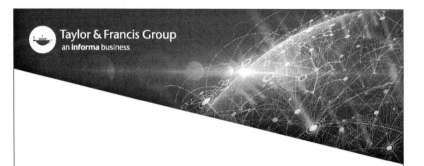